Asphalt Surfacings

Also available from E & FN Spon

Bituminous Binders and Mixers
Edited by L. Francken

Bridge Deck Behaviour
E.C. Hambly

Construction Materials – Their Nature and Behaviour
Edited by J.M. Illston

Continuous and Integral Bridges
Edited B. Pritchard

Dyanmics of Pavement Structures
G. Martincek

Design and Construction of Interlocking Concrete Block Pavements
B. Shackel

Durability of Building Materials and Components
Edited by J.C. Sjöström

Engineering the Channel Tunnel
Edited by C.J. Kirkland

Highway and Traffic Engineering in Developing Countries
Edited by B. Thagesen

Prevention of Reflective Cracking in Pavements
Edited by J.G. Cabrera

Transport, the Environment and Sustainable Development
Edited by D. Banister and K. Button

Asphalt Surfacings

A Guide to Asphalt Surfacings and Treatments Used for the Surface Course of Road Pavements

Edited by

Cliff Nicholls

Transport Research Laboratory

E & FN SPON
An Imprint of Routledge
London and New York

First published 1998
by E & FN Spon, an imprint of Routledge
11 New Fetter Lane, London EC4P 4EE

Simultaneously published in the USA and Canada
by Routledge
29 West 35th Street, New York, NY 10001

© 1998 E & FN Spon

Typeset in 10/12pt Palatino
by Saxon Graphics Ltd, Derby
Printed and bound in Great Britain
by Cambridge University Press

All rights reserved. No part of this book may be reprinted or reproduced or utilized in any form or by any electronic, mechanical, or other means, now known or hereafter invented, including photocopying and recording, or in any information storage or retrieval system, without permission in writing from the publishers.

British Library Cataloguing in Publication Data
A catalogue record for this book is available from the British library

ISBN 0 419 23110 2

Other contributors

Peter Bellin (State Highway Agency of Lower Saxony)

Charles Catt (Consultant)

Stephen Child (Surrey County Council)

Ian Dussek (Wells (Trinidad Lake Asphalt) Limited)

Terry Fabb (Refined Bitumen Association)

Peter Green (BP Research)

Bill Heather (Associated Asphalt)

Jukka Laitinen (Associated Asphalt)

David Laws (W S Atkins East Anglia)

Colin Loveday (Tarmac Quarry Products Limited)

Steven St John (Colas Limited)

Tony Stock (Consultant)

Alan Woodside (University of Ulster)

Contents

Preface   xvii

1 Introduction   1
   1.1   Historical developments   1
   1.2   Terminology   4
   1.3   Current materials and classifications   5
   1.4   Component materials   7
   1.5   Developments   8
       1.5.1   Performance specifications   8
       1.5.2   Comité Européen de Normalisation   10
           1.5.2.1   CEN TC 154, Aggregates   10
           1.5.2.2   CEN TC 19/SC 1, Bitumen   11
           1.5.2.3   CEN TC 317, Tar   12
           1.5.2.4   CEN TC 227, Mixed materials   12
       1.5.3   Technical approval schemes   13
       1.5.4   Strategic Highway Research Program   14
   1.6   References   16

2 Aggregates and fillers   18
   2.1   Rock type classifications   18
   2.2   Properties and methods of test   21
       2.2.1   Dimensions   21
           2.2.1.1   Size   21
           2.2.1.2   Dust content   22
           2.2.1.3   Shape   23
       2.2.2   Relative density, water absorption and porosity   25
       2.2.3   Strength   26
           2.2.3.1   Resistance to impact   26
           2.2.3.2   Resistance to crushing   26
       2.2.4   Durability   27
           2.2.4.1   Resistance to soaking   27
           2.2.4.2   Soundness   27
           2.2.4.3   Resistance to wear   29
           2.2.4.4   Resistance to polishing   30
       2.2.5   Adhesion of bitumen   31

	2.3	Performance and specification		32
		2.3.1	Clean, hard and durable	32
		2.3.2	Cleanliness	33
		2.3.3	Hardness	35
		2.3.4	Durability	37
	2.4	Sources		39
		2.4.1	UK sources	39
		2.4.2	Other European sources	39
	2.5	Filler		40
	2.6	Recycling waste materials		40
		2.6.1	Material types	40
		2.6.2	Factors affecting usage	42
			2.6.2.1 Environment	42
			2.6.2.2 Engineering	42
			2.6.2.3 Economic	43
		2.6.3	Expected development	43
	2.7	Conclusion		43
	2.8	References		44
3	**Binders**			**47**
	3.1	Introduction		47
	3.2	Tar		47
	3.3	Natural asphalts		49
	3.4	Bitumen		50
		3.4.1	Penetration grade bitumen	50
			3.4.1.1 Production	50
			3.4.1.2 Chemical nature	53
			3.4.1.3 Physical properties	56
			3.4.1.4 Bitumen specifications	62
		3.4.2	Bitumen emulsion	63
		3.4.3	Cutback bitumen	67
	3.5	Binder modifiers		68
		3.5.1	Generic types	68
		3.5.2	Polymer modifiers	68
		3.5.3	Chemical modifiers	74
		3.5.4	Adhesion agents	76
		3.5.5	Fibre additives	77
	3.6	References		78
4	**Required characteristics**			**80**
	4.1	Applications		80
	4.2	Monitoring of existing network		81
	4.3	Selection criteria		83
		4.3.1	Structural	83
		4.3.2	Surface properties	84

		4.3.3	Durability	92
		4.3.4	Restrictions imposed by the site	94
	4.4	Uses		95
		4.4.1	New works	95
		4.4.2	Maintenance	97
		4.4.3	Replacement	98
	4.5	Whole-life costing		99
	4.6	Quality systems		102
		4.6.1	Quality control	102
		4.6.2	Quality assurance	103
	4.7	References		106
5	**Rolled asphalt surface courses**			108
	5.1	Concept and history		108
		5.1.1	Concept	108
		5.1.2	History	108
	5.2	Properties and applications		109
		5.2.1	Properties	109
		5.2.2	Applications	111
	5.3	Component materials		114
		5.3.1	Aggregates and filler	114
			5.3.1.1 Coarse aggregates	114
			5.3.1.2 Fine aggregate	116
			5.3.1.3 Filler	117
		5.3.2	Binders	118
		5.3.3	Binder modifiers	120
		5.3.4	Mixed material	124
		5.3.5	Test requirements	129
	5.4	Varieties		129
		5.4.1	Rolled asphalt with pre-coated chippings	129
		5.4.2	High stone-content rolled asphalt	132
		5.4.3	0/3 mm and 15/10 rolled asphalts	132
	5.5	Tack coats		132
	5.6	Maintenance		133
	5.7	Economics		134
	5.8	Conclusion		135
	5.9	References		135
6	**Asphalt concrete (and macadam) surface courses**			139
	6.1	Concept and history		139
	6.2	Properties and applications		141
		6.2.1	Asphalt concrete for UK highways and byways	142
			6.2.1.1 14 mm close-graded surface course	143
			6.2.1.2 10 mm close-graded surface course	144
			6.2.1.3 6 mm dense surface course	144

				6.2.1.4	40 mm single-course 200 pen granite coarse and fine	145
		6.2.2	Asphalt concrete airfield surfacing			148
		6.2.3	International asphalt concrete			151
	6.3	Component materials				152
		6.3.1	Coarse aggregates			153
				6.3.1.1	Adequate skid resistance – for either runway or highway	155
				6.3.1.2	Adequate strength	155
				6.3.1.3	Durability	155
				6.3.1.4	Shape	156
				6.3.1.5	Cleanliness	156
		6.3.2	Fine aggregate			157
		6.3.3	Fillers			158
		6.3.4	Hydrated lime			158
		6.3.5	Binders			160
		6.3.6	Binder modifiers			161
				6.3.6.1	Macadam asphalt concrete	161
				6.3.6.2	Asphalt concretes	163
	6.4	Design				168
		6.4.1	Need for design			168
		6.4.2	Marshall design			170
		6.4.3	Hveem method of design			170
				6.4.3.1	Aggregates	171
				6.4.3.2	Estimation of binder content	171
				6.4.3.3	Specimen preparation	171
				6.4.3.4	Tests	171
				6.4.3.5	Interpretation of results	172
		6.4.4	Mixture approval and verification			172
		6.4.5	Voids			173
		6.4.6	Grading			175
	6.5	Production and paving				176
	6.6	Conclusion				181
	6.7	References				182
7	**Porous asphalt surface courses**					185
	7.1	Terminology				185
	7.2	Concept and history				185
	7.3	Properties				186
		7.3.1	General			186
		7.3.2	Aquaplaning and skidding accidents			187
		7.3.3	Tyre noise			191
		7.3.4	Driver comfort and fatigue			194
		7.3.5	Spray			194

	7.3.6	Glare and illumination	196
7.4	Applications		199
	7.4.1	Reasons for using porous asphalt on highways	199
	7.4.2	Hydraulic applications	199
	7.4.3	Suitability for sites	199
	7.4.4	Use in the United Kingdom	200
		7.4.4.1 Highways	200
		7.4.4.2 Airfields	203
	7.4.5	Overseas	203
7.5	Component materials		204
	7.5.1	Aggregates and filler	204
		7.5.1.1 General	204
		7.5.1.2 Coarse aggregate	205
		7.5.1.3 Fine aggregate	205
		7.5.1.4 Filler	205
	7.5.2	Binders	205
		7.5.2.1 Unmodified binders	205
		7.5.2.2 Modified binders	206
		7.5.2.3 Fibre additives	207
7.6	Manufacture, quality control and laying		208
7.7	Tack coats		208
7.8	Varieties		209
	7.8.1	20 mm porous asphalt	209
	7.8.2	10 mm, 14 mm and 16 mm porous asphalts	210
	7.8.3	Very high void mixtures	211
	7.8.4	Two-layer systems	212
7.9	Performance in service		212
7.10	Recycling		213
7.11	Maintenance		213
	7.11.1	Repairs	213
	7.11.2	Winter maintenance	214
		7.11.2.1 Changes in procedures with porous asphalt	214
		7.11.2.2 Differences in thermal characteristics of porous asphalt and conventional dense surfacings	214
7.12	Economics		216
7.13	References		217

8 Mastic asphalt (and gussasphalt) surface course — 219
8.1	Concept and history		219
8.2	Properties and applications		221
	8.2.1	Waterproofing characteristics	221
	8.2.2	Durability	221

xii   Contents

		8.2.3	Cost	222
		8.2.4	Manufacturing and application equipment and skills	222
		8.2.5	Surface texture	223
	8.3	Component materials		223
		8.3.1	Aggregates and filler	223
			8.3.1.1 Mastic asphalt	223
			8.3.1.2 Gussasphalt	224
		8.3.2	Binders	224
			8.3.2.1 Bitumen	224
			8.3.2.2 Trinidad lake asphalt	225
		8.3.3	Binder modifiers	225
			8.3.3.1 Gilsonite	225
			8.3.3.2 Polymers	225
	8.4	Mastic asphalt		226
		8.4.1	Manufacture	226
		8.4.2	Applications	227
			8.4.2.1 Waterproofing	227
			8.4.2.2 Tunnels and concrete bridge decks	227
			8.4.2.3 Steel bridge decks	228
			8.4.2.4 Other paving	230
			8.4.2.5 Roofing mastic asphalt	230
	8.5	Gussasphalt		231
		8.5.1	Manufacture	231
		8.5.2	Application	232
	8.6	References		237
9	**Stone mastic asphalt surface courses**			238
	9.1	Concept and history		238
		9.1.1	General	238
		9.1.2	History and development	238
		9.1.3	Concept of stone mastic asphalt	240
	9.2	Properties		242
		9.2.1	Deformation resistance	242
		9.2.2	Dynamic stiffness	243
		9.2.3	Flexibility and fatigue life	243
		9.2.4	Durability	244
		9.2.5	Toughness	244
		9.2.6	Skid resistance	245
		9.2.7	Noise generation	246
	9.3	Applications		247
	9.4	Component materials		249
		9.4.1	Aggregates and filler	249
			9.4.1.1 Coarse aggregate	249

			9.4.1.2	Fine aggregate	250
			9.4.1.3	Filler	250
		9.4.2	Binders		250
		9.4.3	Stabilizers, additives and binder modifiers		250
			9.4.3.1	Fibres	250
			9.4.3.2	Polymers	251
			9.4.3.3	Special fillers	251
	9.5	Mixture composition			251
	9.6	Practical aspects			253
		9.6.1	Mixing		253
		9.6.2	Laying		255
	9.7	Cost and maintenance considerations			257
	9.8	References			257
10	**Thin surface course materials**				**259**
	10.1	Concept and history			259
		10.1.1	Definition		259
		10.1.2	History		259
		10.1.3	Categories		261
	10.2	Properties and applications			263
		10.2.1	Properties		263
		10.2.2	Applications		263
	10.3	Component materials			264
		10.3.1	Aggregates and filler		264
		10.3.2	Binders		264
		10.3.3	Binder modifiers		265
	10.4	Varieties			266
		10.4.1	Ultra thin hot mixture asphalt layer – paver-laid surface dressing		266
		10.4.2	Very thin surfacing layer – thin polymer-modified asphalt concrete		266
		10.4.3	UK specification for ultra thin and very thin surface layers		267
		10.4.4	Thin stone mastic asphalt		269
		10.4.5	UK specification for stone mastic asphalt		269
		10.4.6	Eire provisional specification for thin bituminous surfacings		270
		10.4.7	Thoughts for specifiers		271
		10.4.8	Surface finish and texture		272
		10.4.9	Skidding resistance		273
	10.5	Tack coats			274
	10.6	Maintenance			274
	10.7	Economics			274
	10.8	References			275

11 Veneer coats — 277
11.1 Surface dressing — 277
11.1.1 Concept and history — 277
- 11.1.1.1 History — 277
- 11.1.1.2 Basic principles — 278
- 11.1.1.3 Types of specification — 282

11.1.2 Properties and applications — 283
- 11.1.2.1 Preparation — 283
- 11.1.2.2 Types of dressing — 283
- 11.1.2.3 Surface dressing operations — 287

11.1.3 Component materials — 287
- 11.1.3.1 Chippings — 287
- 11.1.3.2 Binders — 288

11.1.4 Equipment — 293
11.1.5 Traffic control and aftercare — 294
11.1.6 Maintenance — 295
11.1.7 Economics — 295

11.2 Slurry surfacing — 297
11.2.1 Concept and history — 297
11.2.2 Properties and applications — 297
11.2.3 Component materials — 299
11.2.4 Contract types — 301
11.2.5 Maintenance — 301
11.2.6 Economics — 302

11.3 High-friction surfacings — 302
11.3.1 Concept and history — 302
11.3.2 Properties and applications — 304
11.3.3 Component materials — 305
- 11.3.3.1 Binders — 305
- 11.3.3.2 Calcined bauxite — 307

11.3.4 Maintenance — 307
11.3.5 Economics — 307

11.4 References — 307

12 Specialized materials — 309
12.1 Introduction — 309
12.2 Cold-laid materials — 309
12.2.1 Fluxed bitumen macadams — 309
12.2.2 Deferred set macadams — 311
12.2.3 Cold mixed materials — 312
- 12.2.3.1 General — 312
- 12.2.3.2 Mixing — 313
- 12.2.3.3 Storage — 314

12.3 Grouted macadam — 316
12.4 Dense Tar Surfacing — 317

12.5	Delugrip	320
12.6	Bitumen bound aggregate	322
12.7	Re-texturing	325
12.8	Coloured materials	328
12.9	Miscellaneous materials	331
	12.9.1 Vehicle noise test track material	331
	12.9.2 55/6F rolled asphalt surface course	332
	12.9.3 Sports playing surfaces	332
12.10	References	332

13 Recycling materials — 335

13.1	Concept and history	335
	13.1.1 Introduction	335
	13.1.2 Terminology	335
	13.1.3 The development of recycling	336
	13.1.3.1 General	336
	13.1.3.2 Hot in-situ recycling	337
	13.1.3.3 Hot in-plant recycling	337
	13.1.3.4 Cold recycling	338
	13.1.3.5 Cold in-plant recycling	338
	13.1.3.6 Comments on recycling developments	338
13.2	Equipment for recycling	339
	13.2.1 In-plant recycling	339
	13.2.2 In-situ recycling	340
13.3	Recycled asphalt as a component material	343
	13.3.1 Recycled asphalt	343
	13.3.2 Reclaimed binder and rejuvenation	344
13.4	Hot recycled mixtures	347
	13.4.1 Performance requirements	347
	13.4.2 Resistance to deformation	348
	13.4.3 Resistance to cracking	348
	13.4.4 Durability	352
	13.4.4.1 Concept	352
	13.4.4.2 RTFOT study	354
	13.4.4.3 Water immersion tests	355
	13.4.4.4 Stiffness	356
	13.4.4.5 Summary	356
13.5	Cold recycled mixtures	357
13.6	Mixture design	358
	13.6.1 Hot mixtures	358
	13.6.2 Cold mixtures	358
13.7	Implementation of recycling	360
	13.7.1 Introduction	360
	13.7.2 Opportunities for recycling	360
	13.7.3 Economics and energy	362

		13.7.4 Implementation of recycling	362
	13.8	References	365
14	**Summary**		366
	14.1	Selection of type of surfacing material or treatment	366
	14.2	Surface treatments	368
	14.3	Surfacing materials	368
		14.3.1 Regulating ability	370
		14.3.2 Skid-resistance	370
		14.3.3 Texture depth	370
		14.3.4 Permeability	370
		14.3.5 Noise	370
		14.3.6 Spray and glare	371
		14.3.7 Physical constraints	371
	14.4	Combination treatments	371
	14.5	Relative economic benefits	372
	14.6	References	372
Appendix A		Draft UK Specification for Stone Mastic Asphalt Surface Course	373
Appendix B		Draft UK Specification for Thin Surface Course Systems	379
Appendix C		Draft Irish Specification for Hot-Laid Thin Bituminous Surfacings	384
Index			393

Preface

This book has been written as a reference book to explain the various asphalt surface course (or wearing course) materials and surfacing treatments that are currently available to engineers. It is not intended to cover the lower, structural layers but reference is made to these layers where appropriate because the properties of all layers interact in producing the required pavement. The book is intended for practising and aspiring maintenance engineers, contracting engineers and consultants who need a clear reference to help them select the materials and/or treatments that are appropriate for use on specific sites. Once it is decided which of them are appropriate, it should then be easier for the engineer to prepare job specifications and/or tender returns for, or to order or apply, a technically correct solution for specific situations.

The contents are intended to explain both the current established position in the United Kingdom, and (to a lesser extent) other parts of Europe, as well the emerging developments that are taking place. Hence, all the main materials and treatments that can be expected to be used for surfacing for the medium term should be covered. The time that this medium term will actually last depends on what new ideas emerge over the next few years, but the industry is relatively conservative and new ideas usually take several years to emerge.

The authors have been selected as acknowledged authorities on their particular topics. They were selected so as to include some from each part of the 'highway engineering industry' – contractors, material suppliers, highway authorities and research establishments. This should achieve a balance between the various approaches required by the different functions they carry out. However, with the moves towards design, build and maintain forms of contract as well as externalization of many local authority functions, the distinctions in those functions are becoming more blurred and all the authors have had to gain an appreciation of the requirements of other professionals in the industry.

The following brief biography of each of the contributors (in alphabetic order) gives an idea of the combined breadth of their experience:

PETER BELLIN, BD, Dipl-Ing, MSc(CE)

Peter is chief engineer for highway pavement engineering, including materials and testing, at the Niedersächsisches Landesamt für Strassenbau (State Highway Agency of Lower Saxony) in Hanover, Germany. In this post, Peter is responsible for many aspects of road pavements, but he has particular interest in specifications, quality assurance, testing asphalt and the construction and maintenance of asphalt pavements.

Peter graduated with a diploma in Civil Engineering, majoring in highway engineering, from the Technical University of Hanover, Germany, in 1961 before working under a training programme with federal and provincial authorities to gain his civil service licence in 1965, a prerequisite for a career as a chartered engineer in the road administration. He then obtained a Master of Science degree from the University of Wisconsin in the United States of America before starting his professional life in the road authority in 1967 as deputy chief engineer of the district office for the reconstruction of the Autobahn near Hamburg. Between 1988 and 1990, he was attached to the Strategic Highway Research Program (SHRP) in Washington, DC, where he fostered the introduction of the stone mastic asphalt technology into the LTPP-Specific Pavement Studies. From then until 1993 he was a member of the SHRP asphalt advisory committee.

Peter is a member of the German Highway and Traffic Research Association and has been active in several working groups concerned with the specification and quality assurance of asphalt; asphalt pavements; aggregates; and waterproofing systems for bridge decks. He has been a member of the CEN TC 227/WG 1, Bituminous Mixtures, since 1990.

CHARLES CATT, MIHT, FIAT

Charles set up as a private consultant on pavement material since taking early retirement as materials engineer for Warwickshire County Council. He has worked on several research projects, mainly on behalf of the Transport Research Laboratory (including research into an alternative 'quiet' material which is durable, analysis of road survey detail and preparation of footway maintenance and design manuals) and of Acland Investments Ltd as well as advising clients on materials problems.

Charles studied at Birmingham University and then joined Tarmac Civil Engineering in 1961 as a site engineer on major road schemes before transferring into the laboratory. In 1967 he moved to Warwickshire County Council as a laboratory engineer, where he remained until his retirement in 1994, rising to materials engineer. As materials engineer, he was responsible for running the materials testing laboratory (which gained NAMAS accreditation in about 1990), for providing advice on various aspects of materials, and for investigating problems with, and failures of, materials. He also led the Mercia Quality Assurance scheme set up by Warwickshire County Council with Staffordshire, Hereford & Worcester, Shropshire and West Midlands Councils.

While with Warwickshire, Charles represented the County Surveyors' Society on the British Standards Institution committee B/510/2 on surface dressing and the Institute of Petroleum panel ST-E-6 on modified bitumen tests as well as several ad hoc task groups. He was Vice-Chairman of the CSS south-west materials group between 1980 and 1994 and the author of the Highways Authorities Standard Tender Document on surface dressing, HASD 4 for several years.

STEPHEN CHILD, BSc, DIP HC&M, MICE, MIHT, MIAT

Stephen is the Group Engineer responsible for managing the Materials and Construction Management Group within the Engineering Consultancy Division of Surrey County Council. The work of his group covers a wide range of materials across routine maintenance, design and construction as well as a site supervision service. He is also the Delegated Engineer responsible for the training of graduates and technicians.

Stephen trained as a Chartered Engineer with Surrey County Council, after which he was appointed Geotechnical and Materials Engineer in the Materials Group. He spent a short time in the private sector with Howard Humphreys as a Principal Engineer in Highway Maintenance. In November 1985, he was awarded the IHT Diploma in Highway Maintenance and Construction, winning the Bomag Prize.

Stephen is involved with the County Surveyors' Society as Secretary to the South East Soils and Materials Engineers' Group, Chairman of the Data Collection Working Party, Skidding Resistance Sub-Group and a member of several other working parties. He represents the County Surveyors' Society on several British Standard Committees including B/502, Aggregates; B/502/2, Aggregates for Concrete; B/502/6, Testing of Aggregates; and B/510/5, Surface Characters. In addition, he is vice-chairman of the South East Branch of the Institute of Asphalt Technology.

IAN DUSSEK, FIHT, FIAT

Ian is a director of Wells (Trinidad Lake Asphalt) Ltd and its associated companies, now part of Associated Aphalt Ltd. His principal role is the marketing of Trinidad lake asphalt and providing technical support for its use, both in the United Kingdom and throughout much of the world. As such, he has been involved with many prestigious highway, bridge and airport projects. Other involvements include plant sales and mastic asphalt contracting in the Caribbean.

On leaving school at St Lawrence College, Ian joined the family business at the laboratory of Dussek Bitumen & Taroleum Ltd. In 1960, he was appointed assistant to the technical director of Previté & Company Ltd and was based at the Fulham laboratories of Limmer & Trinidad Lake Asphalt Company Ltd. In 1962, Ian spent six months in Rhodesia in charge of mastic production and technology, as well as assisting in the development of epoxy resin components for building and civil engineering work. At the end of this time, Ian returned to the family business which became part of the IBE group, including Lion Emulsions and Colas Products, although the group was purchased by Shell International Petroleum Company in 1969. Following two years in the contracting industry, specializing in concrete treatment and texturing, he rejoined the Trinidad lake asphalt industry in 1975.

Ian has taken many posts, including President, of both the Institute of Asphalt Technology and the Institution of Highways & Transportation, as well as of the Worshipful Company of Paviors. He is a member of several British Standards Institution and allied committees, including the Structures Maintenance Committee of the US Transportation Research Board.

TERRY FABB, BSc, FIAT

Terry is the Technical Director of the Refined Bitumen Association (RBA) in which rôle he chairs the RBA's Technical Committee, acts as technical spokesman for the UK bitumen industry and represents the RBA in the European Bitumen Association (Eurobitume).

Terry had a long career at the BP Group Research Centre at Sunbury-on-Thames. His work there covered the product development of petrochemicals, lubricants and fuels and, for the last twenty years of that career, bitumen. During the last eight years of his work for BP, he was the Project Leader responsible for the BP Group's research, development and technical service on bitumen and asphalt. His own R&D efforts included work on low-temperature cracking (for which he received the annual 'Emmons' award of the Association of Asphalt Paving Technologists) and high-temperature rutting (for which he received the annual Argent award of the Institute of Asphalt Technology jointly with Vince Hayes). He also received the Argent award in 1993 for a paper on the case for using porous asphalt in the United Kingdom.

Terry has been secretary of the CEN Working Group responsible for the preparation of harmonized European specifications and test methods for paving grade bitumens since its formation in 1990, and is a member of the corresponding CEN Working Group for asphalts. He is also a member of several British Standards Institution and Institute of Petroleum committees.

PETER GREEN, BSc

Peter is Manager of the Bitumen Technical Unit at the BP Research and Engineering Centre at Sunbury, Middlesex. He has been involved in, and is now responsible for, many projects in support of BP's bituminous activities worldwide. These projects include the development of processing routes for bitumen; bitumen quality; measurement and interpretation of bitumen rheology and its relation to performance; and the development, testing and production of speciality products including modified binders and emulsions.

Peter graduated in 1976 from Kingston University with a chemistry degree and joined BP's Group Research Centre to work on the development of a range of oil products for various applications. In 1984, he transferred to the Bitumen Technical Unit to work on the development of bituminous products. Peter has written and contributed to several papers on bitumen rheology and asphalt properties and sits on the Institute of Petroleum Panels ST-E, Bitumen Tests; ST-E-1/2, Bitumen Rheology; and ST-E-6, Modified Bitumen.

BILL HEATHER, BA, FIHT, MIAT

Bill is the Technical Director for Associated Asphalt Ltd. His responsibilities include the introduction of new special products and processes, which have included Safepave thin surfacing and Glasgrid reinforcement mesh.

Bill started his career as a research assistant for British Drug Houses before moving to the research laboratory of the British Transport Commission. He joined Constable Hart & Company Ltd, where he worked in black-top contracting for eight years before moving to ARC as an area manager. In 1975, he joined Associated Asphalt as an area manager and then regional manager before taking on increasing responsibility for technical matters. He joined the Board of CAMAS Associated Asphalt in the beginning of 1994, which became Associated Asphalt Ltd in 1997.

Bill is a long-standing member of the ACMA Technical Panel; BSI Committee B 510/1, Bituminous Mixtures; CEN TC 227/WG5, Surface Characteristics; and a number of other British Standards and specialist committees concerning the roads industry. Recently, he has joined the British Board of Agrément Highway Authorities Products Approval Scheme (HAPAS) Specialist Group SG3, Thin Surfacing.

JUKKA LAITINEN, MIAT

Jukka is Technical Manager of Associated Asphalt Ltd, where he is responsible for the day-to-day management of Associated Asphalt Pavement Technology. AA Pavement Technology provides comprehensive technical support to the company, including the latest technical facilities for research and development and a fully independent NAMAS laboratory providing extensive services to in-house and external clients, which include consultants, contractors and highway authorities in the United Kingdom and overseas on both highway and airfield projects.

Jukka joined the Central Laboratory of Limmer and Trinidad in 1968, gaining a very broad based knowledge of pavement materials and their testing, including bridge and airfield surfacings. In 1972, he joined Associated Asphalt (subsequently CAMAS Associated Asphalt, now Associated Asphalt Ltd), where he has primarily worked on technical aspects with extensive experience of production management of both static and mobile units. He was closely involved with the development of various products and processes during this period, including the day-to-day production and application of all types of materials. He has a wide range of experience of both highways and airfield contracts in the United Kingdom, Caribbean, Middle and Far East due to his company's involvement in these areas.

Jukka is currently Chairman of the Institute of Asphalt Technology South Eastern Branch and is a member of the IAT Education Committee, which he represents on the Laboratory and Associated Technical Standards Initiative. He is part of a BACMI/LTA working group on the specification for asphalt sports surfaces. Additionally, he has recently joined the British Board of Agrément Highway Authorities Products Approval Scheme (HAPAS) Specialist Group SG2, Overbanding.

DAVID LAWS, CEng, MICE, MIHT, MIAT

David is the Head of Materials Division at Stanton House in Huntingdon, which is part of the East Anglia regional operation of W S Atkins Consultants Plc. The Laboratory undertakes a wide range of testing and supervisory services with full materials consultancy back-up.

David trained as an Engineering Learner with Hertfordshire County Council and completed his training with Ross and Cromarty County Council in their west coast out-office in Dornie. He then worked for Huntingdon and Peterborough County Council, becoming Design Team Leader before being appointed Chief Resident Engineer in Peterborough with Cambridgeshire County Council with the 1974 Local Government reorganization. In 1978 he was appointed Soils and Materials Engineer, became Departmental Quality Control Manager in 1991 and Divisional Manager, Laboratory, in 1992 as part of the Cambridgeshire County Council Engineering Consultancy which, in April 1995, transferred to W S Atkins under a hosting agreement.

David is a member of BSI Committees B/507, B/509/2 and B/510/1, has been an original member of the CSS/TRL Haunching Working Group and is now on the steering group of the Linear Quarry Project. He has also been a member of the HAUC Specifications Working Party since its inception and is one of the Authority nominees as a Certification Panel member for PCSMs. He has been involved with the new developments in thin surfacings since some of the first trials in the United Kingdom during 1991 were on the A1 and A47 in Cambridgeshire; he is currently involved with the monitoring and reporting on the trials covering Safepave, UL-M and stone mastic asphalt on the A1 and A10 together with contractual operation of the performance-based specification of rolled asphalt which involves wheel-track testing of rolled asphalt cores at 60 °C.

COLIN LOVEDAY, FIHT

Colin is Director of Tarmac Heavy Building Materials Ltd, where he has recently completed 25 years working mainly in the fields of asphalt, aggregates and road construction and maintenance.

Colin is Chairman of the Asphalt Technical Committee of the British Aggregates & Construction Materials Industries and a member of the Technical Committee of the European Asphalt Pavement Association, chairing their Task Group on Quality and Certification. He is the chairman or a member of several CEN sub-committees and task groups within TC 154, aggregates, and TC 227, road materials, and of the BSI shadow committees.

CLIFF NICHOLLS, MPhil, BSc(Eng), ACGI, DIC, CEng, MICE, MIStructE, MIAT, MIHT

Cliff is a Project Manager in the Civil Engineering Resource Centre at the Transport Research Laboratory. He is mainly involved in asphalt surface course materials, from rolled asphalt to surface dressing and including porous asphalt, high-friction surfacings and thin surface course materials, but has also carried out research on associated materials such as road markings and trench re-instatement to the HAUC criteria.

Cliff graduated in Civil Engineering from Imperial College, University of London, in 1972 to join Rendel Palmer & Tritton, where he worked on steel bridge design and on site at the Thames Barrier. He moved to the Property Services Agency in 1976 where, after a short period in the design office, he was sent back to IC to get his second degree for research into reliability analysis. In 1980, he moved to the Department of the Environment to act as technical secretary to the various UK Eurocode Technical Panels before returning to research in 1983 at the Civil Engineering Laboratory and then the Concrete Laboratory of the Building Research Establishment. He moved to the Transport and Road Research Laboratory (now the Transport Research Laboratory) in 1986 and took up his present position, and with it an interest in asphalt materials, in 1988.

Cliff represents TRL on several British Standards Institution Committees (B 510/1, Bituminous mixture; B 510/1 WG2, Sampling and Testing of Bituminous Mixtures; B 510/2, Surface Dressing and Slurry Surfacings; and B 510/19, Bitumen), Institute of Petroleum Committees and Panels (ST-E, Bitumen Tests; ST-E-1/2, Bitumen Rheology; and ST-E-6, Modified Bitumen) and British Board of Agrément Highway Authorities Products Approval Scheme Specialist Group (SG1, High-Friction Systems; and SG3, Thin Surfacing) as well as being a BSI delegate on CEN TC227/WG1/TG2 on Test Methods for Bituminous Mixtures. Cliff is also a committee member of the Construction Materials Group of the Society of Chemical Industry.

STEVEN St JOHN, FIHT, MiMgt, MIAT

Steven is Quality Manager for Colas Ltd, in which post he is responsible for ensuring the maintenance and enhancement of Colas' registered quality assurance systems. In addition, he is responsible for supporting business improvement activity within the organization, as well as with its suppliers and customers, and he is also closely involved in international business development.

Steven's background is in specialist contracting, most of it related to highways. His main work experience has been in highway surface treatments, including surface dressing, slurry surfacing and retread. He has been closely involved in surface dressing development with several different companies, working in France, Austria and Scandinavia as well as the United Kingdom. After three years with South Western Tar Distilleries, he joined Colas Roads in 1986. Since then, he has led the team which was responsible for surface dressing the M25 in Kent in 1990, the first motorway surface dressing carried out at night. In 1991, he gained the Professional Diploma in Management from the Open University, which included a dissertation on quality assurance and total quality. He became Quality Manager for Colas Ltd in 1994 and International Business Manager in 1996.

Steven has been a member of the Technical Committee of the Road Surface Dressing Association since 1981. As well as writing some of the RSDA publications, he is a frequent lecturer on their training courses and has been a member of joint RSDA/County Surveyors' Society working parties on traffic signing and on quality assurance. He is a past Chairman of the South East Branch of the Institution of Highways and Transportation.

Dr TONY STOCK

Tony is currently working as a consultant with Stock Tynan Associates, specializing in Pavement Engineering with a global practice, which supports both industry and clients. Dr Stock is also a Visiting Professor at the University of Westminster.

Tony's career in Pavement Engineering started at Surrey University with work on concrete and cement-bound materials. This interest was developed at Nottingham University, developing the first computer program for the design of flexible pavements; the project was sponsored by the Asphalt and Coated Macadam Association and led to the award of a doctorate. Tony then joined the Civil Engineering Department at Dundee University, where he continued with research on Pavement Engineering and on recycling. As a lecturer, he was responsible for research into recycling bituminous materials with support from the Research Council and from the Scottish Division of Tarmac; this work laid some of the technical foundations for the implementation of recycling in the British Isles.

Tony has also been a Visiting Professor at Texas A&M University and has spent a significant period of time working with the Corps of Engineers in their Waterways Experiment Station in Vicksburg, Mississippi. In 1988, Tony joined British Petroleum Research at their research centre in Sunbury-on-Thames as a research associate and leader of the Bitumen R&D team. At this time, he was also a member of the Industrial Advisory Group to the Strategic Highways Research Program. Since leaving BP, Tony was Professor and Head of Civil Engineering at Sheffield Hallam University before setting up his consultancy practice, to which he gives his full attention.

Prof. ALAN WOODSIDE, MPhil, CEng, FICE, FIE Ireland, FIAT, FIQ, FIHT, MCIWEM

Alan is the Director of the Highway Engineering Research Centre in the School of the Built Environment at the University of Ulster as well as holding a personal chair in Highway Engineering since 1994. He leads a staff of researchers and technicians providing a research and consultancy service to the profession as well as providing numerous short courses, both in Northern Ireland and abroad. Alan supervises 15 major research projects which are being sponsored by government and industry from the United Kingdom and abroad.

Alan commenced his training as an Engineering Assistant in 1965 with Larne Borough Council, progressing to Assistant Engineer in 1970. In 1972 he moved to the Joint Engineering Department for Ballymena Borough and Rural District Councils before moving to academia in 1973 when he became a Lecturer in Civil Engineering at the Northern Ireland Polytechnic. Teaching at sub-degree, degree and postgraduate level on many courses, he was promoted through Senior Lecturer to Principal Lecturer by 1982. He was then appointed Senior Lecturer at the University of Ulster and progressed to Reader in Highway Engineering in 1990.

Alan is the Chairman of the CITB (NI) Black Top Committee and the British Standards Committee TG9 on the Durability of Aggregates as well as being a representative on the Comité Européen de Normalisation committee for aggregates. He has also been a Board Member of the Association of Municipal Engineers, a member of the Association of Municipal Engineers Northern Ireland Committee, a member of the Institution of Civil Engineers (Northern Ireland Committee) and a member of the Institution of Highway Engineers (Northern Ireland Committee).

J.C. Nicholls
Editor
3 March 1997

CHAPTER 1

Introduction

J.C. Nicholls, Transport Research Laboratory

1.1 HISTORICAL DEVELOPMENTS

Before discussing the material types currently available and used for surfacing asphalt, or 'blacktop', roads, it is as well to have an appreciation of the historical development of road construction. This is the development of the pavement structure and not the road network.

The first roads were probably developed from animal tracks, where the only construction was route markers to avoid marshes and other inhospitable features (Hindley, 1971). These tended to hold to high ground, such as on the Downs in the United Kingdom, to allow the traveller clear vision and, hence, safety. As human beings became more settled, genuine construction began to be incorporated by clearing the route. Removed stones were often used to build lines of cairns, or even low walls, which served to mark the road. In some parts of the world, the construction even went as far as to include artificially levelling the road, the building of embankments and the digging of ditches.

The earliest paved roads tended to have been block paving in urban areas (which were very limited at that time) and 'corduroy', or log, roads in soft ground conditions. Block paving appears to have started in the Middle East, with bitumen being used for brick and stone bonding as early as 2000 BC, including for the processional roads in Babylon. A typical construction of a corduroy road had a bed of loose branches above which was laid timbers about 100 mm to 150 mm thick parallel to the direction of traffic and finally cross-pieces or sleepers about 75 mm thick. The earliest known examples of corduroy roads in Britain were in Somerset and date back to 4000 BC, in which the surface was obtained by splitting the length of the logs. However, constructed roads are only really necessary when society is based on cities, so fully constructed roads did not really emerge until the ancient empires arose, in particular the Roman Empire.

Roman roads were not all constructed in precisely the same way, the method employed depending on both the local conditions and importance of the route. For the major routes, the ground was usually levelled and aggers, or drainage ditches, dug along both sides of where the road was to be built. Pavement constructions used included:

- placing a foundation of heavy stones over which was laid broken bricks or pottery; and
- excavating the road to a depth of a metre before being filled with loose aggregate and then finished with close-fitting block paving.

When crossing marshy ground, they first laid logs which they covered with rushes to separate the oak from the damaging effect of the lime used as a cement mortar in the road construction. A typical construction is illustrated in Figure 1.1, a cross-section of the Appian Way.

Following the fall of the Roman Empire, the importance of cities diminished and, hence, the need for roads reduced, as well as the loss of an organization capable of constructing and maintaining them. Therefore, there was little true road building during the Middle Ages until the Industrial Revolution and the general movement away from an agrarian economy. In the United Kingdom, the revival of road construction occurred with the advent of toll roads; the leading figures in this era were Thomas Telford (1757–1834) and John Loudon Macadam (1756–1836).

The construction method developed by Telford had a layer of uniformly large stones as the foundation, overlaid by smaller graded aggregate; all the aggregate was carefully washed and sieved prior to use. Macadam also used carefully graded aggregate, but omitted the base layer of large stones. The pavement surface of roads by both were moderately cambered with the gradings being selected so that the surfacings would be sealed with the dust produced by heavy, slow-moving horse-drawn wagons filling any exposed gaps.

The overall structure of Telford's pavements was relatively thick and hence very strong, while that of Macadam's pavements was cheaper, but less robust. The need for the additional strength was widely argued at the time, but both types were better engineered, and hence significantly stronger and more robust, than the typical road pavements previously. The spread of such all-weather roads from being restricted to major conurbations extended the potential for mobility of both people and goods.

While the principal unit of power for vehicles was the horse, the main interest in surface characteristics was in the smoothness of the surface on which the wheels rolled and, to a lesser extent, the grip that the horse could obtain. However, while horse-drawn traffic caused any detritus to help seal the road, the higher speeds that arrived with the

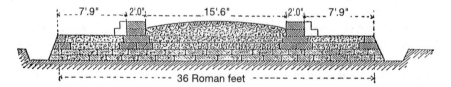

Fig. 1.1 Cross-section of a Roman road (the Appian Way)

automobile meant that these finer particles tended to rise, resulting in great clouds of dust. Initially, measures to reduce the dust included the use of water-bound 'macadams', but the most successful measure was to bind it with tar.

The idea of pre-mixing the aggregate skeleton, or 'macadam', with tar binder before it was laid, so producing a tar-bound macadam, or 'tarmacadam' (section 6.2.1) was first tried in the 1830s before the era of cars. Initially, there was great rivalry between woodblocks, which were quiet, and tarmacadam or asphalt, which were durable and sanitary. However, tarmacadam developed the predominate place in the early twentieth century such that all asphalt is 'tarmac' to the general public. This term has become increasingly inappropriate because bitumen has replaced tar as the most common binder for asphalt paving materials. This change occurred in the United Kingdom because:

- town gas, the main source of coal-tar being a by-product in its manufacture, was replaced by North Sea gas in the 1970s;
- there was increasing concern about the carcinogenic effects of the polycyclic aromatic compounds found in tars; and
- the viscosity of bitumen is less susceptible to changes in temperature than that of tar, resulting in less seasonal variation in performance.

Other asphalt materials were developed which gave impermeable surfacings that would not produce dust as well as resist permanent deformation. The availability of Trinidad lake asphalt (section 3.3) resulted in its being used to produce an asphalt, or mastic, mortar (Chapter 8), generally for use on city streets. The addition of a proportion of larger sized aggregate to bulk up the mortar resulted in the concept of rolled asphalt (Chapter 5), with the predominant binder changing through pitch bitumen to petroleum bitumen. Another method of binding the aggregate that was developed was to spray coal-tar onto the surface, together with an application of some relatively fine aggregate to soak up excessive binder and give grip. This developed into surface dressing (otherwise known as chip seal in the United States of America and some other countries), which is currently used as a maintenance technique for bound, as well as unbound, pavements (section 11.1).

In recent years, communications have fostered a greater interchange of ideas between countries, as well as the increase in the traffic levels imposing greater stresses on pavements, so that there is perceived benefit in developing further material types, and then 'exporting' them. These include stone mastic asphalt (Chapter 9) and the thin surface course systems (Chapter 10). Further developments have occurred with pressures to reduce the environmental impact of roads in terms of less traffic noise and/or spray, which can be achieved using materials such as porous asphalt (Chapter 7).

Some 'imports' to the United Kingdom have had to be modified slightly because of the emphasis placed on the safety of roads by the UK Department of Transport (now the Department of the Environment, Traffic and the Regions). The UK skidding policy has helped to make the United Kingdom among the safest countries in the world in terms of the number of accidents and fatalities per kilometre driven. Despite only the main roads 'belonging' to the Department (or Scottish Office, Welsh Office or Department of the Environment for Northern Ireland in those countries), the other highway authorities in the United Kingdom have tended to follow the lead of these Departments. The main roads belonging to the Department and operated by the Highway Agency, or to the other Government Departments, are the trunk roads; 2537 km of motorways and 8088 km of all-purpose roads which form the strategic network. The remaining roads 'belong' to the relevant local authority, with local authority roads representing 96% of the length of roads in the United Kingdom, but carrying only 69% of the total, and 46% of the commercial, road traffic.

There are, therefore, many different bodies responsible for the public roads in the United Kingdom. Similarly, there are different methods of managing the road network in Europe and elsewhere. Different bodies have different ideas as to what characteristics are required on their roads (Chapter 4). Harmonization through Europe may reduce the variation, but the increased number of different surfacing materials and treatments will allow clients to be more precise when specifying what they require.

1.2 TERMINOLOGY

The development of road pavements (where pavement is the technical term for the road structure and not the commonly used term specifically for footways at the edge of a carriageway) has occurred in many countries with slightly different biases, and hence the terminology used. Hence, the concept that the United Kingdom and United States of America are divided by a common language is possibly nowhere better demonstrated than in the asphalt industry. The UK industry has grown up as two separate industries – macadam and (hot rolled) asphalt – which have since joined up so that there was no common name for mixed materials. To cover all hot mixed materials, the phrases 'macadams and asphalts', 'bituminous materials' or 'blacktop' were used, where 'bituminous' covered tar-bound materials as well as bitumen-bound ones. In the United States of America, the industry was closer to the concrete industry, and the terminology for mixed materials is asphalt (or asphaltic) concrete. Therefore, Portland cement concrete has to be stated in full to avoid confusion with asphalt concrete. Also, bitumen is referred to as asphalt (or asphaltic) cement, so that there is also potential confusion when using the term asphalt on its own.

These differences were confused further by the terms used in 'European' English. However, in borrowing from both traditions, they seem to have produced something better than both sources. They use bitumen (and bituminous) for the hydrocarbon binder and the term asphalt to cover all mixed materials. The European compromise is important not only because it clarifies some situations but also because it is the terminology that is being adopted in CEN standards (section 1.5.2), which will be applied widely across Europe. Therefore, the CEN terminology will generally be used in this book, although often with other names also being given for information.

With all the mixed materials being termed asphalt, several UK terms will be lost. Macadams, widely used on secondary roads, will in future be included under the term asphalt (or asphaltic) concrete; this term was previously considered to refer only to more exactly design materials, generally only used in the United Kingdom on airfields under the title Marshall asphalt. The more open macadam, pervious macadam, has already been re-titled as porous asphalt although the term friction course is still used for the thinner version used on airfields.

Surface dressing, the UK term, is to be used rather than the US term chip seal. The former term arose from the process being used to restore the properties of bound surfaces, while the latter arose from its use to bind the top surface of unbound aggregate. Hence, the former term is more appropriate in the relatively heavily populated European countries with their networks of bound roads.

The layers in a road pavement are also being harmonized. The top running surface, previously known as the wearing course, is now to be called the surface course, and the layer below it, previously the base course in the United Kingdom, is now to be the binder course. The lower, structural asphalt layers are now to be known as the base course rather than roadbase as previously. The base course should not be confused with the previous term for binder course.

1.3 CURRENT MATERIALS AND CLASSIFICATIONS

The traditional surface course materials used in the United Kingdom are rolled asphalt (Chapter 5) and coated macadam (Chapter 6). Both materials are mixtures of the same component materials (generally coarse and fine aggregates, filler and bitumen), but they act in very different ways. Coated macadams rely on aggregate interlock to withstand traffic loading, with the binder just holding the aggregate in place. Rolled asphalt consists of a mortar consisting of binder, filler and fine aggregate in which the coarse aggregate 'floats'. As such, the properties of rolled asphalts are more sensitive to the properties of the binder than a macadam, but the presence of mortar allows chippings, with the specific properties, to be inserted in the surfacing.

In many continental countries, and also in the United States, asphalt concrete (Chapter 6) is very widely used. Asphalt concrete is similar to coated macadam (classified as a type of asphalt concrete by CEN), but the mixture is more designed. The aggregate is designed to have a continuous particle size distribution (PSD) in order that a dense compacted layer should be achieved. The 'ideal' PSD without binder to achieve the maximum compacted density is given by Equation 1.1 (Fuller and Thompson, 1907).

$$P = 100x\left(\frac{d}{D}\right)^{0.5} \qquad 1.1$$

where D = the largest particle used; and
P = total percentage passing sieve size d

The derived curve is often referred to as the 'Fuller' curve, an example of which is shown in Figure 1.2. With binder present, the aggregate grading has to be finer, which can be achieved by reducing the index to less than 0.5 (a range of indexes are shown in Figure 1.2). The optimum binder content for that aggregate skeleton is then found using the 'Marshall' test.

Porous asphalt (Chapter 7) is becoming more popular with the greater emphasis on environmental issues. Porous asphalt is designed to have a large voids content, generally about 20%, originally in order to allow the pavement to drain through, rather than over, the surface course. This reduces spray and minimizes the possibility of aquaplaning; subsequently it has been found to also reduce the noise generated from the traffic.

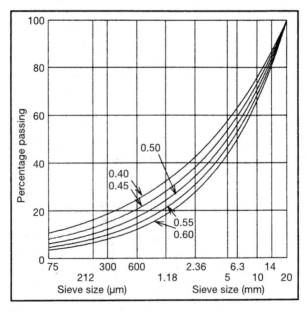

Fig. 1.2 Fuller Curve for a mixture with 20 mm maximum particle size

Gussasphalt (Chapter 8) is a binder-rich mixture that can be machine laid and has been widely used in Germany and surrounding countries. An alternative binder-rich material, mastic asphalt, is commonly used on bridge decks and other localized areas where the high impermeability provided by such binder-rich materials is important to provide the necessary durability.

Stone mastic asphalt (Chapter 9) was developed to be a more abrasion-resistant alternative to gussasphalt, particularly where studded tyres are used. It has been used in continental Europe for some time and is now being introduced into the United Kingdom and America as an alternative to other surfacing materials. Among its advantages are its ability to be laid relatively thinly, as can several other intermediate layer materials (Chapter 10). These materials allow remedial work to include some surface regulation as well as restoring texture depth and skid-resistance without resorting to laying a full surface course.

Veneer coats (Chapter 11) such as surface dressings and slurry surfacings are commonly used surface treatments because they tend to be relatively cheap and so allow the available funds to go further. Surface dressings and slurry surfacings can provide a seal to the surface and the restoration surface characteristics effectively, but cannot provide others that require greater constructional thickness. An alternative surface treatment that is available is re-texturing (Chapter 12). While this treatment can be economic because it does not require the use of additional materials, the properties that can be provided after treatment are totally dependent on the existing surfacing material.

An alternative to the hot mixtures is the use of a cold mixture (Chapter 12). In cold mixtures, the viscosity is achieved during mixing and compaction by the binder being in emulsions rather than being heated; this means that the materials gain their strength when the water is driven off rather than by cooling. This makes these materials more sensitive to the environmental conditions, but does allow material to be kept for some time when used for reinstatement of trenches and other applications requiring small quantities. Cold mixtures are also more environmentally friendly because they do not require the mixture to heated to the temperatures required for hot mixtures.

1.4 COMPONENT MATERIALS

The constituent materials used in all asphalt materials and treatments and aggregate and binder. In addition, many mixtures also include fillers and, for more specialist mixtures, fibres or binder modifiers.

Aggregates (Chapter 2) are split into three size categories, coarse aggregate, fine aggregate and filler. In the United Kingdom, the separation between the coarse and fine categories has traditionally been by the

3.35 mm sieve for macadams and by the 2.36 mm sieve for asphalts while fillers are the very fine aggregates that pass the 75 micron sieve; the upper limit on fine aggregate will be harmonized to 2 mm and that on fillers changed to the 60 micron sieve in European standards (section 1.5.2). Most aggregates are natural, either crushed rock or gravel for coarse aggregate and either crushed rock fines or natural sand for fine aggregate, but artificial aggregates are sometimes used, including slag and, for resin-based high anti-skid surfacing surfaces, calcined bauxite. Commonly used fillers include limestone dust, Portland cement and hydrated lime.

The binder (Chapter 3) is the material that binds the mixture together. Tar used to be used widely (from which we got the term 'tarmac', by which all asphalt mixtures tend to be referred by the general public) but, with the replacement of town gas (of which tar is a by-product) by natural gas as well as for some technical reasons, tar has generally been replaced by bitumen. Bitumen is a hydrocarbon, coming off at the heavy end of the distillation of crude oil. As a natural product, the components in bitumen can vary, with differences in the properties from different areas. Two areas that supply much of the crude used for bitumen in Europe are the Middle East and Venezuela.

Although bitumen has been very successful as a binder, the increase in demands on the pavement surface with ever greater traffic levels on major roads has led to a need to 'improve' the inherent properties of bitumen. Various materials have been used to modify bitumen and tar, some purpose designed and some waste products looking for a use (of which by no means all have actually proved beneficial). The most common categories of binder modifiers (Chapter 3) are natural rubber, synthetic rubber elastomers, plastomers and fibres, even though fibres do not strictly modify the binder, only the mixture.

Rather than using fresh materials, recycling existing asphalt mixtures can be an option (Chapter 13). Recycling can be carried out either on site, rejuvenating the existing material and replacing in roughly the same place, or the old material taken to a coating plant and used as a component in future mixtures. New material is usually mixed with the old in some proportion, particularly in the case of the binder because the old binder will have oxidized and 'aged' in service – possibly being the reason for needing replacement.

1.5 DEVELOPMENTS

1.5.1 Performance specifications

Most countries currently have national standards for pavement materials for use on highways, airfields and other paved areas; the national standards are prepared by the British Standards Institution for the United Kingdom. The primary specifiers then make use of the national standards in their standard specifications, such as the *Specification for Highway Works* (MCHW 1) and the *Highway Authorities Standard Tender Specification* (CSS, 1996) for highways and *Marshall Asphalt for Airfield Pavement Works, Hot Rolled Asphalt and Coated Macadam for Airfield Pavement Works* (DWS 1995) and *Civil Engineering Section 14: Bituminous Materials – Aircraft Pavements* (BAA 1993) for airfields.

Many standards and specifications are based on the concept that the best way to ensure that the client gets what he or she requires is to specify materials and methods that have been found to have acceptable properties in the past. The materials are specified by requiring the use of the components used previously, combined in the same proportions and using the same methods to mix them and apply them. Specifications of this type are generally known as 'recipe' or 'method' specifications.

Recipe specifications can be acceptable when no developments are taking place because there will be information on the performance with particular recipes. However, once changes are occurring, there will not be the historic knowledge of new developments to select what bounds should be placed on component materials, their proportions or the methods of application to ensure satisfactory performance. Even if this knowledge is available, any significant development requires a new specification. Further, it is difficult to compare two materials having different specifications, even if both materials are designed to provide the same properties.

Therefore, there is a general move away from 'recipe' specifications towards performance specifications. The concept is for the client to specify the properties that are required from the treatment, and then allow the contractor to select which material is to be used that will provide those properties. However, the definition of the properties required and the means of measuring them quantitatively is not always possible, particular because one of the properties usually required is 'durability'. The performance has to be able to be assessed at the end of the maintenance period for contractual reasons. Therefore, a hierarchy of methods can be developed to check that the performance is achieved, this hierarchy being as follows:

- performance requirements (where the performance is measured directly, such as initial profile);
- performance-related requirements (where the performance is measured by a test which measures the property on representative samples, such as wheel-tracking to assess rut-resistance); and
- surrogate requirements (where the property is assumed by a surrogate property, such as binder content and/or void content to give assurance of durability).

Care must be taken with the surrogate requirements to avoid the complete 'recipe' being included, as well as the performance requirements, just in case any property has been overlooked.

The change from recipe specifications to performance specifications is occurring in the United States with SUPERPAVE (section 1.5.4) and in many European countries including the United Kingdom. However, the harmonization of European standards (section 1.5.2) has restricted this move because the existing drafts which are being harmonized tend to be recipe specifications; nevertheless, it is anticipated that the next stage of CEN standards will be the development of performance specifications.

1.5.2 Comité Européen de Normalisation

The European Committee for Normalisation, Comité Européen de Normalisation (CEN), have programmes to produce harmonized standards across Europe for many product areas which include asphalt surfacing mixtures and their component materials. This harmonization is important throughout Europe, together with other countries where the standards from a European country are adopted, because, when completed, the standard specifications and associated test methods being developed by the various test committees will be adopted in place of the equivalent national standard by each of the states who subscribe to CEN.

CEN itself will not publish the finished standards itself; the documents will be published separately by each national standardization committee as their national standard. The texts will be identical in most respects (although possibly in different languages) but there may well be significant differences. In developing the CEN standards, while most of the requirements will be mandatory others will be 'optional'. However, optional does not mean that individuals can decide whether to include them, just that the national standardization bodies can decide whether they will be included for use on all occasions in that country.

The relevant CEN committees are TC 227 for mixed materials (including surface dressings, slurry surfacings and (not covered in this book) concrete), TC 154 for aggregates, TC 19/SC 1 for bitumen and TC 317 for tar. Although a great deal of work has been carried out by these committees and their sub-committees and Task Groups, it takes a considerable time to get agreement across Europe with the number of different traditions and languages. The documents for bitumen, which are the most advanced, are not expected to be published until at least 1998. Nevertheless, the changes that are expected to occur to current standards and their implications will be discussed in the relevant chapters, but a list of the documents being prepared by these committees is given below.

1.5.2.1 CEN TC 154, Aggregates

The aggregate gradings that are to be used will be defined by a basic set of sieve sizes plus additional sieve sizes from **either** set 1 or set 2 but with an overriding requirement that the ratio of sieve sizes cannot be less than

1.4. The choice of the set to be used is made by the national standardization body, with the set of sieve sizes that will be used to specify mixed materials in the United Kingdom will be:

- 40 mm
- 31.5 mm
- 20 mm
- 14 mm
- 10 mm
- 6 mm
- 4 mm
- 2 mm
- 1 mm
- 500 μm
- 250 μm
- 125 μm
- 63 μm

There are few changes from the sieve sizes currently used. However, when a particle size distribution is carried out, the following size sieves are required to be included (CEN, 1996) as well as the sieves of the designated product size:

- 125 mm
- 63 mm
- 31.5 mm
- 16 mm
- 8 mm
- 4 mm
- 2 mm
- 1 mm
- 500 μm
- 250 μm
- 125 μm
- 63 μm

The aggregate test methods and procedures that are being developed are:

- methods of sampling
- reduction of laboratory samples
- description and petrography
- common equipment and calibration
- repeatability and reproducibility
- quality control procedures
- particle size
- particle shape
- crushed and broken surfaces
- texture/shape-flow coefficient
- shell content
- fines, sand equivalent method
- methylene-blue test
- grading of fillers
- Micro-Deval resistance to wear
- resistance to fragmentation
- loose bulk density and voids
- compactability
- water content
- particle density
- polished stone value
- studded tyre test
- water suction height
- freeze/thaw test
- magnesium sulphate soundness test
- Sonnenbrand volume stability
- resistance to heat
- resistance to alkali reaction
- chemical analysis
- water leaching test

This list is more extensive than currently required by any individual national specification but it is not intended that all the properties will be mandatory; some will be optional while others will be for information only.

1.5.2.2 CEN TC 19/SC 1, Bitumen

The gradings by which paving grade bitumen will be specified are the permitted range of penetration (in dmm), with the grades being as follows:

- 20–30
- 30–45
- 35–50
- 40–60
- 50–70
- 70–100
- 100–150
- 160–220
- 250–330

The specification requirements for 10–330 penetration grades will be based on the following properties:

- penetration @ 25 °C
- softening point (ring and ball) using the stirred method
- viscosity @ 135 °C
- flash point
- solubility
- resistance to hardening (Rolling Thin-Film Oven Test or Rolling Film Test)
  - change of mass
  - retained penetration
  - minimum softening point

In addition, there will be optional limits on the following:

- wax content (although this is intended to be withdrawn after a limited period)
- penetration @ 5 °C
- Fraass brittle point
- viscosity @ 60 °C
- resistance to hardening (Rolling Thin-Film Oven Test or Rolling Film Test)
  - increase in softening point

1.5.2.3 CEN TC 317, Tar

CEN TC 317 has only recently been set up, and as yet little has been decided.

1.5.2.4 CEN TC 227, Mixed materials

CEN TC 227/WG 1 is producing harmonized specifications and test methods for asphalt mixtures, but with a longer-term intention of achieving performance-related specifications and the necessary test methods to support them. The harmonized specifications being produced initially are to be for:

- hot rolled asphalt;
- asphalt concrete (including both designed 'Marshall' asphalt and 'recipe' macadams);
- porous asphalt (including airfield friction course);
- Mastic asphalt (including gussasphalt);
- stone mastic asphalt;
- very thin surfacings; and
- soft asphalt.

Details of the approach that these specifications will take are given in separate chapters for each of these generic types except for soft asphalt, which is intended primarily for use in Scandinavian countries.

The test methods being developed to support these specifications for mixed materials are:

- sampling
- specimen preparation
- mixing
- binder content
- grading
- binder properties
- mixture density
- binder drainage
- hydraulic conductivity
- indentation of gussasphalt
- 'Marshall' test
- wheel-tracking test
- indirect tensile test
- fatigue test

- specimen density
- voids content
- reference density
- compactability
- water sensitivity
- mixture temperature
- moisture content
- segregation sensitivity
- texture depth
- abrasion to studded tyres
- abrasion of porous asphalt
- creep test
- stiffness
- low temperature cracking
- layer adhesion
- pavement thickness
- specimen dimensions
- impact compactor
- gyratory compactor
- vibratory compactor
- roller compactor
- hot sand test for pre-coated chippings

This list is more extensive than currently required by any individual national specification and it is not intended that all the properties will be mandatory; some will be optional while others will be for information only.

Similarly, CEN TC 227/WG 2 is producing specifications and test methods for surface dressings and slurry surfacings.

CEN TC 227/WG 1 is also developing a system for 'Attestation of Conformity'. This will be based on the 'factory-produced' product (i.e. the as-mixed, rather than the as-laid, material) and, hence, is aimed at the producer/laying contractor interface rather than the contractor/client interface. The latter will be included, but at a lower level. One important aspect of any 'Attestation of Conformity' scheme in which products are given a 'CE' mark is that, if a product is CE marked, it is deemed to be of the relevant standard and cannot be refused by a purchaser. If the purchaser does have reason to doubt the standard, all he or she can do is to use evidence of non-compliance to get the attestation removed from the producer.

1.5.3 Technical approval schemes

In France, there is a system for providing technical advice on products not covered by national standards known as the *Avis Technique*, or Technical Opinion. The Laboratoire Central des Ponts et Chaussées (LCPC), the national highway research laboratory, carries out tests on a material and offers an opinion on the suitability of the product. This is reviewed by a committee involving the specifying authorities, the assessing organization and the manufacturer and, if approved, a document is drawn up setting out its range of application and sites where it has been used. This *Avis Technique* is then used by highway authorities in the procurement of the material with the document being valid for three years. The introduction of the *Avis Technique* procedure has promoted innovation in road materials in France.

In the United Kingdom, the Department of Transport (subsequently the Highways Agency) have promoted, with the assistance of TRL, a less formal five-stage assessment procedure for such materials. The five-stages in the *Highways Agency Procedure for Evaluating New Materials* are:

- Stage 1. Desk Study: assess and evaluate existing information on the material;
- Stage 2. Laboratory Study: test the mechanical properties of materials to allow theoretical predictions to be made of their performance;
- Stage 3. Pilot-Scale Trials: evaluation of construction and performance of materials in small scale trials;
- Stage 4. Full-Scale Trials: full-scale trial on a trunk road to establish whether the previous assessments obtained from Stages 2 and 3 are realized;
- Stage 5. Highways Agency Specification Trials: this stage is necessary to carry out further evaluation of the material and to test the specification under contract conditions.

The procedure is now being developed with specific procedures for various groups of products by the British Board of Agrément on behalf of highway authorities including the Highway Agency, other overseeing organizations and the County Surveyors' Society. The procedure, known as the *Highway Authorities Product Approval Scheme* (HAPAS), will provide a framework for assessing innovative developments and will initially cover:

- high-friction systems;
- overbanding;
- thin surfacings; and
- modified binders for:

It is intended that the various national systems, such as *Avis Technique* and HAPAS, for assessing the more innovative materials and techniques will be harmonized to allow certificates to be acknowledged across Europe as *European Technical Approval* certificates.

1.5.4 Strategic Highway Research Program

In 1987, the Congress of the United States of America allocated $150 million over five years to the Strategic Highway Research Program (SHRP) for research to improve the highway network. The research programme, which also sought to improve road safety for both road users and highway workers, focused on four areas:

- asphalt (the performance of asphalt materials);
- concrete and structures;
- highway operations; and
- long-term pavement performance.

The programme is now complete, with 140 'products' (new specifications, test methods, equipment and reports) having been identified, of which 22 are from the Asphalt Program, which had been allocated about $50 million.

The objective of the Asphalt Program was to produce new specifications for both bitumen and mixed materials which were related to the performance of the materials; however, the short time scale of the project did not allow the development of a rigorous relationship between the material properties and performance under traffic so that, at best, the methods of evaluating materials were performance-related. The list of SHRP products from the Asphalt Program, including their reference numbers, is as follows:

- 1001   Asphalt Binder Specification
- 1002   Bending Beam Test for Asphalt Binders
- 1003   Pressure Ageing Methods for Binders
- 1004   Asphalt Extraction and Recovery Method
- 1005   Direct Tension Test for Low-Temperature Cracking
- 1006   High Temperature Viscosity Test
- 1007   Dynamic Shear Rheometer Test for Binders
- 1009   Binder Chromatographic Test Methods
- 1010   Asphalt Refiners Guide to Binder Performance
- 1011   Asphalt Mixture Specification
- 1012   SUPERPAVE Asphalt Mixture Design System
- 1013   Net Adsorption Test for Asphalt Mixtures
- 1014   Gyratory Compaction for Asphalt Mixtures
- 1015   Rolling Wheel Compaction
- 1017   Shear Test for Asphalt Mixtures
- 1019   Flexural Fatigue Life Test
- 1021   Thermal Stress Restrained Specimen Test
- 1022   Indirect Tensile Creep and Strength Test
- 1024   Environmental Conditioning System
- 1025   Short-Term Asphalt Ageing Method
- 1030   Long-Term Asphalt Ageing Methods

The major 'product' of the Asphalt Program has been the SUperior PERforming PAVEment (SUPERPAVE) system for asphalt mixtures. The asphalt concrete mixtures designed by the programme have aggregate gradings passed on the 'Fuller' curve (Figure 1.2) with an index of 0.45 in Equation 1.1 (section 1.3).

One important aspect that has been introduced into the structure of the specifications is that the properties are considered in terms of climatic zones. It is no longer necessary to have every binder able to withstand loads at +60 °C and –20 °C; where the climate is more temperate, less onerous requirements may be stipulated. Hence, the binder specification is an attempt to define the behaviour of materials under specific traffic and environmental conditions. A set of material tests have been developed to measure the physical properties of both modified and unmodified binders. These tests are much more sophisticated than the penetration and softening point, which are widely used to specify binders.

To support the mixture specification, a suite of test methods have been developed to assess the resistance of mixtures to:

- rutting;
- fatigue cracking; and
- low temperature cracking,

taking into account the effects of:

- ageing;
- moisture susceptibility; and
- loss of adhesion.

This suite of tests is currently being used in the United States in pilot trials of the SUPERPAVE specification.

In addition, the US Federal Highway Administration (FHWA) is taking the lead in helping US highway agencies with the implementation of the research outputs. FHWA is also continuing the work of evaluating the long-term pavement performance experiments. With this commitment to improve and rationalize their specifications, the specification approach and test methods will obviously become a strong influence in Europe and throughout other areas of the world, including through CEN in the further development of their 'harmonized' CEN specifications now being prepared.

However, before the SHRP approach is widely adopted by others, the 'products' from SHRP need to be validated. The major concern is the robustness of the relationship between the results of the new tests and the performance of the materials in service. Other concerns are the cost of the equipment (about $100k for the binder tests and about $300k for the mixed materials) and the time scale for completing the design of a mixture. Nevertheless, the evaluation of SUPERPAVE and the associated specifications is taking place and some aspects of them will be introduced into CEN standards at some time in the future.

1.6 REFERENCES

Manual of Contract Documents for Highway Works. Her Majesty's Stationery Office, London.
  Volume 1: Specification for Highway Works (MCHW 1).
British Airports Authority (1993) *Civil Engineering Section 14: Bituminous Materials – Aircraft Pavements. Quality System Reference Specification SP-CE-014-04-0A,* British Airports Authority Technical Service Division.

References

Comité Européen de Normalisation (1996) *Tests for Geometrical Properties of Aggregates; Part 2. Determination of Particle Size Distribution – Test Sieves, Nominal Size of Apertures. BS EN 933-2: 1996.* British Standards Institution, London.

County Surveyors' Society (1996) *Highway Authorities Standard Tender Specification.* Cheshire County Council.

Defence Works Services (1995) *Marshall Asphalt for Airfield Pavement Works; Hot Rolled Asphalt and Coated Macadam for Airfield Pavement Works. Defence Works Functional Standards.* Her Majesty's Stationery Office, London.

Fuller, W.B. and S.E. Thompson (1907) The laws of proportioning concrete, *Trans. American Society of Civil Engineers* **59**, 67–172.

Hindley, G. (1971) *A History of Roads*, Peter Davies, London.

CHAPTER 2

Aggregates and fillers

Prof. A. Woodside, University of Ulster

2.1 ROCK TYPE CLASSIFICATIONS

The classification of aggregates by means of a British Standard Specification was first published as BS 63 in 1913 and was primarily concerned with road aggregates. The trade names had been proposed by the Geological Survey in order to provide a 'simple' classification. It was considered unnecessary to name all the rocks by their precise geological or traditional name and so similar rocks were grouped together and named after the predominate rock type within that group.

When BS 812 was issued in 1943, a shorter list of 11 trade groups was published. This reduction was considered to be a more scientific classification and was based mainly on origin, as had been recommended by Knight in 1935. The fundamental change was to divide road-stones into rock types with all artificial/synthetic materials grouped together. The revision of BS 812 in 1984 provided a list of rock types used for aggregates, as follows with descriptions for some of the more commonly used aggregate types:

- artificial
  - crushed brick
  - slags
  - calcined bauxite
  - synthetic aggregate
- basalt
  - andesite[a] (a fine-grained, usually volcanic, variety of diorite)
  - basalt (a fine-grained basic rock, similar in composition to gabbro, usually volcanic)
  - basic prophite
  - diabase
  - dolerite (a basic rock, with grain size intermediate between that of gabbro and basalt) (including theralite and teschenite)
  - epiorite
  - lamprophyyre
  - quartz-dolerite
  - spilite

- flint
    - chert (cryptocrystalline[b] silica)
    - flint (cryptocrystalline[b] silica originating as nodules or layers in chalk)
- gabbro
    - basic diorite (diorite is an intermediate plutonic rock, consisting mainly of plagioclase, with hornblende, augite or biotite)
    - basic gneiss
    - gabbro (a coarse-grained, basic, plutonic rock consisting essentially of calcic plagioclase and pyroxene, sometimes with olivine)
    - hornblende-rock
    - norite
    - peridotile
    - picrite
    - serpentine
- granite
    - gneiss (a banded rock, produced by intense metamorphic conditions)
    - granite (an acidic, plutonic rock consisting essentially of alkali feldspars and quartz)
    - granodiorite
    - granulite (a metamorphic rock with granular texture and no preferred orientation of the minerals)
    - pegmatite
    - quartz-diorite
    - syenite (an intermediate plutonic rock consisting mainly of alkali feldspar with plagioclase, hornblende, biotite or augite)
- gritstone (a sandstone, with coarse and usually angular grains; including fragmental volcanic rocks)
    - arkose (a type of sandstone or gritstone containing over 25% feldspar)
    - greywacke (an impure type of sandstone composed of poorly sorted fragments of quartz, other minerals and rock; the larger grains are usually strongly cemented in a fine matrix)
    - grit
    - sandstone (a sedimentary rock composed of sand grains naturally cemented together)
    - tuff (consolidated volcanic ash)
- hornfels (a thermally metamorphosed rock containing substantial amounts of rock-forming silicate minerals)
- limestone
    - dolomite (a rock or mineral composed of calcium magnesium carbonate)
    - limestone (a sedimentary rock consisting predominantly of calcium carbonate)
        - marble (a metamorphosed limestone)

- porphyry
  - aplite
  - dacite
  - felsite
  - granophyre
  - keratophyre
  - microgranite[a] (an acidic rock with grain size intermediate between that of granite and rhyolite)
  - porphyry
  - quartz-porphyrite
  - rhyolite[a] (a fine-grained or glassy acidic rock, usually volcanic)
  - trachyte[a] (a fine-grained, usually volcanic, variety of syenite)
- quartzite
  - ganister
  - quartzitic sandstone
  - recrystallized quartzite
- schist
  - phyllite
  - schist (a metamorphic rock in which the minerals are arranged in nearly parallel bands or layers; platy or elongated minerals such as mica or hornblende cause fissility in the rock which distinguishes it from gneiss)
  - slate (a rock derived from argillaceous sediments or volcanic ash by metamorphism, characterized by cleavage planes independent of the original stratification)
  - all severely sheared rocks

Notes
a The terms microgranite, rhyolite, andesite or trachyte, as appropriate, are preferred for rocks alternatively described as porphyry or felsite.
b Composed of crystals so fine that they can be resolved only with the aid of a high-power microscope.

Although surfacing aggregates account for only a small proportion of the total road construction, they form a significant part of the total cost. Should they fail prematurely, the engineer is left with a road which, although it may be structurally sound, has a surfacing not capable of providing the necessary level of performance to ensure adequate safety for the public. To limit their premature failure, specifications impose more stringent requirements than for any other layer.

Generally, surfacing aggregates are required to be 'clean, hard and durable', possess properties which provide the minimum levels of skid- and abrasion-resistance and provide an adhesive bond with bitumen. With the imminent inclusion of European CEN test methods into British specification (section 1.5.2), it is advisable to consider how aggregates are likely to be specified in the future.

2.2 PROPERTIES AND METHODS OF TEST

2.2.1 Dimensions

2.2.1.1 Size

Particle distribution, as measured by BS 812: Part 103 (BSI, 1985a), describes the size of individual particles and is sub-divided into two sections:

- **Sieve tests**. Using a nest of sieves, it is possible to classify the material by means of a grading analysis. However, Figure 2.1 shows a comparison of dry and wet gradings and demonstrates how routine dry sieving analysis can substantially underestimate the amount of material passing the 75 μm sieve. Where not appreciated, substantial errors can be made in a number of areas, such as the ability to meet the required specification grading limits or the asphalt mixture design.

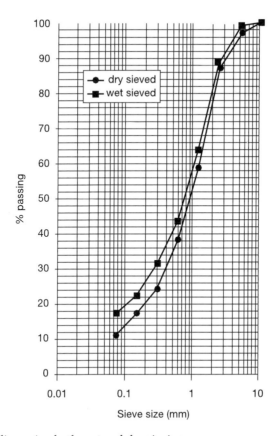

Fig. 2.1 Grading using both wet and dry sieving

- **Sedimentation test**. A gravimetric sedimentation type test which determines the proportion of material finer than 20 μm in particle size in aggregates.

The grading analysis may enable the sample to be described as:

- gap-graded, where some particles are retained on the larger sieves and the remainder on smaller sieve sizes with little material being retained in the middle sieves within the nest;
- continuously- or dense-graded, where a significant number of particles are retained on each sieve size within the range; or
- single-size grading, where the majority of particles are retained on one sieve having passed the next larger sieve.

Aggregate retained on a 3.35 mm test sieve is known as **coarse** aggregate, while **fine** aggregate is that material which substantially passes a 3.35 mm test sieve and can be:

- natural sand;
- produced by crushing rock (crushed rock fines); or
- a mixture of sand and crushed rock fines.

Research has shown that particle size has a direct relationship to skid resistance and the grading of aggregates is a fundamental factor in the design of asphalt mixtures.

2.2.1.2 Dust content

The degree of cleanliness of a roadstone is usually regarded as being defined by the amount of clay, silt and dust which is present on the fine and coarse fractions. In the current version of BS 812: Part 103 (BSI, 1985a), washing and sieving is the preferred method for particle size distribution analysis and should always be used for aggregates which may contain clay or other materials likely to cause agglomeration of particles (Woodside, Given et al., 1993). BS 812: Part 103 also states that 'dry sieving will give inaccurate results for aggregates containing clay but is quicker and less laborious'.

Dust is generally regarded as material which is less than 75 μm (63 μm with the harmonized European standards) in size. However, within this fraction there may exist a large variation in size, type and shape of constituent particles. Therefore, a further sub-division is required when attempting to assess the properties of dust; for example, a very fine grading will increase the surface area available to bitumen and, consequently, may:

- cause serious problems with the binder content in asphalt mixtures; and/or

- influence the ability of the bitumen to bond with the aggregate surface if a sandwich of numerous small particles are present rather than a single layer of coarser-sized dust.

Generally, the coarser sizes within the dust grading (those larger than 1 μm) consist of micro-particles of the parent rock whereas those less than 1 μm in size may be either unreactive rock-flour or be water-active clay minerals, the latter being potentially detrimental to performance.

Current specifications tend to recommend the use of Plasticity Index to assess aggregate fines. This measure involves the testing of all material which is less than 425 μm in size and is liable to produce different results from different operators. The clay-sized material of less than 2 μm has the greatest potential to cause failure, and the ideal test is the methylene blue absorption (MBV) test (CEN, 1995; Woodside and Woodward, 1984). A simplified version of this test involves adding 30 mL of water to 1 g samples of dust and observing the quantity of methylene blue dye that is absorbed. The simplified method is primarily of use for testing basalts and other basic igneous rocks which contain expansive clays and, if other rock-types are to be assessed, alternative pass limits must be applied; for example, tentative limits of <1.0 for Tertiary basalts and <0.6 for Silurian greywackes are used in Northern Ireland.

It is important to understand the properties of the dust content as well as assessing the total dust present on the stone; the latter can be easily performed using an ultrasonic water-bath. Research (Craig, 1991) has shown that 1.0% by mass of dust on particles is acceptable if the dust is sound whereas 0.5% by mass is the maximum recommended if the dust is unsound in nature; these figures are contrary to many current specifications.

2.2.1.3 Shape

The need to classify aggregates in terms of their shape has long been of interest, even as long ago as 1880 when Sorby attempted to subjectively classify sand grains into different groups according to their physical dimensions and surface characteristics. Further development by Wentworth led to the roundness and flatness ratio being derived in 1922. However, a roundness value was developed to quantify shape in terms of one of five categories: rounded, sub-rounded, curvilinear, sub-angular and angular. Knight (1935) reported the work and suggested Table 2.1.

A major advance occurred when Zingg (1935) proposed shape classifications based on the measurement of length, breadth and thickness. This gave:

- the thickness to breadth ratio – a flatness ratio; and
- the breadth to length ratio – an elongation ratio.

Table 2.1 Classification of aggregate shape

Classification	Description	Examples
Rounded	Fully water worn or completely shaped by attrition	River or sea gravels, seashore and wind-blown sand
Irregular	Naturally irregular or partly shaped by attrition with rounded edges	Pit sand, pit gravels, land and dug flints
Angular	Possessing well-defined edge intersection of roughly planer faces	Crushed rocks of all types, talus or screes
Flaky	Material, usually angular of which the thickness is small relative to the width and/or the length	Laminated rocks

Source: Knight (1935).

When using ratio values of 0.66, it was possible to classify the particles as either discs, equi-dimensional, blades or rods. One of the benefits of this method proposed by Zingg was that each individual particle shape could be plotted to form what is known as a Zingg graph. If particles have their thickness almost equal to their breadth, they will have a flatness ratio value of >0.9; similarly, if their breadth is almost equivalent to their length, the elongation ratio might exceed 0.9. Consequently, the plot of such values would clearly fall within the equi-dimensional area in a Zingg graph.

Aggregate particles may be classified in terms of their physical dimensions and how they relate to the mean size. An example is the flakiness index to BS 812: Section 105.1 (BSI, 1985b), which is determined by separating the flaky particles and expressing their mass as a proportion of the mass of the total sample; the test must be carried out on all particle sizes in the sample. Aggregate particles are deemed to be flaky when they have a thickness (smallest dimension) less than 0.6 of their mean sieve size. This shape is taken as the mean of limiting sieve sizes used for determining the size fraction in which the particle occurs. The test is not applicable to material passing a 6.3 mm test sieve or retained on a 63 mm test sieve.

The determination of particle shape by shape index has been proposed as a European Standard (section 1.5.2.1) with individual particles in a sample of coarse aggregate being classified on the basis of the ratio of their length (L) to thickness (E) using a calliper. The shape index is calculated as the mass of particles with a ratio of dimensions L/E more than 3 expressed as a percentage of the total dry mass of particles tested. It is anticipated that the calliper system will replace the present British system of determining flakiness index by the use of slotted sieves.

Rather than using callipers to determine the various particle measurements, McCool (1990) developed a contact sensor to measure the greatest, intermediate and least dimensions. Linked to a personal computer, the results could be calculated in the form of ratios and plotted onto a Zingg graph. Further automation has been developed by the Laboratoire Regional des Ponts et Chausées at Nantes (Lebas and Maldonado, 1989) using computer analysis of video images to calculate the greatest and intermediate dimensions; further work is on-going to measure the third dimension.

Shape has an influence on the performance of an aggregate, increasing the proportion of flaky and elongated particles which cause the overall strength of the aggregate (and the resultant asphalt mixture) to decrease. The 10% fines value will decrease and the aggregate impact and crushing values increase. This could be as much as 50% for certain aggregates. Furthermore, flaky and elongated particles will tend to break under the roller and provide poor embedment and minimal texture depth. However, it has been shown that a cubic particle may not be the best shape with which to produce porous asphalt (Chapter 7); high flakiness index values will improve the permeability of the mix (Woodside, Woodward et al., 1993).

2.2.2 Relative density, water absorption and porosity

BS 812: Part 2 (BSI, 1995) outlines three different methods of expressing the relative density of an aggregate; it may be expressed as the oven-dried value (odRD), the saturated surface-dry value (ssdRD) or the apparent value (appRD), whereas water absorption (WA) is expressed as the difference in mass before and after drying at $(105 \pm 5)\,°C$ for 24 hours. Several methods may be employed, using either a wire basket, gas jar or pycnometer to hold the sample. However, the grading of the test sample would influence the choice.

In the assessment of 10/14 mm crushed rock aggregate, it is advisable to use the gas-jar method. This involves duplicate 1000 g samples being washed to remove dust and then immersed in a gas-jar at 20 °C for 24 hours, then weighed (mass B), the container being emptied and refilled with water only and weighed (mass C). The aggregate is surface dried and weighed (mass A), then placed in an oven at 105 °C for 24 hours, cooled in an airtight container and weighed (mass D). The various values of relative density can then be determined from Equations 2.1–2.5.

$$\text{Relative density on an oven-dried basis (odRD)} = \frac{D}{A-(B-C)} \quad (2.1)$$

$$\text{Relative density on a saturated surface-dry basis (ssdRD)} = \frac{A}{A-(B-C)} \quad (2.2)$$

$$\text{Apparent relative density (appRD)} = \frac{D}{D-(B-C)} \quad (2.3)$$

$$\text{Water absorption (WA)} = \frac{100(A-D)}{D} \quad (2.4)$$

$$Porosity = \frac{100WA\ ssdRD}{WA + 100}\ \% \tag{2.5}$$

Relative density is directly related to an aggregate's volume and, therefore, necessary when designing an asphalt mixture, as well as being referred to when specifying the rate of spread of chippings for surface dressing. Furthermore, the saturated surface-dry relative density is also used in the calculation of aggregate abrasion values. The water absorption is an excellent means of indicating potential durability problems such as frost susceptibility.

2.2.3 Strength

2.2.3.1 Resistance to impact

Stewart of Cape Town University, South Africa, developed a test in the mid-1940s which became the aggregate impact value test now found in BS 812: Part 112 (BSI, 1990a). The aggregate impact value (AIV) gives a relative measure of the resistance of an aggregate to suddenly crack. The test enables samples of 10/14 mm particles to undergo 15 blows of a standard weight of 13.5 kg with the material passing 2.36 mm being expressed as a proportion by mass in % of the original. Aggregates with an AIV less than 10 may be considered strong whereas those with an AIV greater than 30 should be considered weak.

It is envisaged that the British impact test will be replaced by the German Schlagversuch impact test, which is likely to be the CEN method of assessing the strength under impact (CEN, 1991b). Samples of 8/12.5 mm aggregate are impacted ten times and then graded using 8 mm, 5 mm, 2 m, 0.63 mm and 0.2 mm sieves. The impact value (Sz) is determined by adding the proportion passing values (in %) and dividing the sum by 5. High quality aggregate should have a value less than 18. It has been found that a reasonable correlation exists between the AIV and Sz test methods (Woodside and Woodward, 1993).

2.2.3.2 Resistance to crushing

The aggregate crushing value (ACV) to BS 812: Part 110 (BSI, 1990b) gives a relative measure of the resistance of an aggregate to crushing under a gradually applied compressive load. The test is carried out using 10/14 mm particles which are placed into a steel cylinder fitted with a freely moving plunger and subjected to a slowly applied compressive load of 400 kN over a period of 10 minutes. This action causes the aggregate to degrade and is dependent upon the crushing resistance of the material. The degree of crushing is assessed by sieving the crushed aggregate on a 2.36 mm sieve and expressing the fine material passing this

Properties and methods of test

sieve as a proportion of the original in %. Aggregates with an ACV of less than 10 may be considered to be strong whereas those greater than 25 should be regarded as weak.

The 10% fines value (TFV) to BS 812: Part 111 (BSI, 1990c) is the load (in kN) required to produce 10% of fine material passing a 2.36 mm sieve from an original 10/14 mm sample. The test specimen is placed into a steel cylinder fitted with a freely moving plunger as for the ACV test. The specimen is then subjected to a slowly applied compressive load over a period of 10 minutes, the load for the first test being estimated by Equation 2.6.

$$Required\ force = \frac{4000}{AIV}\ kN \qquad (2.6)$$

The maximum force is then recorded. The whole specimen is then sieved on a 2.36 mm sieve in order to determine the proportion of fines. If the proportion does not fall within the range 7.5% to 12.5%, another specimen should be tested using an adjusted maximum loading to bring the proportion of fines within this range. The 10% fines value can be determined from the maximum force to produce a proportion within the required range using Equation 2.7.

$$TFV = \frac{14\ Force}{Fines + 4}\ kN \qquad (2.7)$$

where $Force$ = Maximum force applied (kN);
         $Fines$ = Proportion of fines produced (%).

2.2.4 Durability

2.2.4.1 Resistance to soaking.

Many aggregates lose their inherent strength when saturated. The soaked 10% fines value is one method of assessing the reduction from the value obtained using the standard method (section 2.2.3.2). The test is basically the same except that the aggregate is soaked in water for 24 hours at 20 °C before being surface dried and crushed in the normal manner. Finally, the total crushed sample is dried for 12 hours at (105 ± 5) °C and allowed to cool before sieving. The result is then reported as the 10% fines value carried out on soaked aggregates.

2.2.4.2 Soundness

The magnesium sulphate soundness value (MSSV) to BS 812: Part 121 (BSI, 1989a) is a good method for determining the soundness of an aggregate. Not having been included in earlier editions of BS 812, it was introduced to identify certain aggregates which, when tested by the mechanical strength tests, are apparently suitable for use but which tend to fail in service. The test was first reported in France around 1828 as a method to assess the frost resistance of building stone. It was accepted as the American Society for Testing and Materials (ASTM) Standard C88 in 1931 and has been used widely throughout the world. Although a modified version has been used for UK runway specifications (PSA, 1987; DWS, 1995), it use has been minimal in the UK highway industry until very recently.

The BS 812: Part 121 test comprises of subjecting samples of 10/14 mm aggregate (for airfields, all fractions are tested) to five 24-hour cycles of 17 hours immersion in a saturated solution of magnesium sulphate solution followed by 7 hours drying at 110 °C. The samples are then sieved on the 10 mm sieve (ASTM use the next smaller sieve) and the magnesium sulphate soundness value (in %) for each specimen is calculated by Equation 2.8.

$$MSSV = 100 \frac{Mass\ retained}{Initial\ mass} \% \qquad (2.8)$$

For UK airfields, the results have previously been reported as the proportion being broken down (that is, unsound) rather than, as in BS 812: Part 121, the proportion of sound aggregate.

Failure of an aggregate particle occurs due to repetitive stressing caused by the growth of salt crystals in any weaknesses present. A result of 90% indicates a sound aggregate whereas a result less than 70% would be an unsound stone, but acceptable pass limits vary depending on use. The *Specification for Highway Works* (MCHW 1) specifies a minimum of 75% of an original 10/14 mm sample remaining after the five cycles of immersion.

Many Europeans do not favour this test on the grounds that the introduction of the chemical process of crystallization is not representative of in-service performance. Also, it has been replaced, or there is a strong lobby wishing to replace it, by other test methods in many parts of the world including America, Australia, New Zealand and, more recently, Canada. Nevertheless, the test is considered by many, including the major UK civil and military airport authorities, as an ideal method of ranking aggregates.

Resistance to freeze/thaw and thermal expansion/contraction is an equally important aspect of the durability of aggregates and has promoted the German freeze/thaw (F/T) test (CEN, 1991d) to be recommended as a European Standard. However, experience of this test would suggest that the method is limited and merely causes mild mechanical stressing. Work carried out in Iceland (The Aggregates Committee, 1992) has shown that a better indication of freeze/thaw resistance can be achieved by:

- increasing the number of cycles to 70;
- reducing the temperature to ± 4 °C; and
- using a 1% solution of sodium chloride.

Aggregates which have passed the German method have failed this Icelandic test. A logical use of increasing the number of freeze/thaw cycles is to specify aggregate capable of surviving 10, 50, 70 or 100 freeze/thaw cycles depending on the extent to which the situation is demanding.

2.2.4.3 Resistance to wear

The aggregate abrasion value (AAV) to BS 812: Part 113 (BSI, 1990d) gives a measure of the resistance of an aggregate to surface wear by abrasion. Two flat specimens are made from approximately 23 selected and correctly oriented particles embedded in a resin base. The two samples are fixed in contact with a steel horizontal grinding lap and then loaded while an abrasive sand (Leighton Buzzard) is continually fed onto the lap surface for 500 revolutions. The aggregate abrasion value is determined from the difference in mass before and after abrasion using Equation 2.9.

$$AAV = \frac{3000 \text{ (Initial mass–Mass after abrasion)}}{\text{Saturated surface–Dry particle density}} \qquad (2.9)$$

An aggregate with an AAV greater than 10 is usually unsuitable as a surfacing material whereas one with a value less than 4 would be regarded as an extremely hard-wearing aggregate.

In France and in some other countries, the wet micro-Deval (MDE) test (CEN, 1991a) has been used to assess and specify the frictional wear or attrition properties necessary for surfacing aggregate. The standard test consists of a 500 g sample of 10/14 mm particles placed in a steel cylinder along with 2500 mL of water and 5000 g of 10 mm diameter ball bearings. The sealed container is rolled at 100 rpm for 2 h and then sieved over a 1.6 mm sieve to determine the proportional weight loss as a percentage

The two tests, AAV and MDE, appear to be measuring different characteristics. The Department of the Environment published the findings of a study of the sources of high-specification aggregate for road surfacing materials (Thompson *et al.*, 1993) which also included an evaluation of the probable effects of the proposed CEN Standards in the current specifications. A comparison of 52 samples of aggregate indicated an apparent lack of correlation between AAV and wet micro-Deval results. Thompson explains that

> the fact that samples with aggregate abrasion values of less than 10 can have micro-Deval coefficients ranging from 12 to 40 has disturbing implications for the compliance of certain existing HSA sources with any future European Specification requirements that may be based on the micro-Deval test.

None of the sources had a micro-Deval coefficient (MDE) within the French Class A specification requirement of less than 10. However, extensive research work in the United Kingdom (Woodward, 1995; Perry, 1996) has failed to establish a reliable correlation between AAV and MDE for aggregate used in road surfacing. This lack of correlation has made it impossible to set limiting values of MDE which would correspond to AAV limits established over many years.

However, the mechanisms of the two tests are completely different. The AAV test complements the PSV test (section 2.2.4.4), with the abrasion resistance ensuring that the macro-texture of the surface is maintained while the PSV limits help to ensure that the micro-texture of the particles is maintained. Together, they ensure a satisfactory resistance to skidding and, hence, the safety of the road user. The MDE is an assessment of the overall resistance to wear of an aggregate, and thus of its durability. Therefore, the AAV test should not be used as a substitute for MDE, and it has been suggested that the AAV be renamed the surface aggregate abrasion value (SAAV) to avoid any confusion.

2.2.4.4 Resistance to polishing

The polished stone value (PSV) to BS 812: Part 114 (BSI, 1989b) gives a measure of the resistance of a roadstone to the polishing action of a vehicle tyre under conditions similar to those occurring on the road surface. The test is in two parts:

- samples of stone are subjected to a polishing action in an accelerated polishing machine; and then
- the state of polishing reached by each sample is determined by measuring the skid-resistance using the portable pendulum skid-tester.

The test is carried out on aggregate particles which pass the 10 mm sieve and are retained on the 10/14 mm flake sorting sieve. Four curved specimens of each sample and four specimens of a control stone are made, each consisting of between 35 and 50 particles embedded in resin. The polishing process is carried out on 14 specimens (6 pairs of test samples plus a pair of control samples) and lasts for 6 hours, using corn emery for the first 3 hours and emery flour for the remaining 3 hours. The test is repeated and the polished stone value for each aggregate is calculated from Equation 2.10 using the average of the 4 samples tested.

$$PSV = S + 52.5 - C \qquad (2.10)$$

where S = mean of the four test specimens (two from each run) recorded to 0.1; and
C = mean of the four control specimens (two from each run) recorded to 0.1.

This test enables the highway engineer to rank the aggregates and, hence, to select those which will perform adequately on the selected site. Research (Woodside, 1981) has shown that other factors will induce a much higher and quicker rate of polish on many roadstones. Factors such as stress, speed, loading and even the use of salt in winter maintenance may cause the particles to polish much quicker, thus reducing the micro-texture of the surface and, consequently, diminishing the overall skid-resistance of the surfacing.

However, it has been shown (Perry, 1996) that extended polishing may cause 'plucking' in some of the gritstones and, thus, produce a regenerated surface with enhanced PSV results. Furthermore, the test does not reflect the environmental effect on the roadstone during its lifetime of service and, again, it has been shown (Woodside and Woodward, 1984) that the surface of many aggregate particles is enhanced by winter conditions.

High PSV aggregates are those with values exceeding 68 whereas low PSV aggregates have values less than 50; it is very difficult to enhance the value obtained from any source. However, research (Woodward, 1995) has shown that aggregates with a low 10% fines value (section 2.2.3.2) tend to produce higher PSV results. Similarly, the higher the aggregate abrasion value (section 2.2.4.3), the higher the PSV. Consequently, the highway engineer is faced with the difficult task of balancing AAV against PSV.

2.2.5 Adhesion of bitumen

When assessing the performance characteristics of an aggregate, the highway engineer relies heavily on the properties already covered in the section. However, very little is known about the inter-relationship between bitumen and aggregate in terms of how different bitumens perform in relation to different aggregates and in-service conditions. The ability to quantify this property in the laboratory might ensure that premature failures could be avoided, and many researchers have sought means of achieving this goal. Hence, numerous tests are now available, of which three of the most relevant are:

- Vialit plate test (LCPC, 1963), which measures the aggregate particle/bitumen adhesion properties when subjected to sudden impact;
- Instron adhesion pull-off test (INAPOT) (Craig, 1991), which measures the adhesion characteristics of aggregate particles embedded in bitumen and then subjected to a gradual extractive force; and
- Net adsorption test (NET), which was initially developed in 1986 as part of the American Strategic Highway Research Program (section 1.5.4) as a method which could be used to quantify the adsorptive nature and water sensitivity of bitumen/aggregate combinations, assesses a fundamental measure of aggregate/bitumen adhesion and moisture sensitivity.

Curtis *et al.* (1992) attempted to answer the basic questions relating to the mechanism of stripping and the possibility of relating simple adsorption and desorption measurements, made with aged and unaged bitumen on different aggregate, to the stripping mechanisms. The method developed was the NET and involves the preparation of 1 g/L solution of bitumen to toluene, the removal of 4 mL of solution, further diluting it to

25 mL with toluene and then determining its adsorption at 410 mm with a spectrophotometer which had been standardized with pure toluene. A sample of aggregate weighing 50 g and passing the No. 4 American Society for Testing & Materials (ASTM) sieve is placed in 50 mL Erlenmeyer flasks with 140 mL of the bitumen/toluene solution. The flasks are agitated for 6 hours on a shaker table before 4 mL of solution is removed, diluted to 25 mL with toluene and a 6-hour adsorption obtained. Two millilitres of water is then added to the Erlenmeyer flask, which is shaken for a further 12 hours and a final adsorption determined. The % net adsorption can then be determined by Equations 2.11 to 2.13.

$$\text{Initial Adsorption } (A) = \frac{V_a C(A_i - A_a)}{WA_i} \quad (2.11)$$

$$\text{Net Adsorption } (A_n) = \frac{V_r C(A_i - A_r)}{WA_i} \quad (2.12)$$

$$\% \text{ Net Adsorption } (\% A_n) = 100 \frac{A_n}{A} \% \quad (2.13)$$

where A_i = initial adsorption (mg/g);
A_a = adsorption reading after 6 hours (mg/g);
A_r = adsorption reading after addition of water (mg/g);
C = initial concentration of bitumen in solution (g/L);
V_a = volume of solution in the flask when the 6-hour sample is taken (normally 140 mL);
V_r = volume of solution in the flask when the water sample is taken (normally 136 mL); and
W = mass of aggregate (g).

The higher % net adsorption indicates the greater resistance to stripping of bitumen from aggregate by water.

Alterations to this method of presentation of results (Woodside et al., 1996) have provided a more realistic description of the aggregate/binder adhesion characteristics using Equations 2.14 to 2.15.

$$\text{Modified Initial Adsorption } (\% A_i^*) = 100 \frac{A_i}{A_{max}} \% \quad (2.14)$$

where A_{max} = maximum adsorption reading (mg/g).

$$\text{Modified Net Adsorption } (\% A_n^*) = 100 \frac{A_n}{A_{max}} \% \quad (2.15)$$

2.3 PERFORMANCE AND SPECIFICATION

2.3.1 Clean, hard and durable

The revision of the *Specification for Highway Works* (MCHW 1) in 1991 specified, for the first time, the mechanical properties required for asphalt materials based on research at TRL (Bullas and West, 1991). Previous editions had used the term clean, hard and durable without defining what this meant. For aggregates to be used for surfacing, the following minimum properties, when tested to BS 812, are required:

- **Clean**. The fraction less than 75 μm (section 2.2.1.2) shall not exceed the limits stated in BS 594: Part 1 (BSI, 1992) and BS 4987: Part 1 (BSI, 1993) when determined by wet sieving.
- **Hard**. Coarse aggregates shall have:
  - a 10% fines value (TFV) (section 2.2.3.2) of greater than 140 kN for natural crushed and uncrushed rocks and greater than 85 kN for blastfurnace slag when tested in a dry condition.
  - an aggregate impact value (AIV) (section 2.2.3.1) of less than 30% for natural crushed and uncrushed aggregates and 35% for blastfurnace slag when tested in a dry condition.
- **Durable**. When required, the aggregate source shall have a magnesium sulphate soundness value (MSSV) (section 2.2.4.2) of greater than 75%.

Certain other properties are required and include resistance to abrasion as measured by the aggregate abrasion value (AAV) (section 2.2.4.3), polishing as measured by the polished stone value (PSV) (section 2.2.4.4) and an affinity to bitumen (section 2.2.5).

In order to obtain more durable and cost-effective pavements, there is now a gradual move from what have been, essentially, recipe-based specifications to those which are end-performance based. Although the changes to the *Specification for Highway Works* (MCHW 1) have defined the basic aggregate properties required, the document is still very much a national recipe-type specification; aggregate selection could be based on performance so that a successful end-product can be predicted more accurately. To further the use of end-performance specifications, this section proposes tests that have been selected on their ability to provide a quick and cheap indication of in-service performance while using the minimum of specialized equipment.

2.3.2 Cleanliness

Cleanliness is a problem that is often not fully appreciated, but it only takes one sizeable failure to have justified investigation of an aggregate's dust. All dusts do not behave in a similar manner; some may be inert whereas others may expand in the presence of water and cause adhesion problems (section 2.2.1.2).

A quick means of assessing dust quality is a simplified version of the methylene blue (MBV) test (section 2.2.1.2). Another simple test is the French Vialit Plate (section 2.2.5), which can be used to evaluate the effect of variables such as the amount and type of dust, the type of binder, the use of adhesion agents, the temperature and the moisture content.

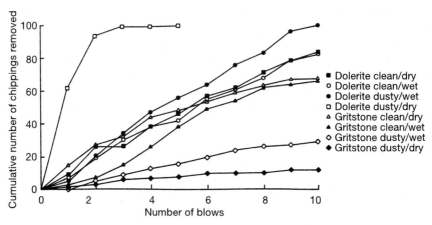

Fig. 2.2 Vialit plate results for two aggregates under differing test conditions

Figure 2.2 shows the effect that dust and moisture can have on the adhesion characteristics for two aggregates. Although the data obtained are limited, they can provide the engineer with a quick means of highlighting a future problem.

The test parameters mentioned may be further quantified by the Instron adhesion pull-off test (INAPOT) (section 2.2.5), which involves the extraction of aggregate prisms from a layer of bitumen. Figure 2.3 shows a range in performance for different combinations of rock-type and proportion of dust.

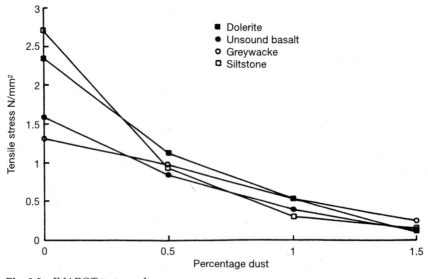

Fig. 2.3 INAPOT test results

As most high-quality aggregates are used at considerable distances from their source, some indication of the quantity of dust produced during transport and handling should be known. Dry tumbling test methods, such as the dry micro-Deval test (section 2.2.4.3), are a means to indicate an aggregate's susceptibility to producing dust.

Overall, the following procedures should be considered by the engineer to assess the cleanliness of aggregates and to supplement a simple indication of the quantities present:

- wet sieving to determine the proportion of dust present;
- the determination of the grading below 75 μm;
- an assessment of the quality of dust by the methylene blue test;
- an assessment of the potential to produce fines during transport and handling;
- the introduction of localized limits depending on the rock-type; and
- the use of the Vialit plate and/or INAPOT test methods to assess the performance of the aggregate interaction with the proposed binder.

2.3.3 Hardness

When assessing aggregate hardness in terms of performance, there are many factors which influence testing methods. Aggregate tests use sizes and shapes not used in practice – rarely does 100% single-sized 10/14 mm aggregate arrive on site.

Some methods, such as aggregate abrasion value (AAV) (section 2.2.4.3), use samples from which flaky particles have been removed, an operation which may improve the aggregate quality. For example, the flaky portion of a gritstone may be predominantly siltstone or shale in composition and the testing of individual constituents have shown differences of up to 20 units of AAV to occur. Removal of the flaky portion during the testing process may mask an aggregate susceptible to premature failure. Shape also influences an aggregate's strength characteristic with the strength decreasing as a chipping becomes less of a cube and increasingly blade-, rod- or disc-shaped.

The tests usually specified, such as aggregate impact value (AIV) (section 2.2.3.1), 10% fines value (TFV) (section 2.2.3.2) and AAV, all assess the dry aggregate. However, noticeable decreases in strength may occur if the aggregates are tested wet. This phenomenon has been recognized and, in the 1990 revision of the relevant parts of BS 812, alternative soaked versions of these tests are given. Certain aggregates which meet the dry specification limits may have their strengths reduced by over 50%, which clearly has considerable consequences given the wet climate of the United Kingdom.

If a source produces a geologically uniform aggregate, the engineer can design and specify with a certain degree of confidence. However:

- gritstones may contain siltstone and shale; and
- basalts may contain vesicular particles and/or be affected to different degrees of weathering.

Often, as little as 5% of such material can initiate failure but, during testing, their presence is frequently masked. Figure 2.4 shows an example where different ratios of a hard, sound aggregate and a soft, unsound aggregate were subjected to the AAV test. The two aggregates had the following properties:

Aggregate	AAV (section 2.2.4.3)	MSSV (section 2.2.4.2)
Hard, sound	3.4	98
Soft, unsound	18	60

Based on these values, the soft, unsound aggregate would quickly wear away if used as a surfacing aggregate but the AAV results indicate that up to 80% of this material would meet the normally specified limit of AAV < 10. This example also illustrates a problem with many test methods in that the stronger aggregates protect the presence of much weaker material. Figure 2.4 also shows the results for the wet micro-Deval (MDE) test (section 2.2.4.3) which illustrate the greater ability of this method to differentiate the presence of weak material in a composite mixture.

The European test methods which are likely to be incorporated into specifications in the future include the German Schlagversuch impact (Sz) test (section 2.2.3.1) and the Los Angeles (LA) test (CEN, 1991c). The Los Angeles test assesses an aggregate's resistance to fragmentation and

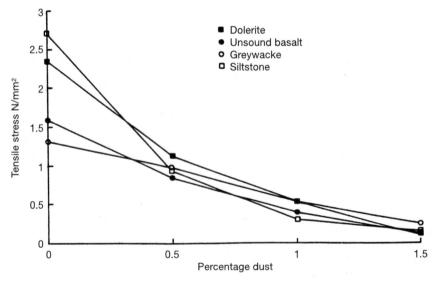

Fig. 2.4 Ability of aggregate abrasion value and Micro-Deval coefficient test methods to indicate the presence of soft and unsound aggregate particles

has been preferred to existing UK test methods as the CEN test because it is dynamic in action, tests a larger sample and exposes the presence of weaker constituents. A comparative study has been carried out (Woodside and Woodward, 1993) testing a selection of 9 basalt and 4 greywacke aggregates to BS and proposed CEN test procedures; the results are given in Table 2.2.

Therefore, the following approach should be considered by the engineer to assess the hardness of aggregates:

- dry strength testing should be continued;
- wet strength tests are to be preferred;
- if a source produces only one discernible rock-type, further testing can be restricted;
- if a source contains a number of rock-types, it should be assessed by methods which highlight the existence of weaker constituents; and
- individual constituents should be tested and reference examples kept on site.

2.3.4 Durability

Traditionally, durability was left to the discretion of the engineer and often not considered a problem. However, the magnesium sulphate soundness value (MSSV) test (section 2.2.4.2) was introduced into BS 812 as Part 121 (BSI, 1989a) in order to assess the soundness of aggregate. In the forward, it states that:

Table 2.2 Comparative study of BS and proposed CEN test methods

Rock type		WA	AIV	TFV	AAV	PSV	MSSV	MDE	LA	Sz	F/T
Basalt	1	1.7	12.6	298	7.81	55	90.6	40.2	10.0	12.8	1.9
	2	2.16	10.3	298	7.85	57	82.5	30.9	12.8	10.2	1.7
	3	1.35	14.6	324	5.56	57	92.0	34.4	14.9	12.0	1.7
	4	1.1	10.7	276	8.9	59	94.7	33.0	10.2	13.5	n/a
	5	0.7	8.2	438	4.6	54	98.8	22.3	11.0	10.6	0.3
	6	1.3	13.2	308	5.3	54	96.6	24.5	16.3	14.6	0.3
	7	1.07	12.0	294	8.5	56	96.3	32.0	14.7	10.6	0.4
	8	2.15	12.9	320	6.64	56	91.7	32.2	13.9	11.8	0.9
	9	2.86	15.3	223	7.61	59	74.3	36.6	16.0	13.4	1.9
Greywacke	1	0.6	8.3	403	5.99	67	98.5	20.0	11.8	9.8	0.4
	2	0.65	12.1	375	2.59	65	98.0	6.9	11.2	11.2	0.3
	3	0.43	12.4	383	5.62	63	99.2	19.0	13.0	9.4	0.3
	4	0.45	8.1	394	5.00	61	98.9	11.5	10.3	9.6	0.2

BS tests WA Water Absorption; AIV Aggregate Impact Value; TFV Ten per cent Fines Value; AAV Aggregate Abrasion Value; PSV Polished Stone Value; MSSV Magnesium Sulphate Soundness Value.
Proposed CEN tests MDE Micro-Deval co-Efficient; LA Los Angeles; Sz Schlagversuch impact value; F/T Freeze/Thaw.

there is the need in some circumstances, for a test to identify certain aggregates which are apparently suitable for use, when tested by other Parts of this Standard (e.g. the mechanical strength tests described in BS 812: Parts 110, 111, 112 and 113) but which fail in service.

However, engineers may not fully grasp what the term durable encapsulates. Consideration should be given to not only the soundness of an aggregate but also any factor which may affect its ability to remain serviceable in use. One of the simplest factors is loss of strength (section 2.3.3) if wet, while resistance to freeze/thaw (section 2.2.4.2) is another. Resistance to frictional wear or attrition can be assessed by the wet micro-Deval (MDE) test (section 2.2.4.3). In France, road aggregate is classified by combining the MDE and Los Angeles (LA) values (AFNOR, 1982). The limits for the different classes are shown in Table 2.3. If this practice were adopted in the United Kingdom, many high PSV aggregates would not be able to meet the Category A or B requirements.

Although there is now a British Standard to assess soundness, durability still remains a difficult property for the UK highway engineer to assess and requires research into the many different test methods available so that reliable indications of quality depending on rock-type and in-service conditions are possible. Nevertheless, the following approach should be considered by the engineer to assess the durability of aggregates:

- specifications should be rock-type based and/or allow localized limits rather than nationally or internationally specified acceptance limits because durability-related failure affects different aggregates in different ways;
- the continued use of the magnesium sulphate soundness value test;
- the use of wet strength tests;
- the use of the 'new European tests' such as Los Angeles, wet micro-Deval and freeze/thaw; and
- the long-term implications of using surfacing aggregates and how the environment will affect their performance.

Table 2.3 French aggregate classification

Category	Los Angeles value (LA)	Micro-Deval coefficient (MDE)	Polished stone value (PSV)
A	≤ 15	≤ 10	≥ 55
B	≤ 20	≤ 15	≥ 50
C	≤ 25	≤ 20	≥ 50
D	≤ 30	≤ 25	≥ 50
E	≤ 40	≤ 30	≥ 50

2.4 SOURCES

2.4.1 UK sources

There are currently 1300 quarries being worked in the United Kingdom (BACMI, 1994), collectively supplying more than 300 million tonnes of aggregate each year. Within the United Kingdom, the aggregate wealth is not evenly distributed and is divided by a line drawn through England from Flamborough Head in East Yorkshire to Portland Bill in Dorset. The area to the east of this divide is well endowed by many sand and gravel pits and a bountiful supply of marine aggregates, but there are very few sources of hard rock; whereas, to the north and west of the dividing line, there are large deposits of igneous, sedimentary and carboniferous rocks in addition to sand and gravel sources.

Furthermore, due to the population density and the high demand for construction materials, the existing reserves of sand and gravel within the south east are rapidly being depleted, consequently this area is supplied from the Mendips in the west and Leicestershire in the Midlands and, to a lesser extent, from Scotland, Ireland, Wales and Cornwall. Similarly, East Anglia is provided from a variety of regions, with the Midlands and North Wales quarries supplying a major portion of the shortfall in crushed rock.

2.4.2 Other European sources

A survey of the aggregates industry in 16 European countries (Anon, 1996), carried out in 1995, estimated that RMC were the largest company, having the market leadership in Germany and a principal company in France, Spain and the United Kingdom. Redland were considered to be the second largest company, with interests in France and the United Kingdom, third place going to Tarmac[1] and the next largest were Holderbank, Lafarge, Italcementi, Hanson, Minorco, Pioneer and Heidelberger. The total aggregate production in the 16 countries of Europe was estimated to exceed 2088 million tonnes in 1995 with Germany producing more than one-third of the total; the other major contributors were the United Kingdom, France and Spain. However, Norway has specialized in the production of high-grade rock-armour stone and exports considerable quantities to the United Kingdom and the Benelux countries.

The general trend appears to be towards the construction of well-managed super-quarries similar to the Foster Yeoman development at

Note

1. The survey was prior to the asset swap between Tarmac and Wimpey, giving Tarmac all the quarries, including those purchased from Alfred McAlpine by Wimpey the previous year.

Glensanda in Scotland, where huge volumes of rock can be directly loaded by a conveyor-belt system onto a waiting ship, thus reducing the expensive on-shore handling costs. Aggregate sources in the upper Rhine region tend to use barges as an economic means of transporting their product down the river to service the lowlands of Europe, whereas other suppliers prefer to use rail for moving aggregates.

2.5 FILLER

Filler is an aggregate of which at least 75% passes a 75 μm test sieve; the sieve will become 60 μm when the CEN become operative. Care should be taken to ensure that filler obtained from the fine aggregate in a mixture is not more than 8% by mass of that fine aggregate. The filler within an asphalt mixture may take the form of crushed rock (from the parent rock in the mixture), crushed slag, hydrated lime, Portland cement or other material approved by the purchaser. It is worthy of note that the use of 2%, by mass of the total aggregate, of hydrated lime or Portland cement filler will reduce the risk of water stripping the binder from some of the aggregates. Furthermore, the use of adhesion agents to the bitumen may also achieve the same result.

In order to determine whether clay minerals are present in the fine aggregate or filler, the methylene blue test (section 2.2.1.2) or the sedimentation test (section 2.2.1.1.) may be employed. Clay particles containing secondary minerals are capable of expanding to a size more than 300% of the original, thus causing premature failure in an asphalt mixture.

2.6 RECYCLING WASTE MATERIALS

2.6.1 Material types

The national resources of the world are diminishing with use. It is increasingly important for industry and the public to recognize, and act on, the concept of sustainable development so that the needs of the present are met without compromising those of the future. In the light of this, recycling (that is, using again that which has already been processed) can play an essential role. The forecast for future demand for aggregates is set to increase, as our infrastructure and environment generally are further improved and maintained. Therefore, it is right that the sources of raw materials are accomplished thoughtfully and responsibly, taking fully into account all the various environmental, economic and social factors involved (BACMI, 1996). Recycled aggregates fall into two broad categories:

- waste and demolition materials from the construction industry; and

- secondary materials, a term used to collectively refer to a whole range of industrial by-products and wastes of varying use and value, ranging from premium products such as slags to variable products such as colliery spoil.

Throughout Europe, there is a good deal of recycling of aggregates: 40% of all aggregates (5% from asphalt surface courses and 15% from asphalt binder courses and bases) used in Germany and 80% of all used in the Netherlands are considered to be recycled. Even within the United Kingdom, a good deal of aggregate recycling already takes place: secondary materials account for around 10% of total aggregate usage:

- colliery spoil (comprising minestones, siltstones, shales, earths, sandstones and sometimes limestones) is used for bulk fill but its variable quality usually prohibits other applications;
- China clay sand is used in certain parts of the country for mortar and as concreting sand and has potential for bulk fill, but it is generally found in the South-West of England, a location considered by many to be remote from most construction markets;
- pulverized fuel ash (PFA) is used as a cement substitute and for lightweight blocks as well as having potential for bulk fill, although the decline in coal-fired electricity generating plants is steadily reducing the production;
- most slags derived from the iron and steel industry are already used as aggregates and cementitious materials;
- the majority of asphalt planings are re-used on low-volume roads or recycled within road construction as hot or cold mixtures (Chapter 13);
- slate tips are a potentially useful source of secondary aggregate but are often in relatively inaccessible areas of the country and do have handling problems;

Table 2.4 Uses of secondary materials in 1990

Materials	Annual utilization
Colliery spoil	2.8 mt
China clay waste	1.5 mt
Slate waste	0.5 mt
Power station ash PFA & FBA	5.7 mt
Blastfurnace slag	4.0 mt
Steel slag BOS	0.2 mt
EAF	0.2 mt
Demolition and construction wastes	11.0 mt (of which 1 mt is graded aggregates)
Asphalt road planings	6.0 mt
Others	0.1 mt

- scalpings (the material left from the operation of rock quarries that is not suitable for the production of graded aggregates) are often re-used in road construction and maintenance; and
- at least 60% of all demolition materials and construction wastes find their way to secondary usage, which research suggests is a higher proportion than that in many other European countries.

Table 2.4 shows the relative extent of each of these secondary materials. A variety of means to increase further the use of aggregate recycling is currently being examined, including the introduction of land-fill taxes. It should be noted that:

- secondary materials account for 10% of total aggregate usage;
- at least 60% of all demolition materials and construction wastes are recycled;
- many quarrying companies produce secondary materials;
- the Department of the Environment, Transport and the Regions has called for a doubling of the use of recycled materials; and
- the quarrying industry is actively working towards this increase.

2.6.2 Factors affecting usage

Over the next decade, the extent to which recycling can be increased will depend upon three principal factors: environmental, engineering and economic.

2.6.2.1 Environment

The whole environment cycle has to be taken into account, in particular the impact on the environment today, because of the ever-increasing importance that is attached to the preservation of our surroundings and the heritage we leave to future generations. Using secondary materials means that less virgin land is disturbed, and this is a fundamental argument in its favour. However, some deposits of waste products suitable for recycling have been landscaped, or are regarded as part of the landscape, so that their removal would be indistinguishable from quarrying; in recent years, the necessary permissions to rework some tips have been refused by local authorities.

Transport is another crucial component within the environmental, as well as part of the economic equation. Transport over long distances, shipping out by sea is a possibility, even though road transport is first necessary. However, this would frequently entail the building of substantial extra port facilities, with all the environmental upheaval involved.

2.6.2.2 Engineering

Engineering considerations are factors that have a bearing on recycling, with some such constraints being based on practical application while others are less soundly based. Restrictions inhibiting the current use of secondary materials are being actively addressed by the Building Research Establishment and others. There is a very real possibility that the current research into the performance of secondary materials will result in some clarification of technical specifications. This would almost certainly open the way for wider secondary usage, but not before perhaps the greatest hurdle – the perception of suitability by construction specifiers and clients – is overcome.

The builders of roads, airfields, schools, housing, hospitals, airports, factories, offices and all the other structures which rely upon aggregates for their construction are responsible for ensuring the long-term effectiveness and safety of their work. Therefore, it is hardly surprising that the ultimate clients, their insurers and those specifying the materials to be used will take a conservative approach in order to minimize the risk of later problems for which they would be liable.

This conservatism emphasizes the urgent need for research to improve the quality of information on the performance of recycled and construction aggregates. The trend in many parts of the construction industry towards higher quality standards and greater durability (which also has implications for sustainability) further underlines this requirement.

2.6.2.3 Economic

The popular perception of the economics of recycling is often that it is cheaper but, as industries as diverse as paper manufacture and car production will testify, this is not necessarily the case. With aggregates, transportation costs are a major factor in the overall price. The general availability of sources of aggregate near to the user site almost always means that they are cheaper than supplies brought in from further afield. In road building especially, where aggregate costs account for some 30% of project costs, such price differentials can be very significant. However, in some areas when used in really substantial quantities, recycled materials can begin to show cost advantages. Lack of confidence has so far precluded secondary usage at levels where cost-savings through economies of scale would apply.

2.6.3 Expected development

The UK Department of the Environment has put its very substantial weight behind the use of secondary aggregates and, in its *Guideline for Aggregates* (DOE, 1994), it is relying on greatly increased use of secondary aggregates as part of its long-term strategy to meet aggregate demand. However, the actual amount of secondary materials employed in the future will depend on:

- further research;

- improved data on the specifications required for, and the suitability of, the product for differing applications;
- promotion of the qualities of the various materials; and
- possibly, an acceptance by the end user of an increased degree of risk.

2.7 CONCLUSION

This chapter has considered the problem faced by the Engineer of trying to predict aggregate performance in order to provide the public with a safe and longer-lasting road surface. Rather than supply a set of answers, its aim has been to increase awareness of the complexities of aggregate performance and the limitations of current selection methods. It has also introduced future European test methods which will cause fundamental changes to British specification requirements. Finally, as aggregate is a naturally occurring product and prone to change, can its future assessment against a recipe specification be allowed to continue?

2.8 REFERENCES

Manual of Contract Documents for Highway Works. Her Majesty's Stationery Office, London.
  Volume 1: Specification for Highway Works (MCHW 1).
Association Française de Normalisation (1982) *Aggregates – Characteristics of Aggregates Intended for Road Works. Norme Française Homo-Logueé NF P18-321, May 1982*. Association Française de Normalisation, Paris.
Aggregates Committee, The (1992) *Aggregates for Pavements. Frost-Resistance Test. V-209*,Reykavik, Iceland.
Anon (1996) Analysis of the markets and prospects for aggregates, ready-mixed concrete, asphalt and lime in 16 European cuntries, *BDS Marketing & Research*, May.
British Aggregates & Construction Materials Industries (1994) *Why Quarry, Quarrying: An Introduction. Information Sheet 1*, British Aggregates & Construction Materials Industries, London.
British Aggregates & Construction Materials Industries (1996) *A Question of Balance – Recycling*, British Aggregates & Construction Materials Industries, London.
British Standards Institution (1985a) *Testing Aggregates. Part 103: Methods for Determination of Particle Size Distribution, Section 103.1: Sieve Tests. BS 812: Part 103.1: 1985*, British Standards Institution, London.
British Standards Institution (1985b) *Testing Aggregates. Part 105: Methods for Determination of Particle Shape, Section 105.1: Flakiness Index. BS 812: Part 105.1: 1985*, British Standards Institution, London.
British Standards Institution (1989a) *Testing Aggregates. Part 121: Methods for Determination of Soundness. BS 812: Part 121: 1989*, British Standards Institution, London.
British Standards Institution (1989b) *Testing Aggregates. Part 114: Methods for Determination of the Polished-Stone Value. BS 812: Part 114: 1989*, British Standards Institution, London.

References

British Standards Institution (1990a) *Testing Aggregates. Part 112: Methods for Determination of Aggregate Impact Value (AIV). BS 812: Part 112: 1990*, British Standards Institution, London.
British Standards Institution (1990b) *Testing Aggregates. Part 110: Methods for Determination of Aggregate Crushing Value (ACV). BS 812: Part 110: 1990*, British Standards Institution, London.
British Standards Institution (1990c) *Testing Aggregates. Part 111: Methods for Determination of 10% Fines Value (TFV). BS 812: Part 111: 1990*, British Standards Institution, London.
British Standards Institution (1990d) *Testing Aggregates. Part 113: Methods for Determination of Aggregate Abrasion Value (AAV). BS 812: Part 113: 1990*, British Standards Institution, London.
British Standards Institution (1992) *Hot Rolled Asphalts for Roads and Other Paved Areas; Part 1, Specification for Constituent Materials and Asphalt Mixtures; Part 2, Specification for Transport, Laying and Compaction of Rolled Asphalt. BS 594: Part 1: 1992, BS 594: Part 2: 1992*, British Standards Institution, London.
British Standards Institution (1993) *Coated Macadam for Roads and Other Paved Areas; Part 1, Specification for Constituent Materials and for Mixtures; Part 2, Specification for Transport, Laying and Compaction. BS 4987: Part 1: 1993, BS 4987: Part 2: 1993*, British Standards Institution, London.
British Standards Institution (1995) *Testing Aggregates. Part 2: Methods for Determination of Density. BS 812: Part 2: 1995*, British Standards Institution, London.
Bullas, J.C. and G. West (1991) *Specifying Clean, Hard and Durable Aggregate for Bitumen Macadam Roadbase*, Department of Transport TRRL Research Report 284, Transport and Road Research Laboratory, Crowthorne.
Comité Européen de Normalisation (1991a) *Method of Test for the Determination of Resistance to Wear of Aggregates: Micro-Deval Test. Draft European Standard CEN/TC 154/SC6/N136E*, Comité Européen de Normalisation TC 154/TC 154.
Comité Européen de Normalisation (1991b) *Method of Test for the Determination of Resistance to Impact of Aggregates: Schlagversuch, Draft European Standard CEN/TC 154/SC6/N135E*, Comité Européen de Normalisation TC 154/TC 154.
Comité Européen de Normalisation (1991c) *Method of Test for the Determination of Resistance to Fragmentation of Aggregates: Los Angeles Test. Draft European Standard CEN/TC 154/SC6/N134E*, Comité Européen de Normalisation TC 154/TC 154.
Comité Européen de Normalisation (1991d) *Method of Test for the Determination of Resistance to Freeze/Thaw. Draft European Standard CEN/TC 154/SC6/N148E*, Comité Européen de Normalisation TC 154/TC 154.
Comité Européen de Normalisation (1995) *Tests for Geometric Properties of Aggregates, Part 12: Assessment of Fines – Methylene Blue Test. Draft European Standard prEN 933–12*, Comité Européen de Normalisation TC 154/WG.
Craig, C. (1991) A Study of the Characteristics and Role of Aggregate Dust on the Performance of Bituminous Materials, DPhil Thesis, Department of Civil Engineering and Transport, University of Ulster.
Curtis, C.W., R.L. Lytton and C.J. Brannan (1992) *Influence of Aggregate Chemistry on Absorption and Desorption of Asphalt*, 71st Annual Meeting, Transportation Research Board, Washington DC.
Defence Works Services (1995) *Marshall Asphalt for Airfield Pavement Works; Hot Rolled Asphalt and Coated Macadam for Airfield Pavement Works*, Defence Works Functional Standards, Her Majesty's Stationery Office, London.
Department of the Environment (1994) *Guidelines for Aggregates*, MPG6, Her Majesty's Stationery Office, London.
Knight, B.H. (1935) *Road Aggregates, Their Use and Testing*, The Roadworkers' Library, Volume III, Edward Arnold & Company, London.

Laboratoire Central des Ponts et Chausées (1963) Adhesion test on vialit plate, *Bulletin de Liaison Ponts et Chausées* 4, Laboratoire Central des Ponts et Chausées, Paris.

Lebas, A. and A Maldonado (1989) Un Centre d'Essais Lourds au Service de la Recherche et de la Promotion dans l'Industrie des Carriers, *Bulletin Liaison Laboratoire Ponts et Chaussées* 160, Feb.–Mar., Laboratoire Central des Ponts et Chausées, Paris, 37–42.

McCool, P.D. (1990) Rational Design Method for Surface Dressing, DPhil Thesis, University of Ulster.

Perry, M. (1996) A Study of the Factors Affecting the Polishing of Gritstone Aggregate, DPhil Thesis, University of Ulster.

Property Services Agency (1987) *Standard Specification Clauses for Airfield Pavement Works; Part 4. Bituminous Surfacing,* Property Services Agency, Croydon.

Sorby, H.C. (1880) On the Structure and Origin of Non-Calcareous Rocks, *Proceedings of the Geological Society*, **36**, Geological Society, London, 46–96.

Thompson, A., J.R. Greig and J. Shaw (1993) *High Specification Aggregates for Road Surfacing Materials,* Technical Report, Department of the Environment, London.

Wentworth, C.K. (1922) The shape of beach pebbles, *Geological Survey, Professional Paper* 131, Geological Society, London, 75–102.

Woodside, A.R. (1981) A Study of the Characteristics of Roadstones with Particular Reference to Polishing and Skidding Resistance, MPhil Thesis, University of Ulster.

Woodside, A.R. and W.D.H. Woodward (1984) *The Methylene Blue Absorption Soundness Test,* Internal Report, Department of Civil Engineering, University of Ulster.

Woodside, A.R. and W.D.H. Woodward (1993) Assessing surfacing aggregate performance – is clean, hard and durable enough? *Municipal Engineer, Proceedings of the Institution of Civil Engineers*, Institution of Civil Engineers, London, 151–5.

Woodside, A.R., W. Given, W.D.H. Woodward and M. Megaw (1993) The relationship of unbound aggregate fines to pavement performance, *Proceedings of Euroflex 1993; European Symposium on Flexible Pavements*, Lisboa, Portugal.

Woodside, A.R., W.D.H. Woodward, B. Kelly and C. Lycett (1993) Assessing the factors which influence the permeability and strength of a porous asphalt, *Proceedings of 5th Eurobitume Congress*, Stockholm, 623–6.

Woodside, A.R., W.D.H. Woodward, T.E.I. Russell and R.A. Peden (1996) The relationship between aggregate mineralogy and adhesion to aggregate, *Performance and Durability of Bituminous Materials, Proceedings of Symposium, University of Leeds*, Mar., E & F N Spon, London.

Woodward, W.D.H. (1995) Laboratory Prediction of Surfacing Aggregate Performance, DPhil Thesis, University of Ulster.

Zingg, T.H. (1935) A contribution to the analysis of coarse gravel, *Schweizerische Mineralogische und Petrographische Mitteilungen*, **15**, 133–40.

CHAPTER 3

Binders

P.J. Green, BP Research

3.1 INTRODUCTION

By the middle of the nineteenth century, the problem of dust generated by the horse-drawn traffic in towns was causing much concern (section 1.1). However, the change from horse-drawn to motor vehicles not only greatly exacerbated the dust problem but also placed new demands on the smoothness and durability of the road surface. This led to serious attention being given to the materials used in the construction of streets and roads.

The availability of a regular supply of tar from local coal gas works provided one of the first materials for treating the previously unbound aggregates in streets and pavements. Later, rock asphalt and Trinidad lake asphalt were the sources of bitumen first used to bind mixtures of aggregates in road building in the modern style. By the turn of the century the potential of petroleum as a readily available source of low-cost, high-quality bitumen was being exploited on a small scale. From then to the present day, a range of petroleum bitumens have been developed and used in materials for use in road construction.

A general classification of binders for asphalt materials and surface treatments is shown in Figure 3.1. The most widely used of these in increasing importance are coal tars, natural asphalts and petroleum bitumens.

3.2 TAR

Coal tar was one of the first materials tried for suppressing the dust generated from the traffic on unbound roads. It was originally the by-product of the carbonization of coal during the production of town gas and coke. However, the production and widespread use of natural North Sea gas since the 1970s has meant that the main source of coal tar at the present time is from coke production for the steel and smokeless fuel industries.

Coke is produced either by the destructive distillation of coal at about 1000 °C or carbonization at lower temperatures (600 °C to 750 °C) for smokeless fuel production. The vapours from these processes are condensed and become the crude tars which must be refined further to obtain products suitable for use as road binders. Fractional distillation is

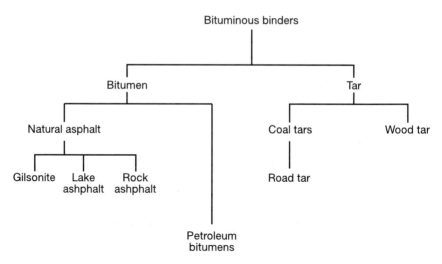

Fig. 3.1 General classification of binders for asphalt materials and surface treatments

used to remove volatile oils such as creosotes and the residue or pitch is used as the main component for road tars. A range of grades of road tar, varying in consistency, are produced by blending together various proportions of the residue pitch and the distillate oils. Tars from the two carbonization processes are termed 'high-temperature' and 'low-temperature' tar.

The properties of road tars are defined by BS 76 (BSI, 1974). The principal property used for the characterization of road tars is viscosity, which is indirectly measured by determining the time taken for a fixed quantity of tar to flow through a standard orifice in a container known as the 'Standard Tar Viscometer' (STV) shown in Figure 3.2. The higher the viscosity of the tar, the longer the efflux time and vice versa. The viscosity of the tar is expressed as an 'equiviscous temperature' (EVT), which is defined as the temperature at which the tar will have a flow time of 50 seconds from the STV. Having determined a flow time at a particular temperature, factors are provided in BS 76 for converting the flow time obtained at the flow temperature to an EVT.

The main grades of road tar are from 34 °C to 54 °C EVT in 4 °C grade intervals. The more viscous grades (higher EVT) are used for more heavily trafficked roads with the lower viscosity grades (lower EVT) being used where traffic levels are lower.

In recent years, the production and usage of coal tar for road purposes has dramatically declined. The two main reasons for this reduction are:

- the advent of a ready supply of North Sea gas resulted in the total disappearance of town gas and with it the supply of coal tar from this source; and
- there is a far greater awareness of the effects of materials on health and the environment.

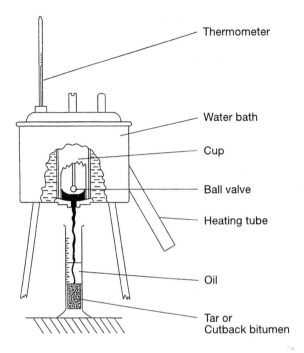

Fig. 3.2 Standard tar viscometer

It has been well established that tars and tar oils contain high proportions of polycyclic aromatic compounds (PCAs) (Table 3.1) which are known to be carcinogenic (Wallcave *et al.*, 1971) and concerns of such high levels of toxic chemicals on workers have restricted the use of tar even further. Indeed, in Germany, although tar is no longer used, there is much concern over the disposal of tar-based asphalt waste from worn out roads.

3.3 NATURAL ASPHALTS

In some parts of the world, bitumens occur naturally. In such cases, they are normally found mixed with inorganic material and are known as 'natural asphalts'. Although there are small deposits of natural asphalt in various parts of the world, the largest and best known is the 'Lake' in Trinidad. The lake covers an area of some 100 acres and is sufficiently hard to permit men to walk on it and even to support wheeled vehicles for short periods of time. The asphalt in the lake is moving to the extent that holes dug in the surface will slowly disappear over a period of about 24 hours. Despite the large quantity of asphalt which has been removed from the lake over the last century, the level of the lake is reported not to have dropped significantly.

Table 3.1 Polycyclic aromatic compounds (PCAs) in different bitumens and coal tar pitches

Polycyclic aromatic compound (mg/kg)	Bitumen			Coal tar pitch	
	A	B	C	A	B
Benzo (a) anthracene	2.1	0.7	35	8900	12500
Benzo (a) pyrene	1.7	2.5	27	8400	12500
Benzo (k) fluoranthrene	ND	+	ND	7100	9000
Indeno (1,2,3-cd) pyrene	ND	Tr	1.0	7300	9300

ND Not detected   + Not determined but present in small amount   Tr Trace

Trinidad lake asphalt (TLA) is dug from the lake and initially contains a mixture of bitumen, water and very finely divided mineral matter. The raw lake asphalt is purified locally by heating to about 160 °C in open stills to melt it and to drive off the water. Coarse foreign bodies such as plants, the remains of fallen trees and the occasional long lost item of equipment are removed by passing the liquid asphalt through a screen. Having gone through this process, the 'refined' TLA, or 'Trinidad Epuré' as it is known, typically has a penetration at 25 °C of about 2 dmm and a softening point of about 95 °C. The composition is typically as follows:

Bitumen   55% by weight
Mineral matter   35% by weight
Organic matter   10% by weight

The mineral matter is extremely fine, with nearly half of it being less than 10 microns in diameter. Being of such a small size, it is extremely difficult to remove from the asphalt and, therefore, must be taken into account when designing the composition of asphalt mixtures.

TLA is too hard to be used as binder for road surfacing materials and it is normally softened by blending with 100, 200 or 300 penetration grade bitumen to produce a binder with the required properties. TLA/bitumen blends are sometimes used in the United Kingdom for rolled asphalt surfacing courses where improved resistance to deformation is required. They are also used in Europe in various asphalt mixtures.

3.4 BITUMEN

3.4.1 Penetration grade bitumen

3.4.1.1 Production

By far the most commonly used binder in the world today is petroleum bitumen. This is manufactured by subjecting selected crude oils to atmos-

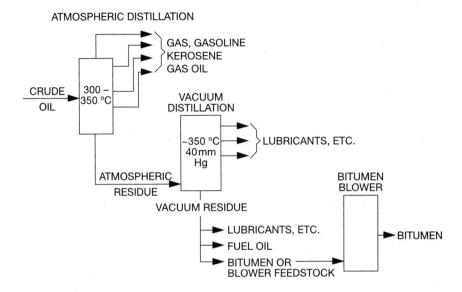

Fig. 3.3 Typical bitumen production scheme

pheric distillation followed by vacuum distillation (Figure 3.3). The vacuum distillation residue, also called 'vacuum residue', is the basic bitumen produced in the refining process. It is known as 'residual' or 'straight-run' bitumen.

For some crude oils, the whole range of paving grades can be produced simply by varying the temperature and/or vacuum conditions inside the vacuum distillation unit and separating more or less distillate from the residue. The vacuum residue is run out from the vacuum unit at typically 350 °C to 360 °C and passed, via a heat exchanger, to storage at about 180 °C to 200 °C. However, bitumen cannot be made by this route in all cases and further processing is required.

From over 1500 different crude oils which are available throughout the world, less than 100 are suitable on their own for bitumen manufacture. The two reasons for this are:

- the bitumen content of crude oil is very variable; and
- some crudes are not suitable for bitumen due to the basic chemical composition of the crude.

Figure 3.4 shows the relative amounts of various petroleum fractions obtained by distilling a range of crudes, from very heavy to very light. At the very light extreme some crudes, such as Gippsland from Australia, contain less than 1% bitumen while, at the very heavy extreme, Vittorio from Sicily contains over 90% bitumen. Most bitumen producers want to manufacture other products in addition to bitumen and for this reason Vittorio would not be preferred as a bitumen crude. At the other end of the scale Gippsland contains so little bitumen that it would be uneconomic to use it for bitumen production. Crude oils are normally selected to give a bitumen yield of between 15% and 60% by weight of crude. On this basis, the most suitable crudes for bitumen production are those in the range from heavy to medium.

As an example of the unsuitability of some crudes for bitumen, nearly all crudes from the Far East such as Indonesia and Australia contain a lot of wax. This wax is not completely removed by the vacuum distillation process and would be present in significant amounts (up to 50% by weight) in the final bitumen. The result would be a poor quality bitumen with poor adhesion to aggregate. For this reason, Middle East crudes, from which bitumens with much lower wax contents (generally up to 6%) can be produced, have to be imported into the Far East and Australia.

One way of looking at the differences between heavy and light crudes is the amount of 200 pen bitumen which may be obtained. An alternative is the effective distillation temperature, or true boiling point, required to produce a 200 pen bitumen. Figure 3.4 also shows how the effective distillation temperature (cut point) varies considerably from crude to crude, with the heavier crudes producing 200 pen bitumen at lower temperatures than lighter crudes.

Although the vacuum residue may be used directly as bitumen, with some crude oils further processing is necessary. The two main reasons for this are that:

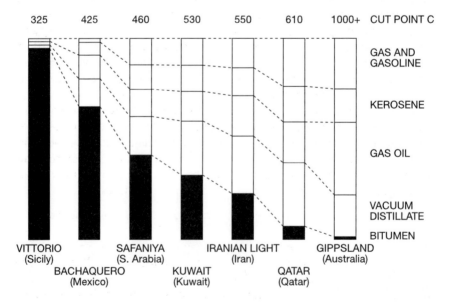

Fig. 3.4 Different crude oil compositions

- there are physical limits to the equivalent distillation temperature which can be achieved due to the design of the vacuum unit and vacuum pumps such that, in practice, 560 °C to 570 °C is about the maximum equivalent distillation temperature (cut point) that is used. While it may not be possible to produce the harder grades even at the maximum cut point of the vacuum distillation unit; and
- further processing may be required in that the properties of the vacuum residue may not be entirely suitable for a finished bitumen such as, for example, a certain vacuum residue may have a low softening point for the penetration grade or too low a viscosity at mixing and laying temperatures.

In such cases, the normal course is to blow air through the vacuum residue. Air blowing is a controlled process which may be carried out in a batch or continuous mode using a blowing tower typically about 12 m (40 feet) high and about 3.65 m (12 feet) in diameter. In a continuous operation vacuum residue, typically at 160 °C to 180 °C passes through a heat exchanger and enters the blowing tower. Air is pumped into the base of the tower via spargers. The chemical reaction which takes place between the oxygen in the air and the bitumen generates heat and raises the temperature of the product in the blower to about 200 °C to 260 °C. Product is drawn off from the base of the blower and passed through a heat exchanger to pre-heat the vacuum residue going into the tower and the product then passes to storage at about 180 °C to 200 °C. Because the blowing process generates a lot of heat, it is necessary to inject steam and water into the top of the tower above the level of bitumen to stop the temperature rising too high and prevent a runaway condition. The off-gas from the blowing tower passes out through the top of the unit, through a knock out drum to separate gas from liquid droplets and the gases are then passed on to an incinerator via a flame arrestor which prevents blow back of the incinerator flame. The process is quite simple in principle but must be carried out under carefully controlled conditions.

Hence, it should be clear that bitumen is definitely not a by-product of the petroleum industry. The selection of suitable crude oils for bitumen manufacture is extremely important if bitumen is to be produced both economically and within specification. It is the skill of the bitumen refiner in making that selection which allows the economic production of bitumens with suitable properties for their designed application.

3.4.1.2 *Chemical nature*

Bitumens are complex mixtures of very many chemical compounds of high molecular weight, typically between 500 and 50 000. Because of its complexity, a complete analysis of the composition of bitumen would be almost impossible. However, it is possible to characterize the constituents by a wide variety of methods such as solubility in different solvents, by molecular weight or by broad chemical type. Chromatography, in conjunction with solvent extraction, is commonly used to separate the components of bitumen into the following four broad chemical groups:

- **Asphaltenes**: These are brown/black amorphous solids of high molecular weight, typically 1000–50 000. They are precipitated from bitumen by dissolving in a paraffinic solvent such as n-heptane. Asphaltenes generally comprise 5–25% by weight of the bitumen.
- **Resins**: Dark brown solids or semi-solids which are soluble in heptane. They are adhesive and very polar in nature. Molecular weights are typically 900–1300. Resins may comprise 5–50% by weight of the bitumen.
- **Aromatics**: These are generally dark brown viscous liquids. Molecular weights are typically 500–900. Usually present at 40–60% by weight in the bitumen.
- **Saturates**: Solids or viscous liquids, light coloured with molecular weights in the range 500–800. This fraction may be present from 1 to 25% by weight in the bitumen.

The resins, aromatics and saturates fractions are often known collectively as the 'maltenes' fraction.

These four general compound types are regarded as making up bitumen in a colloidal system in which the asphaltenes are present as 'micelles' dispersed in the lower molecular weight maltenes. It is generally accepted that the asphaltene micelles are stabilized by a sheath of compounds predominantly found in the resins fraction. The sheathed asphaltenes are dispersed in the 'oily' medium of aromatics and saturates, as shown in Figure 3.5.

When a harder bitumen is produced by increasing the temperature and/or the vacuum conditions inside the distillation column, it is the more volatile saturates and aromatics fractions which are removed from the residue, thus concentrating the asphaltenes and resins. Therefore, a hard bitumen will have a higher asphaltenes content than a softer grade from the same crude oil (Figure 3.5). In the production of harder bitumens by air blowing, the concentration of the saturates, being relatively inert to oxidation, remains largely unchanged. However, the reactivity of the aromatics and resins fractions to air and the resulting formation of asphaltenes from this reaction, increases the concentration of the asphaltenes in the bitumen more so than by distillation alone. Thus, a bitumen produced by air blowing will have a higher asphaltenes content than an equi-penetration bitumen produced by vacuum distillation from the same crude oil.

At ambient temperatures, bitumen is a very stable, inert material, especially when it is in bulk. However, when exposed to the atmosphere in very thin films, as in macadams, it slowly hardens until eventually it becomes brittle. The mechanism of this hardening is complex, as would

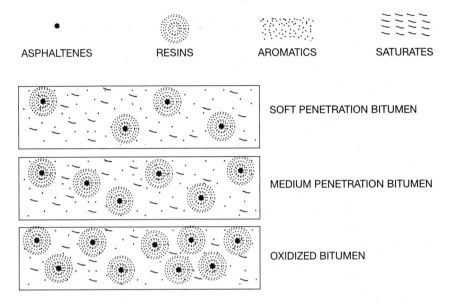

Fig. 3.5 Colloidal nature of bitumen

be expected from the complex nature of bitumen. Also the mechanism is difficult to study under real application conditions due to the contamination from traffic such as droppings of lubricating oil and diesel fuels and waxes from tyres. However, studies under controlled conditions in the laboratory have shown that oxidation is the main reaction occurring and this results in similar changes in the broad chemical composition which occur at high temperatures.

At normal bitumen storage temperatures of 130 °C to 180 °C, bitumen is fairly stable in bulk and may be stored hot for quite long periods without significant changes occurring. This is due to the very restricted access of oxygen to bitumen inside a storage tank. However, if the bitumen is pumped to and from storage and bitumen is allowed to return to the tank via a 'splash-back' system through the vapour space above the bitumen in the tank, significant hardening can occur due to greater access of oxygen.

During mixing, storage, transport and laying of asphalts and macadams, thin films of bitumen are exposed at high temperatures to atmospheric oxygen. Under these conditions, significant chemical changes can occur which result in significant physical changes. As in the production of bitumen by air blowing, the main chemical changes which are believed to take place are oxidation, condensation and polymerization. In consequence, the net changes in composition as measured by the broad chemical composition described above are substantial reductions in the aromatics fraction, together with increases in the resins and asphaltene contents. The saturates contents are usually little changed, due to the relatively low reactivity of the compounds in this fraction. These changes in the chemical composition result in an increase in the average size of the molecules present (the molecular weight increases) and this in turn is accompanied by a hardening of the bitumen. Once asphalts are compacted, providing the mixture has sufficiently low air voids, analysis of bitumens recovered from road trial sections has shown that relatively few further changes occur in the chemical composition as a result of oxidation. However, with permeable mixtures, such as macadams and porous asphalt, the bitumen will continue to harden and eventually become brittle and will crack either under traffic or thermally-induced stresses or a combination of both.

3.4.1.3 Physical properties

The main physical properties which are relevant to the performance of a bitumen as a road binder are its rheological (elastic stiffness and flow) properties and its mechanical strength.

The mechanical strength of bitumen is important at low temperatures when it becomes progressively harder and eventually brittle. In this brittle state, the fracture strength is of the order of 4×10^6 Pa, which is low compared to other engineering materials such as steel and concrete. The temperature at which bitumen becomes brittle is dictated by its rheological properties, principally the grade or hardness of the bitumen.

The rheological properties of bitumen may be expressed in both fundamental and empirical properties. Traditionally, empirical properties have been used to characterize and specify bitumens throughout the world. The properties mostly used for this characterization are as follows.

Needle penetration (BS 2000: Part 49 (BSI, 1993a), IP 49)
This test measures the consistency of bitumen by measuring the depth to which a standard (but arbitrary) needle will penetrate into a sample under a specified load at a specified temperature (Figure 3.6). Normally, the load is 100 g and the temperature is 25 °C. Clearly, the higher the penetration the softer the bitumen and vice versa.

Softening point (BS 2000: Part 58 (BSI, 1993b), IP 58)
The temperature at which a disc of bitumen is unable to support a standard metal ball (Figure 3.7). The softening point is approximately an equiviscous temperature for unmodified bitumens, when such bitumens have a viscosity of approximately 1200 Pa.s.

Fraass breaking point (IP 80)
A measure of the brittleness expressed as the temperature at which cracks appear in a thin film of bitumen on a thin metal plate when the

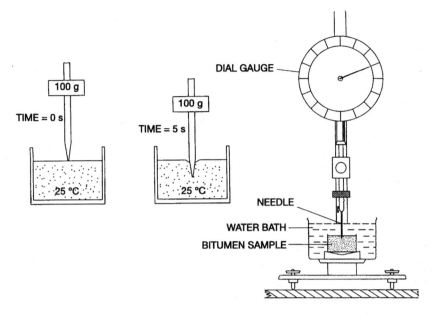

Fig. 3.6 Bitumen penetrometer and penetration test

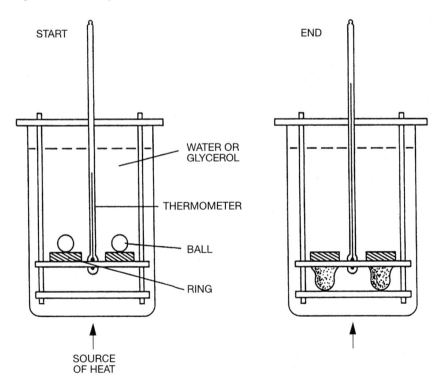

Fig. 3.7 Softening point (ring and ball) test

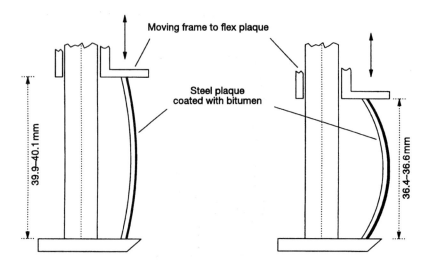

Fig. 3.8 Fraass breaking point test

latter is simultaneously cooled and flexed under standard conditions (Figure 3.8). It has been shown that the Fraass breaking point is approximately an equi-stiffness temperature for unmodified bitumens, the 'equal' stiffness being about 100 MPa at a loading time of 11 seconds.

While the above empirical tests have served very well over the years for specifying bitumens, increasing demands on road-building materials as a result of increasing traffic densities and axle loadings have led to the development of a range of high performance road binders containing various additives. The properties of these high-performance modified binders can be quite complex compared to the traditional unmodified bitumens and it has become clear that the empirical methods of specification are not necessarily adequate in ensuring satisfactory performance. As a result of this, there has been a strong move, particularly in the United States of America following the Strategic Highways Research Program (SHRP) (Section 1.5.4), towards specifications which define the relevant properties of bitumens in more fundamental terms (Anderson, 1994).

Before describing these performance-based specifications, it is useful to consider the fundamental properties of bitumen. Bitumen is a 'visco-elastic' material in that it exhibits either purely viscous properties or purely elastic properties or (most usually) a combination of viscous and elastic properties, depending on the temperature and rate of loading.

At high temperatures, above about 100 °C, bitumen behaves as Newtonian liquid in that its viscosity is not dependent on the stress applied but simply on the temperature. Several viscometers are commercially available which are suitable for measuring the viscosity of bitumen at these temperatures. Many rely on the principle of applying a shear stress or shear strain to a sample of bitumen contained between two

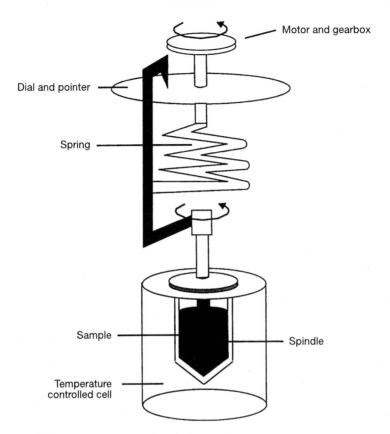

Fig. 3.9 High-temperature rotational viscometer

surfaces moving relative to each other and measuring the resulting strain rate or shear stress respectively. The viscosity is defined as the ratio of the shear stress to the shear strain rate and the units are Pascal seconds (Pa.s). A schematic of a high-temperature rotational viscometer is shown in Figure 3.9.

An alternative to the 'shear' viscometers is the capillary viscometer. This is essentially a narrow bore tube through which the sample of bitumen flows either under its own weight or under the influence of a known pressure, as shown in Figure 3.10. The time required for the bitumen to flow between two marks on the tube is converted to a viscosity using calibration factors.

For a large number of bitumens, it has been found that the change in viscosity with temperature is described by Equation 3.1.

$$\text{Log}_{10}[\log_{10}(\eta)] = A + B\log_{10}(T + 273) \qquad (3.1)$$

where η = the viscosity (MPa.s);
T = the temperature (°C); and
A and B are positive and negative constants, respectively.

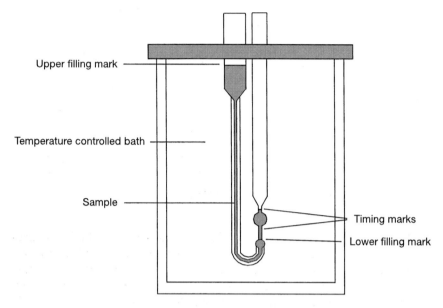

Fig. 3.10 Capillary viscometer

By plotting the left hand side of Equation 3.1 against $\log_{10}(T + 273)$, viscosity–temperature charts may be constructed which provide linear plots over a reasonably wide temperature range.

As the temperature is reduced, the viscosity of bitumen increases and, at very low temperatures below about −10 °C, bitumens almost behave as elastic solids and will instantly deform when a stress is applied and instantly recover when the stress is removed. However, at these low temperatures, bitumen is in a very vulnerable condition due to its low mechanical strength and, if the applied stress is too high, fracture will occur.

At intermediate temperatures, bitumen exhibits 'visco-elastic' behaviour in that it deforms as both an elastic solid (instantaneously) and as a liquid (over period of time). This combination of viscous and elastic behaviour is illustrated in the creep test depicted in Figure 3.11. When subjected to a stress for a period of time, the bitumen will deform in the manner depicted. This deformation is composed of the following three elements:

- instant elastic (recoverable) strain;
- delayed elastic (recoverable) strain; and
- permanent (non-recoverable) strain or viscous flow.

The actual relationship will depend on the temperature; viscous flow will dominate at high temperatures while, at low temperatures, the amount of viscous flow will be much reduced.

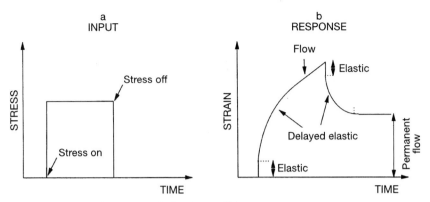

Fig. 3.11 Visco-elastic creep behaviour of bitumen

When the applied stress is removed from the bitumen, the initial elastic strain is recovered instantly and the delayed elastic strain is recovered over time but the permanent strain (or viscous flow) is not recovered. It is the viscous flow which, after repeated loading from traffic, eventually results in permanent deformation of asphalt mixtures. Creep tests of this type are useful in evaluating the properties of bitumens at long loading times and/or low temperatures.

To evaluate the behaviour of bitumens under conditions which more accurately simulate traffic movement, oscillation tests are used. It is usual in such tests to subject the bitumen to an oscillating sinusoidal stress and measure the resulting strain (Figure 3.12). The magnitude of the resulting sinusoidal strain which is induced in the bitumen is directly related to the

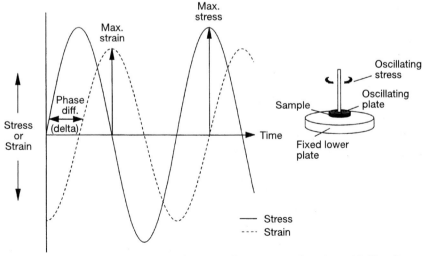

Fig. 3.12 Deformation response of a visco-elastic material to sinusoidal loading

complex stiffness modulus, G^*, of the bitumen where G^* is the peak applied stress divided by the peak applied strain. However, due to its visco-elastic nature, the strain induced in the bitumen will be out of 'phase' with the applied stress. The degree to which the strain is out of phase with the stress is known as the phase angle, and is a measure of the visco-elastic character of the bitumen under the specific conditions of temperature and frequency of loading. This visco-elastic character is usually described by the elastic and viscous moduli, G' and G'' respectively, which are related to the phase angle and the complex modulus by Equations 3.2 and 3.3.

$$[G^*]^2 = [G']^2 + [G'']^2 \quad (3.2)$$

$$\tan(\delta) = \frac{G''}{G'} \quad (3.3)$$

By varying the temperature at which the stress is applied to the bitumen, it is possible to simulate a wide range of traffic loading and climatic conditions and, hence, obtain much information about the performance of binders.

3.4.1.4 Bitumen specifications

Specifications represent primarily the attempts by highway authorities to ensure that bitumens used in the region for which they are responsible

Table 3.2 Properties included in National Specifications for Road Bitumens

Property	Country				
	UK	France	Germany	USA	Australia
Penetration at 25 °C	✓*	✓*	✓	✓†	
Penetration at 15 °C					✓
Softening point	✓	✓	✓		
Fraass point		✓	✓		
Ductility			✓	✓†	✓‡
Wax content		✓	✓		
Ash content			✓		
Solubility	#✓	#✓	#✓	#✓	#✓
Density		✓	✓		
Flash point		✓		✓	✓
Durability					✓
Viscosity at 60 °C				✓	✓‡
Viscosity at 135 °C				✓	✓
Rolling Thin Film Oven Test (RTFOT)					✓
Thin Film Oven Test (TFOT)				✓	
Loss on heating	✓	✓			

* Also after loss on heating ‡ Also after RTFOT
† Also after TFOT # Various solvents are specified

Table 3.3 SHRP binder specifications

Temperature	Functional property	Measurement	Parameter specified	Method/ equipment
High	Pumping/ mixing/laying	Viscosity	Newtonian viscosity	Brookfield Viscometer
Intermediate	Rutting Fatigue cracking	Binder stiffness, elastic and viscous components	$G^*/\sin(\delta)$ $G^*.\sin(\delta)$	Dynamic Shear Rheometer
Low	Thermal cracking	Creep stiffness	Creep stiffness after 60 seconds loading	Bending Beam Rheometer
		Strain at break	Failure strain for 1 mm/min elongation	Tensile Test

are suitable for the purposes intended. The specifications attempt, or should attempt, to allow for significant variations in climatic and traffic conditions from country to country and within a country.

Bitumen specifications have developed over the years on a virtually national basis. This is not entirely due to variations in climate or traffic but also to different perceptions of which properties are important for good performance. Table 3.2 summarizes the main features of some national bitumen specifications. Bitumen consistency or grade is generally specified either by penetration (as in Europe) or by viscosity (as in United States of America and Australia). Additional properties are also specified to try to control the rheological behaviour and chemical and ageing characteristics. Bitumen specifications in Europe are being collated and revised under the Comité Européen de Normalisation (CEN) to standardize on one set of specifications for all the member states (section 1.5.2).

The specifications shown in Table 3.2 are mainly based on empirical test methods such as penetration, softening point and ductility. While specifications based on such methods have proved adequate in the past and will probably continue to do so for unmodified binders, the advent of high performance, modified binders has exposed deficiencies in empirically based specifications for this type of binder.

In an attempt to improve the condition of the road network in the United States of America, SHRP (section 1.5.4) investigated many aspects of bitumen and asphalt performance. One of the results of this five-year programme was a set of specifications for binders for use in asphalt concrete which were based on fundamental rheological and engineering properties rather than those measured by empirical tests. The main elements of the specification address the performance properties of the bitumen at application temperatures and also at the service temperatures encountered under particular climatic conditions (Table 3.3). For performance at service temperatures, the specification defines limits for the rheological properties of the binder which studies have shown to relate to the pavement distress modes of permanent deformation, fatigue cracking and low temperature cracking. Limiting values of the performance parameters must be met at the extremes of pavement temperature expected in service. In addition, the Rolling Thin Film Oven Test (RTFOT) is specified to simulate the ageing which a binder undergoes during mixing in a conventional asphalt pug-mill and a pressure ageing vessel (PAV) has been developed to simulate ageing after 5 to 10 years' service (Anderson, 1994).

3.4.2 Bitumen emulsion

An emulsion is a dispersion of minute globules of one liquid throughout another liquid with which it is not miscible. Bitumen emulsions are examples of oil-in-water emulsions. The bitumen content usually lies between 30% and 70% depending on the application for the emulsion. The primary object of emulsifying the bitumen is to obtain a low viscosity product which can be used without the heating which is normally required for using bitumens. When an emulsion is used, the water is readily lost by evaporation or it may separate from the bitumen because of the chemical nature of the surface to which the emulsion was applied. This process is known as breaking.

Bitumen emulsions are normally produced by dispersing hot bitumen in water containing an emulsifying agent (soap). Dispersion is achieved by feeding the bitumen and water phases via metering pumps into a colloid mill (Figure 3.13). The mill essentially comprises a high speed rotor revolving inside a stator. The high shear inside the mill produces the very small particles of bitumen (the dispersed phase) dispersed in the water (the continuous phase). Bitumen droplet sizes are typically in the range <1 to 20 microns diameter. Because of absorption of the emulsifying agent onto the surface of the bitumen droplets, they remain in suspension and do not readily coalesce.

Emulsifying agents (emulsifiers) are chemicals whose molecules have different affinities for water and oil at each end of the molecule and, as such, are only partially soluble in both. The hydrophobic (water insoluble, oil soluble) end of the emulsifier molecule is usually a hydrocarbon chain which will orientate itself into the bitumen droplets (Figure 3.14). The hydrophillic (water soluble, oil insoluble) end usually has an electrostatic charge and orientates itself into the water phase. In this way, the emulsifier gives the bitumen droplets an electrostatic charge which causes mutual repulsion and gives the emulsion its stability. The type or polarity of the electrostatic charge depends on the emulsifier used. Soaps

Bitumen 65

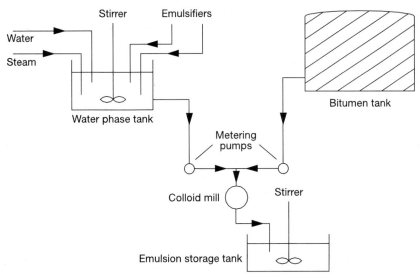

Fig. 3.13 Schematic diagram of bitumen emulsion manufacture

of fatty acids, such as sodium or potassium oleate, provide a negative charge to the bitumen droplets (Equation 3.4).

$$RCOOH + NaOH \rightarrow RCOO^- + Na^+ + H_2O \qquad (3.4)$$

where R is a hydrocarbon chain, typically C_{14} to C_{20};
       RCOOH is a fatty acid;
       NaOH is sodium hydroxide; and
       $RCOO^- + Na^+$ is a fatty acid soap.

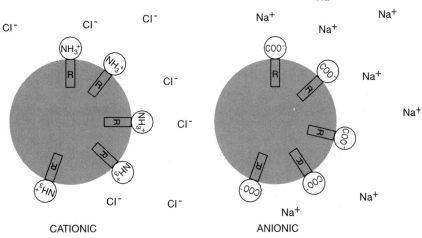

Fig. 3.14 Stabilization of bitumen droplets by emulsifiers

Where the bitumen droplets are negatively charged in this way, the emulsions are called 'anionic' emulsions. Other types of emulsifier can impart a positive charge to the bitumen globules. Examples of such emulsifiers are amines which have been neutralized with hydrochloric or acetic acid (Equation 3.5).

$$RNH_2 + HCl \rightarrow RNH_3^+ + Cl^- \qquad (3.5)$$

where R is a hydrocarbon chain, typically C_{14} to C_{20};
        RNH_2 is an amine;
        HCl is hydrochloric acid; and
        $RNH_3^+ + Cl^-$ is amine hydrochloride.

Emulsions containing positively charged bitumen droplets are called 'cationic' emulsions.

The main characteristics of bitumen emulsions which are important for their use are viscosity, stability and rate of breaking.

The viscosity of an emulsion is important in determining the ease with which the emulsion can be handled and applied. The viscosity depends on the following two main features of the emulsion:

- bitumen content; and
- the size of the bitumen droplets.

For a given droplet size distribution, bitumen contents up to about 60% by weight bitumen have relatively little effect on emulsion viscosity (Figure 3.15). However, above about 65% by weight bitumen content, the viscosity of the emulsion rises significantly. For this reason, emulsions for surface dressing which usually contain about 70% by weight bitumen are normally applied hot to reduce the viscosity of the emulsion for spraying.

The electrostatic forces between the charged droplets of bitumen in the emulsion cause an increase in the emulsion viscosity ('electroviscous

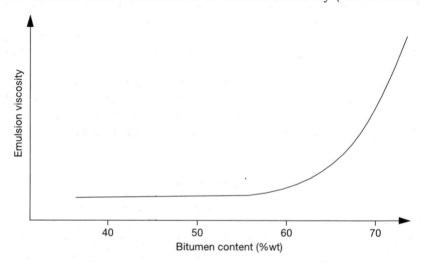

Fig. 3.15 Viscosity of emulsion as a function of bitumen content

effect'). For a given bitumen content, increasing the emulsifier content increases the electrostatic forces and, thereby, increases the emulsion viscosity.

The stability of an emulsion is important not only to ensure satisfactory storage times but also the rate of breaking. The stability can be regulated by the amount and/or the type of emulsifier used and controls the ability not only to store and transport the emulsion but also its breaking properties. For both anionic and cationic emulsions, there are several grades of different breaking characteristics. Rapid-setting emulsions are used for surface dressing, while medium or slow-setting emulsions are used for 'mixtures', that is mixed with aggregate either in concrete-type mixers or in situ. Rapid-setting emulsions are not used for mixtures because they would tend to set during the mixing process and clog the mixer. Generally in making mixtures, the finer the aggregate used, the slower setting the emulsion that has to be used. Therefore, stone mixtures require medium-setting emulsions and sand mixtures require a slow-breaking emulsion.

With anionic emulsions, the breaking process is predominantly by evaporation of the water in the emulsion continuous phase. Because of this, anionic emulsions are susceptible to temperature and humidity in terms of their breaking properties. In contrast, cationic emulsions break mainly by chemical coagulation. When the emulsion is brought into contact with most types of stone (particularly those with an appreciable silica content such as granite, quartzite and basalt), there is an interaction of charges which breaks the emulsion. Thus, a more rapid break is possible with cationic emulsions because they are less susceptible to weather conditions than anionic emulsions. Because of this, they are being increasingly used in place of anionic emulsions.

Anionic and cationic emulsions must not be mixed together because the opposite charges on the bitumen droplets in each type of emulsion will cause rapid coalescence of the dispersed bitumen.

Emulsions must be protected from frost during storage. If freezing occurs, there is a risk of premature breaking of the emulsion in the storage container.

In Europe, the vast majority of bitumen emulsions are used for surface dressing. Slurry sealing is also an important technique in many parts of the world which have long distances of lightly trafficked roads and, in this application, bitumen emulsions are used almost exclusively.

3.4.3 Cutback bitumen

Cutback bitumens are blends of penetration grade bitumen (usually 100 penetration) with a solvent or flux. The lower viscosity of cutback bitumens allows lower working temperatures than would be required for unmodified penetration grades. In this respect, they are similar to emulsions. The main applications of cutback bitumens are for surface dressing and, to a lesser extent, for macadam mixtures. Whether the application is in surface dressing or macadam mixtures, the performance of the cutback relies on the evaporation of the flux during application and service at a predictable rate, resulting in a residual binder with the appropriate properties.

A variety of flux oils are used throughout the world, including white spirit, kerosene, gas oil, coal tar oils (creosotes) and mixtures of these. The choice of flux used is determined by the rate at which the cutback 'cures', or hardens in use, due to evaporation of the flux. This evaporation is related to the volatility of the flux oil and climatic conditions. In addition to these 'commodity' distillates, proprietary fluxes with carefully tailored evaporation characteristics are also used.

Cutback bitumens are commonly classified and specified by viscosity, rate of cure and solvent distillation characteristics and penetration of the base bitumen. Viscosity depends on the amount of flux present in the cutback and is commonly specified and measured using an efflux viscometer. In the United Kingdom, the Standard Tar Viscometer (STV) is used at 40 °C. Three grades of cutback bitumens are recognized by BS 3690: Part 1 (BSI, 1989), these being 50 seconds, 100 seconds and 200 seconds.

In addition to viscosity, cutback bitumens are commonly described by their curing or setting rates as one of the following:

RC   Rapid curing (commonly white spirit based)
MC   Medium curing (commonly kerosene based)
SC   Slow curing (commonly gas oil based)

In the United Kingdom, British Standard cutback bitumens are of the kerosene based, medium curing type.

3.5 BINDER MODIFIERS

3.5.1 Generic types

The increasing demands of traffic on road building materials in recent years has resulted in a search for binders with improved performance relative to normal penetration grade bitumens. This effort to obtain improved binder characteristics has led to the evaluation, development and use of a wide range of bitumen modifiers which enhance the performance of the basic bitumen and hence the asphalt on the road. Table 3.4 lists the generic types of bitumen modifiers and additives, together with examples of each type.

3.5.2 Polymer modifiers

Polymers are playing an increasingly important role in the asphalt industry and are the most technically advanced bitumen modifiers currently available. As early as 1873, a patent was granted for an asphalt paving

Table 3.4 Types and examples of binder modifiers and additives

Modifier type	Example
Thermosetting polymers	Epoxy resin Polyurethane resin Acrylic resin Phenolic resin
Thermoplastic rubbers	Natural rubber (crumb or latex) Vulcanized rubber (e.g. tyre) Styrene-butadiene-styrene block copolymer (SBS) Styrene-butadiene rubber (SBR) Ethylene-propylene-diene terpolymer (EPDM) Isobutene-isoprene copolymer (IIR) Nitrile rubber Butyl rubber Styrene-isoprene-styrene block copolymer (SIS) Styrene-ethylene-butadiene-styrene block copolymer (SEBS) Polybutadiene (PBD) Polyisoprene
Thermoplastic crystalline polymers	Ethylene vinyl acetate (EVA) Ethylene methyl acrylate (EMA) Ethylene butyl acrylate (EBA) Polyethylene (PE) Polypropylene (PP) Polyvinylchloride (PVC) Polystyrene
Chemical modifiers and extenders	Organo-metallic compounds (e.g. manganese) Sulphur Lignin
Fibres	Cellulose Alumino-magnesium silicate Glass Asbestos Polyester Polypropylene
Adhesion agents	Amines Amides
Antioxidants	Amines Phenols Organo-zinc/organo-lead compounds
Natural asphalts	Trinidad lake asphalt Gilsonite

mixture containing rubber latex and, in 1902, rubber-modified asphalt was being laid in France. While there is a variety of polymers currently being used worldwide in the asphalt paving industry, they generally fall in one of the following two major polymer families:

- thermoplastic, crystalline polymers; and
- thermoplastic rubbers.

Crystalline polymers, often termed 'plastomers', include such materials as polyethylene, polypropylene, polyvinyl chloride (PVC), polystyrene, ethylene vinyl acetate (EVA) and ethylene methyl acrylate (EMA). The group of thermoplastic rubbers, often called 'elastomers', consists of polymers such as natural rubber, styrene-butadiene rubber (SBR), styrene-butadiene-styrene (SBS), styrene-isoprene-styrene (SIS), polybutadiene (PBD) and polyisoprene.

An important effect of both types of polymer modifiers is on the temperature susceptibility of the stiffness of bitumen. Compared with many materials, bitumen viscosity or stiffness is highly dependent on temperature. Indeed, it is this characteristic which makes bitumen an ideal binder for many applications. However, where extreme conditions of high (say 45 °C to 60 °C) or low (say below −10 °C) pavement temperatures are encountered, the viscosity or stiffness of bitumen can lead to premature failure of a surfacing by rutting or cracking, particularly if traffic demands are also unusually high. In contrast to bitumen, organic polymers are generally far less susceptible to changes in temperature in terms of stiffness. Providing both the type of base bitumen and polymer modifier are carefully chosen to ensure compatibility, incorporating the polymer modifier into bitumen can significantly reduce the temperature susceptibility of the binder in the service temperature range. In order to obtain maximum benefit from the use of polymer modifiers and to ensure ease of application, it is essential to ensure that the polymer modifier systems used do not result in unduly high viscosities at elevated temperatures. This may be achieved by the use of polymer modifiers which, in combination with the chosen type and grade of base bitumen, exist within the bitumen as a polymer network through thermally reversible bonds in the service temperature range but which, at application temperatures, are dissociated, thereby reducing binder viscosity (Figure 3.16). Both crystalline polymers (such as EVA) and thermoplastic rubbers (such as SBS) are able to form similar polymeric networks in bitumen.

As a result of the reduced temperature susceptibility of polymer modified binders (PMBs), it is possible to increase the stiffness of the binder at high pavement service temperatures in order to reduce rutting and, at the same time, to reduce the stiffness of the binder at low pavement temperatures in order to reduce brittleness and cracking.

The improvement in deformation resistance obtained by the use of a PMB in asphalt can be illustrated in a laboratory wheel tracking test (Bouldin and Collins, 1992). In the test, which is typically carried out at high pavement service temperatures of 45 °C to 60 °C, either the depth of the rut developed or the rate of development of the rut is determined as a function of the number of passes of a standard loaded wheel over the asphalt specimen. Figure 3.17 shows results obtained in the test which

Binder modifiers

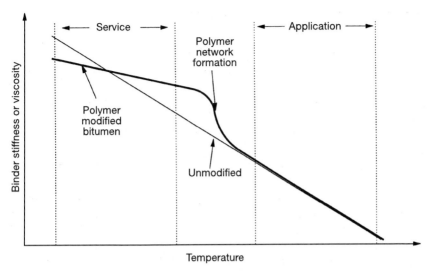

Fig. 3.16 Binder stiffness/viscosity as a function of temperature

show the much slower development of the rut for an asphalt containing the PMB.

The benefit of lower stiffness of PMBs at low pavement service temperatures can be demonstrated in the laboratory using tensile tests on asphalt (Shahmohammadi, 1994; Stock and Arand, 1993). In one type of test, an asphalt sample is cooled but not allowed to thermally contract, thus causing a thermally induced stress to build up as the temperature is reduced. In a separate experiment, the tensile strength of the asphalt is

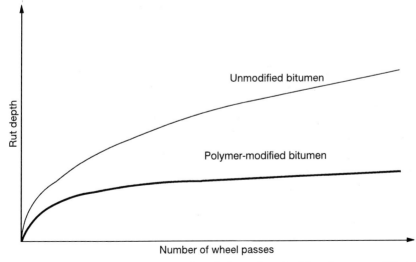

Fig. 3.17 Improved rut resistance of asphalt produced with polymer-modified bitumen

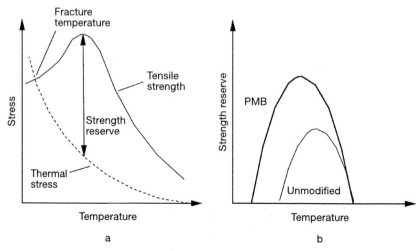

Fig. 3.18 Improved low temperature properties of asphalt produced with polymer-modified bitumen

determined as a function of temperature. The difference between the tensile strength at any temperature and the tensile stress at that temperature is known as the 'strength reserve' (Figure 3.18a), and is a measure of the load bearing capacity of the asphalt. At some temperature, the thermal stress in the asphalt will equal the tensile strength and the asphalt will fail due to cracking. This temperature is known as the 'fracture temperature'. Both strength reserve and fracture temperature are important performance characteristics of asphalts at low pavement temperatures. Figure 3.18b shows typical curves of strength reserve as a function of temperature for an unmodified bitumen and a PMB. It can be seen that the modified binder not only increases the strength reserve relative to the unmodified bitumen but also reduces the fracture temperature.

The reduced temperature susceptibility resulting from polymer modification can be a key benefit in improving binder performance at both high and low service temperatures. However, it has been found that other properties of the binder may also be improved by the use of suitable polymer modifiers. Fatigue lives of dense asphalt mixtures and also thin layers may be significantly increased when PMBs are used (Figure 3.19) (Whiteoak, 1990).

Better adhesion to aggregates, increased tensile strength and better cohesion of PMBs have been found to result in significant improvements in the performance of surface dressings and thin layers. The enhanced performance of PMBs in surface dressings may be demonstrated in the laboratory using the Vialit pendulum impact test. In this test, a thin film of the binder is sandwiched between two blocks of metal or aggregate and the energy to remove the upper block using a standard sideways

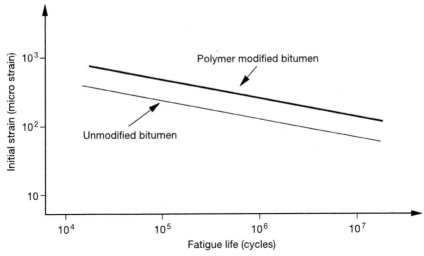

Fig. 3.19 Relationship between fatigue life and strain for bitumen and polymer-modified bitumen

impact is measured. This energy is plotted as a function of temperature (Figure 3.20). The benefits of increased cohesion and adhesion of the PMB over unmodified bitumen can clearly be seen, the modified binder requiring higher impact energies to remove the aggregate block.

In some European countries such as France, Austria and Spain, the Cantabro abrasion test is used to design porous asphalts because it is

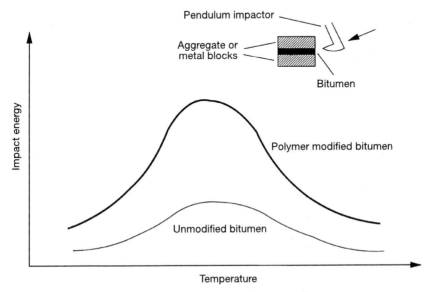

Fig. 3.20 Impact performance of binders as a function of temperature

believed that the test reflects the mechanism of deterioration on the road. In this test, a 100 mm diameter cylindrical specimen of asphalt (such as used for the Marshall Test) is tumbled in a rotating steel drum for 5 minutes and the loss in mass due to abrasion is measured. The test is often carried out before and after a period of immersion in water to try to estimate the tendency of the binder to strip from the aggregate. Figure 3.21 shows typical results using unmodified bitumen and PMB where the improved performance of the latter is evident.

Polymeric modifiers for bitumen are becoming increasingly used worldwide in applications where enhanced performance of the binder is required to counteract adverse conditions of climate and/or traffic. However, the use of PMBs does not necessarily guarantee satisfactory performance in all situations. Due to its complex chemical nature and the interactions between different chemical species in bitumen, there is almost invariably a delicate balance in terms of compatibility between any polymer modifier and bitumen. Achieving this balance depends not only on the accurate selection of grade and chemical composition of base bitumen and polymer modifier but also on the processing conditions used for the production of the PMB. If the correct chemical balance is not accurately achieved, the performance of the PMB will not only be impaired but could actually be inferior to that of an unmodified bitumen.

3.5.3 Chemical modifiers

Much work has been carried out, particularly in the United States of America and Canada, on the use of sulphur in asphalt mixtures by either

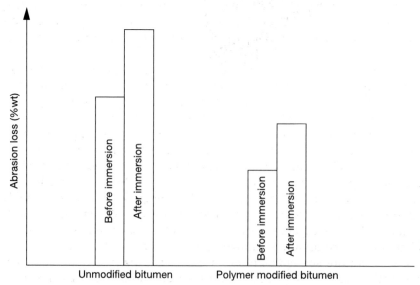

Fig. 3.21 Cantabro abrasion test results on porous asphalts with different binders

pre-blending with bitumen or adding directly to the asphalt during the coating process. This work was largely prompted by the low cost of sulphur and its availability to a large excess.

The interaction of sulphur with bitumen is not simple and depends not only on the chemical composition of the latter but also on the temperature. Between about 120 °C and 150 °C, a combination of chemical reactions and physical alterations occurs which leads to changes in the rheological properties. Up to about 20% by weight of sulphur can be 'dissolved' into bitumen. Part of this sulphur reacts chemically with components in the bitumen while the excess constitutes a separate phase. At asphalt mixing and laying temperatures, the excess molten sulphur has a low viscosity, about 0.03 Pa.s, and can significantly improve the workability of asphalt mixtures. As the binder cools, the sulphur slowly re-crystallizes resulting in a hardening of the mixture. This re-crystallization process can take up to two weeks, after which time the increased stiffness of the binder results in improved deformation resistance of the asphalt mixture. However, due to their lower thermal susceptibility compared with unmodified bitumens, sulphur-bitumen blends are more flexible and less brittle at low pavement temperatures.

Despite the successful results obtained using sulphur to modify bitumen and asphalt, it is little used. The main reason for this is that, above 150 °C, bitumen undergoes a vigorous dehydrogenation reaction with sulphur, producing large quantities of toxic hydrogen sulphide gas and an effect on binder properties similar to that obtained by blowing with air at high temperatures. Although attempts have been made to control the bitumen and asphalt temperatures to below 150 °C in order to avoid the production of hydrogen sulphide gas, the risk to asphalt workers during laying and compaction would appear too great. Potential problems with hydrogen sulphide emissions may also have to be addressed if the pavement is to be recycled at the end of its life.

A number of inorganic compounds have also been found to modify favourably the properties of bitumen. Copper sulphate is reported to reduce the tendency to flow at high pavement temperatures and, at the same time, reduce brittleness at low temperatures. However, the use of copper sulphate lost favour due to difficulties in controlling the continued reaction with the bitumen and hence ultimate binder properties.

Other metallic compounds have also been used for chemically modifying the properties of bitumen. In both the United States of America and Canada, patents have been granted describing the use of oil soluble organo-manganese compounds, either alone or in combination with organo-cobalt or organo-copper compounds. A commercially available blend of such compounds, known as 'Chemcrete', is claimed to improve the temperature susceptibility of bitumen, thereby improving deformation resistance at high pavement temperatures and reducing cracking at low pavement temperatures. It is thought that the mechanism by which the organo-metallic compounds have this effect is that they catalyse the reaction between thin films of bitumen and atmospheric oxygen to form diketone compounds which, in turn, form stable complexes with the metal. These complexes are then able to form a structure involving other bitumen molecules which increase the viscosity and strength of the binder. The magnitude of the strength and viscosity increase depends not only on the level of manganese concentration but also on the source and chemical composition of the bitumen and availability of oxygen to the binder mixture. Because of the sensitivity of the process to these variables, the benefits of such modification in UK trials have been extremely variable. It has been found that the degree of hardening obtained was sometimes insufficient to prevent deformation at high pavement temperatures while, at the other extreme, the hardening reaction proceeded too far and resulted in very brittle binders which caused cracking of the pavements. 'Chemcrete' is no longer marketed in the United Kingdom.

3.5.4 Adhesion agents

One of the principal functions of an asphalt binder, whether tar or bitumen, is to act as an adhesive either between road stones or between road stone and the underlying surface. The adhesion of asphalt binders to road stone presents few problems in the absence of water, although excessive dust may lead to failures. However, because road stones are more easily wetted by water than by tar or bitumen, the presence of water can lead to difficulties, either in the initial coating of damp or wet aggregate or in maintaining an adequate bond between the binder and stone. Failure to maintain the bond between the binder and stone in the presence of water in known as 'stripping'.

A binder which is able to coat wet aggregate is said to have 'active' adhesion with that aggregate. Where a binder has coated dry aggregate and does not undergo stripping when immersed in water, adhesion is said to be 'passive'.

The coating of aggregate is greatly facilitated by using a low viscosity binder. On the other hand, stripping of binder from aggregate in the presence of water is resisted better by binders of higher viscosity. Thus, although advantage can be taken of the viscosity/temperature characteristics of the asphalt binders by obtaining initial adhesion with a hot binder, the viscosity of which will markedly increase on cooling to road temperature, it is usually necessary to compromise between the lower viscosity needed to give the best initial coating on the stone and the higher viscosity desirable to give better protection against stripping.

Addition of filler to the asphalt mixture increases the viscosity of the binder and, therefore, is of assistance in controlling the rate at which stripping occurs. However, although the rate at which stripping progresses may be controlled by increasing the binder viscosity, stripping is not eliminated entirely.

There are certain fillers which have more than a physical action. Examples of these are hydrated lime and Portland cement which, when added to an asphalt mixture in concentrations of 1–2% by weight, may very much reduce, or even completely eliminate, stripping. With these particular fillers, there is a chemical action as well as the normal physical action of increasing the viscosity of the binder. It is believed that these fillers strongly adsorb the polar constituents of bitumen and can even combine with weakly acidic compounds, such as carboxylic acids, to produce bitumen-compatible compounds. Such compounds will be strongly adsorbed at negatively charged aggregate surfaces, thereby giving improved adhesion.

The use of cationic surface active agents is now common where a bitumen–aggregate adhesion problem potentially could exist. These compounds are usually organic amines of relatively high molecular weight ('fatty amines'). When added to bitumen at levels of about 0.2–1.5% by weight, such amines orientate themselves such that the hydrocarbon organic chain is retained in the bitumen while the amine group remains at the surface or bitumen–aggregate interface. The net effect is that the long hydrocarbon chain acts as a 'bridge' between the bitumen and aggregate, thus ensuring a strong bond.

Unfortunately, although fatty amines are very effective as adhesion agents for bitumen, they are relatively unstable at the high temperatures normally used for storing bitumen. It is possible with some of these adhesion agents that 50% of the amine can be inactivated at 120 °C and, at 180 °C, the doped bitumen can be stored for only a few hours before the amine loses all its activity. Because it is desirable to be able to store doped bitumens at elevated temperatures without the amine losing its adhesion activity, a range of adhesion/anti-stripping agents were developed based on other organic nitrogen-containing compounds such as amidoamines and imidazolines. These compounds are often used in conjunction with fatty amines, combining their improved heat stability with the generally superior adhesion characteristics at lower dosage levels of the fatty amine.

3.5.5 Fibre additives

Fibres are added to mixtures and are commonly referred to as binder modifiers. However, they do not affect the rheological properties of the binder, although they do modify certain properties of the mixture. Therefore, fibres are considered as additives and not binder modifiers. This is reflected in the draft CEN terminology.

Mixture segregation (binder drainage) during transportation and laying is recognized as a potential problem with certain gap-graded asphalt mixtures, such as stone mastic asphalt (SMA) and porous asphalt (PA). It has been found that, depending on their type and nature, the use of fibrous materials in these asphalt mixtures can significantly reduce or even eliminate the problem. The fibres used in such mixtures may be either

Table 3.5 Characteristics of different fibres used in asphalt

	Diameter (microns)	Length (mm)	Density (g/ml)
Chrysolite	0.1–1	0.5–1.0	2.7
Rock wool	3–7	0.2–0.8	2.7
Glass wool	5–6	0.2–1.0	2.5
Cellulose	20–40	0.9–1.5	0.9

organic (cellulose) or inorganic (such as asbestos, rockwool and glass). The stabilizing effect of the fibre modifier depends not only on the nature of the material but also on fibre shape in terms of length and diameter. Table 3.5 summarizes the main characteristics of the fibres currently used.

Various grades of fibres are available within each type, varying not only in fibre dimension but also in the type of surface treatment used during production. For example, different grades of cellulose fibre are available with different densities due to different amounts of added filler. Fibres may also be treated chemically to impart surface active properties, thereby improving dispersion within the binder in the asphalt.

Without exception, fibre modifiers are currently incorporated into the asphalt at the mixing plant. Due to density differences between all types of fibre and bitumen and consequent problems of segregation, a stable pre-blend of hot bitumen and fibres is not yet available. However, certain grades of fibre may be supplied blended with a hard bitumen in pellet form and these may be used at the mixing plant, suitable adjustment being made where necessary to the grade of bitumen being used to coat the aggregate.

3.6 REFERENCES

Standard Methods for Analysis and Testing of Petroleum and Related Products, Institute of Petroleum, London.
  Standard Method of Test for Penetration of Bituminous Materials (IP 49)
  Softening Point of Bitumen Ring and Ball (IP 58)
  Breaking Point of Bitumen Fraass Method (IP 80)
Anderson, D. (1994) The SHRP binder test methods and specification, *Revue Generale des Routes et des Aerodromes*.
Bouldin, G. and J.H. Collins (1992) Wheel tracking experiments with polymer and unmodified hot-mix asphalt, *Polymer Modified Binders, ASTM Spec. Tech. Publ. No. 1108*, American Society for Testing and Materials, Philadelphia, 50–60.
British Standards Institution (1974) *Specifications for Tars for Road Purposes. BS 76: 1974*, British Standards Institution, London.
British Standards Institution (1989) *Bitumens for Building and Civil Engineering; Part 1, Specification for Bitumens for Roads and Other Paved Areas. BS 3690: Part 1: 1989*, British Standards Institution, London.

References

British Standards Institution (1993a) *Petroleum and its Products; Part 49, Penetration of Bituminous Materials. BS 2000: Part 49: 1993*, British Standards Institution, London.

British Standards Institution (1993b) *Petroleum and its Products; Part 58, Softening Point of Bitumen (Ring and Ball). BS 2000: Part 58: 1993*, British Standards Institution, London.

Shahmohammadi, R. (1994) *Highways Surfacings*, Jonestown University.

Stock, A.F. and W. Arand (1993) Low temperature cracking in polymer modified binders, *Proceedings of the Association of Asphalt Paving Technologists*, Austin, Texas, 23–53.

Wallcave, L.H., Garcia, R. Feldman, W. Lijinsky and P. Shubik (1971) Skin tumourigenesis in mice by petroleum asphalts and coal-tar pitches of known polynuclear aromatic hydrocarbon content, *Toxicology and Applied Pharmacology* **18**(1), 41–52.

Whiteoak, D. (1990) *The Shell Bitumen Handbook*, Shell Bitumen, Chertsey, Surrey, 164–5.

CHAPTER 4

Required characteristics

W.P.F. Heather, Associated Asphalt Ltd

4.1 APPLICATIONS

Asphalt surfacings are flexible both in performance and in application providing safe, durable, cost-effective surfacings in uses that range from main runways of international airports, motorways and trunk roads through minor roads, housing estates, industrial areas and car parks to footways, sports tracks and tennis courts; the list of applications is almost endless. Any area that is trafficked by vehicles or pedestrians can be provided with a suitably designed asphalt surface. Although examples of bitumen usage in roads can be found in biblical times (section 1.1), it is the twentieth century that has seen the transition worldwide from unbound roads, that were dusty in summer and impassable in adverse conditions, to an infrastructure of clean and safe paved roads, over 90% of which are bitumen bound and which sustain traffic in all weather conditions.

This revolution in mobility and access is taken for granted, certainly in all developed nations and, increasingly, in Second and Third World countries. However, it is only in the last few years that the roads engineer has had the ability to assess the complex performance model of asphalt surfacings predictively from laboratory tests with any degree of certainty and, even now, we are only on the bottom rungs of the ladder leading to reliable and accurate end-performance tests. When achieved, the need for experienced engineers to select the criteria relevant to the end usage requirements will still be paramount.

This book assumes an adequate pavement structure for the road (or other paved area) and concentrates on the only visible part of the structure, the asphalt surface course and its characteristics; this is the most important aspect from a user's point of view. A successfully designed pavement should be like a building in taking the adequacy of performance of the foundation for granted and should be concerned with maintaining the fabric, using the building analogy, or the surfacing, in a pavement context. To achieve that end, the correct choice of asphalt surfacing for each application is essential to optimize and maintain the overall performance of the trafficked structure.

For the many applications previously noted, varying the proportions of aggregates, fillers and binder will provide satisfactory surfacings for nearly all of them. The exception is for very heavy-duty pavements, that is those where heavy, slow-moving or stationary loads impart the greatest stress on the surfacings, where modification of the binder, mixture or both may be required. Certain specialist sports surfacings, like professional running tracks, are outside the scope of this book because they incorporate little or no mineral aggregate, generally cement, rubber and wholly synthetic binders such as polyurethane and epoxy resins.

Asphalt surfacings are dynamic systems in performance terms, varying in accordance with temperature and rate of loading. The United Kingdom is fortunate in that there is a temperate climate and, therefore, does not have to cope with the extremes of climatic conditions that exist in Continental climatic zones. However, the relatively small annual temperature range is more than offset by the other factor in the performance equation, the rate of loading, because of the density of traffic, 66 vehicles per kilometre in 1993 in the United Kingdom, which is 50% above the average in the European Union and which results in a motorway network which is one of the most heavily trafficked anywhere in the world. Therefore, this high trafficking reinforces the need to choose the most appropriate surfacings to cope with the traffic loadings.

A further consequence of the density of traffic is a need for the UK economy to have its infrastructure functioning at all times because failures, and the cost of failures in terms of congestion and delays, have to be minimized. Therefore, there is a need to constantly monitor the roads network and to devise maintenance strategies to minimize the risk of failure in order to ensure that timely maintenance can prolong indefinitely the many billions of pounds invested in the infrastructure. It is fortunate in one sense that it is the surface course that is the prime indicator of the state of the road structure, because distress signs on the surface, such as cracking and rutting, are reliable in pointing the engineer to the need to investigate the causal elements of such distress and whether it is confined to the asphalt surfacing or the structure beneath.

The characteristics of the various surfacing applications will be discussed in this chapter, together with the reasons why the United Kingdom has evolved a coherent policy for these characteristics and how they can be achieved, maintained and, indeed, improved.

4.2 MONITORING OF EXISTING NETWORK

Given an existing network, a method has to be devised to assess the condition of the network in order to plan a maintenance schedule that keeps the network in satisfactory condition such that it achieves or exceeds its design life. As the network is in use at all times, a methodology that can assess its condition without impeding traffic flows and is cost effective and accurate is clearly desirable. The problem with all monitoring systems is the need to build a data bank from its early life so that the deterioration of the system over time can be assessed. It is unfortunately the case that reliable as-built

Table 4.1 Maximum permitted number of surface irregularities

Irregularity	Surfaces of carriageways, hard strips and hard shoulders				Surfaces of lay-bys, service areas, all asphalt binder course and upper base courses in pavements without binder courses			
	4 mm		7 mm		4 mm		7 mm	
Length (m)	300	75	300	75	300	75	300	75
Category A* roads	20	9	2	1	40	18	4	2
Category B* roads	40	18	4	2	60	27	6	3

* Category B is generally for low-speed (under 50 km/h) roads.

information on roads is rarely available; also, to date, the *Specification for Highway Works* (MCHW 1), used for trunk roads, and British Standards require acceptance criteria that are not readily translatable into usable data for network monitoring equipment. Default criteria, such as specified in Clause 701 of the SHW at the time of construction, are described as irregularities, that is variations from the profile of the road surface as measured by a rolling straight edge, that exceed 4 mm or 7 mm which, for Class A trunk roads, must not exceed 20 or 2, respectively, in longitudinal profile. The criteria are reproduced in Table 4.1.

Although mandatory for acceptance criteria, these restricted wavelengths (section 4.3.2) do not measure the true profile and only give a limited approximation to the 'rideability', that is the evenness experienced by the road user. Work is currently progressing in the Comité Européen de Normalisation (CEN) in the drive to harmonize European Standards (section 1.5.2) to devise test methods that assess the true profile so that the starting point data, when new construction is opened to traffic, can be transferred to the monitoring systems as an accurate base from which to monitor deterioration over time. It is the deterioration of the longitudinal and transverse profiles that is one of the prime indicators of the structural condition. The more uneven the road:

- the greater the vehicle operating cost;
- the more uncomfortable the ride; and
- the greater the rate of deterioration to failure.

Conversely, the more even the ride, the more the improvement in these negative factors. Too little regard has been given to making improvements in the field of evenness in new construction (that is, in laying techniques) which would have great benefits for road users and the economy as a whole. For example, research has shown that, for heavy goods vehicles (HGVs), road damage can be related to unevenness in profile in the two wavelength ranges of 1.3 m to 2.5 m and 5 m to 18 m (Williams, 1994). No specification in use anywhere (to the author's knowledge) requires compliance to this sort of wavelength criterion (section 4.3.2).

The High-Speed Road Monitor (HRM) has been developed over time by the TRL by combining the high-speed profilometer with a sensor texture meter; the former to measure profile and rut depth and the latter to measure macro-texture, or roughness of the surface. The macro-texture of a surfacing is currently usually defined contractually as the profile depth, that is the depth beneath a plane connecting the tips of the asperities of a road surface and measured by a fixed volume of sand or glass beads divided by the diameter of the area on a dry surface over which they are spread by a standard rubber puck to BS 598: Part 105 (BSI, 1990). Sensor-measured texture-depth, as measured by the HRM and the mini-texture meter (BSI, 1990), differs in that it measures the deviation in depth from a nominal plane, i.e. sensor-measured texture depth is a measure of the standard deviation whereas the sand-patch texture-depth is a measure of the mean. Hence, although the relationship between the two types of measure will be consistent if the surface 'shape' remains the same, a different relationship can be expected for a different shape of surfacing. This has become important with the growth of different surfacing types.

Macro-texture lies in the range of 0.30 mm to 50 mm, after which the term mega-texture has been used to describe part of the profile, but for practical purposes a range of 0.3 to 5 mm would be more appropriate because Clause 921 of the *Specification for Highway Works* (MCHW 1) requires a minimum initial texture depth, as measured by sand patch, of 1.5 mm for UK high-speed (>90 km/h) trunk roads. This compares with the minimum legal tyre tread depth of 1.6 mm. The need for macro-texture has long been established (Sabey, 1966) in determining high-speed wet skidding-resistance and, more importantly, the relationship between texture and accidents (Roe *et al.*, 1991). The role of micro-texture will be described in section 4.3.2.

The HRM can operate at speeds up to 95 km/h and, therefore, without impeding traffic flow which makes it ideally suited to assess completed sections of the network. The intent is that the entire trunk road network is surveyed every two years and can be viewed as a system to highlight lengths of the network in most need of maintenance; if, say, structural deterioration is identified by proportional change in longitudinal profile, particular lengths of the road can be assessed by more discriminating methods such as the falling-weight deflectometer (FWD).

4.3 SELECTION CRITERIA

4.3.1 Structural

The surfacing is an integral part of the structural design, whether of a multi-layer asphalt pavement, a flexible composite (cementitious road base) pavement or a rigid composite carriageway, because, apart from trial sections of exposed aggregate concrete (popularly dubbed 'whisper concrete'), an asphalt surfacing is at present mandatory on the most heavily trafficked trunk roads. The structural contribution of a surfacing, apart from surface dressing and other veneer treatments (Chapter 11), is always positive for a asphalt mixture but, in terms of performance, is outweighed by the characteristics demanded of it. This is not the case for all applications because combination surfacings consisting of cement grouted macadam can provide very significant contributions to the overall structural design with stiffness values in excess of the capability of the Nottingham Asphalt Tester to assess where the surfacings have to cope with high stress, slow moving heavy loading or static loading (section 4.3.3). The effect can be likened to case-hardening in metallurgy because, with this construction, the surfacing has a much higher stiffness modulus than the underlying asphalt base and road base materials, a reversal of the normal situation in terms of structural contribution. Most surfacing materials, including rolled asphalts (Chapter 5) to BS 594 (BSI, 1992), asphalt concrete macadams (Chapter 6) to BS 4987 (BSI, 1993), design asphalt concretes (Chapter 6) and thin surfacings (Chapter 10), fall in the stiffness modulus range of 1 GPa to 3 GPa; however, in HD 26/94 (DMRB 7.2.3), porous asphalt (Chapter 7) at 50 mm laid thickness is stated to contribute only 20 mm in stiffness value to the combined asphalt layer. Because the structural contribution of surfacings lies within a relatively narrow range compared to base and road base materials, other criteria are necessary to make an informed choice.

4.3.2 Surface properties

It is the surface characteristics which determine selection of asphalt surfacings, and the starting point is what characteristics might be considered to constitute the ideal surfacing. It should be:

- smooth for comfort and low vehicle operating costs, minimizing rolling resistance and therefore reducing fuel costs, tyre wear and vehicle wear;
- able to provide a surface of sufficient friction between the tyre contact patch and the surface to maintain control of the vehicle at all speeds and in all weather conditions;
- quiet, glare-free and able to create no spray under wet conditions;
- durable, resisting wear and deterioration of its required characteristics;
- economical to place and economical to maintain; and
- (preferably) capable of being re-used or recycled at the end of its useful life span.

It is clear from these desirable characteristics that an ideal surfacing does not exist. Therefore, it is necessary to compromise, and selection has to be made as to which characteristics are most needed in a specific application.

In analysing the ideal characteristics listed, smoothness is a function of the initial construction process and also of mixture design. A dense graded asphalt concrete, for example Marshall asphalt (Chapter 6) commonly used as a runway surfacing material in the United Kingdom, gives a smooth, low texture, high stiffness, durable surfacing, resistant to deformation in aircraft usage but would not be allowable for trunk road applications in the United Kingdom because of low macro-texture and consequently poor wet skidding-resistance. The majority of us have had experience of autoroute driving in France with kilometres of smooth asphalt concrete with little texture and which are hazardous in the wet.

The application of coated chippings in rolled asphalt surface course (Chapter 5) to provide the 1.5 mm texture depth specified does little for smoothness in profile or ride quality but is a necessary additional process to give the safety factors required at high speed and adverse weather. The components of friction are complex and it is beyond the scope of this publication to discuss the current state of understanding in any detail. As mentioned previously, texture has long been a concern of specifications in the United Kingdom and it is only relatively recently that friction and related safety are gaining in the national road agendas of countries in the European Union and elsewhere.

Macro-texture can be provided by:

- the process, as in rolled asphalt (Chapter 5) where a 35% stone content dense gap graded mixture has effectively no texture before the chippings are rolled into the surface; or
- the mixture design, as in porous asphalt (Chapter 7) where the grading envelope is such that, after compaction, it has approximately a 20–25% void content which ensures deep texture between the surface aggregate particles.

Macro-texture is one component of overall friction while micro-texture is another which provides friction at low speeds; both are necessary for overall friction. Micro-texture is a natural property of the aggregate, assessed by a relative measure of its resistance to polishing under traffic against an aggregate of known polishing characteristics and codified in HD 28/94 (DMRB 7.3.1) and BS 812: Part 114 (BSI, 1989). The value obtained is the polished stone value (PSV) (section 2.2.4.4), which is an early example of an end-performance test, an indirect measure of micro-texture. For practical purposes in road usage, micro-texture lies in the range of 0–0.3 mm.

The skid-resistance of a surface can be measured by a coefficient of friction, the sideway force coefficient (SFC). The equipment used to measure the coefficient on highways is the SFC routine investigation machine (SCRIM). It consists of a freely rotating, standardized smooth rubber tyred wheel, loaded to 200 kg, mounted in the vehicle at a 20° angle to the direction of travel. Water is sprayed onto the surface in front of the wheel at approximately 0.95 L/s and the resistance to sliding, the sideway force, is measured by a load cell. This coefficient is not a constant and varies seasonally and with speed and traffic, especially HGV traffic. Therefore, although SFC can be measured at speeds up to 100 km/h, the results are normally standardized to 50 km/h. This equipment has been in use for many years, following on from early models developed by the TRL.

Figure 4.1 shows the seasonal variation that occurs in the SFC of exposed aggregate in the road. This variation may be due to:

- the self-cleansing action of the higher rainfall in winter which causes the larger gritty particles and de-icing compounds to abrade the exposed aggregate surface and restore the frictional characteristics; whereas
- the finer dustier particles in summer polish the aggregate surface, thus reducing the frictional characteristics.

The surface temperature of the road may also affect the resistance of the rubber in the test tyre, contributing further to the seasonal difference. There is a relationship between the PSV of the aggregate in the surfacing and its SFC which equates roughly to Equation 4.1, which is valid for traffic up to 750 commercial vehicles per lane per day (cv/l/d). As friction drops with increasing commercial vehicle traffic, compensation has to be made to the required PSV to achieve SFC.

$$\frac{PSV}{100} = \text{Mean Summer SCRIM Coefficient} \qquad (4.1)$$

For the safety factor to be maintained, the friction values as measured by SCRIM are required to be reported as the mean summer values because friction is seasonal; preferably from at least three measurements between May and September when the friction is normally at its lowest.

Table 4.2 PSV of aggregates necessary to achieve the required skidding resistance in asphalt surfacings under different traffic conditions

Required mean summer SCRIM Coefficient at 50 km/h	PSV of aggregate necessary					
	Traffic (in commercial vehicles per lane per day)					
	250 or under	1000	1750	2500	3250	4000
0.30	30	35	40	45	50	55
0.35	35	40	45	50	55	60
0.40	40	45	50	55	60	65
0.45	45	50	55	60	65	70
0.50	50	55	60	65	70	75
0.55	55	60	65	70	75	
0.60	60	65	70	75		
0.65	65	70	75			
0.70	70	75				
0.75	75					

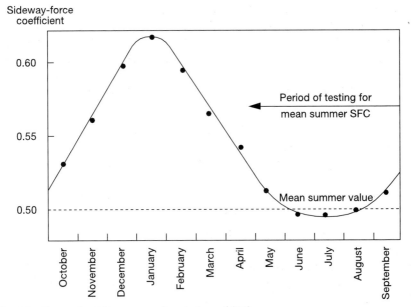

Fig. 4.1 Example of the seasonal variation of SFC

Table 4.3 Sideway-force coefficient and risk of skidding accidents in wet conditions

Sideway-force coefficient at 50 km/h	Relative risk of skidding accidents
0.6 and greater	1
0.5–0.6	1.7
0.4–0.5	1.9
0.3–0.4	15
0.2–0.3	25

Table 4.3 (Hosking, 1993) illustrates the risk of wet skidding accidents with falling SFC values and, for this reason, minimum investigatory levels exist to ensure an adequate level of friction, as reproduced from HD 28/94 (DMRB 7.3.1) in Table 4.4.

There is a requirement that, for trunk roads, the SCRIM values are taken at not more than three-year intervals. It has been found that, when the binder film wears off the aggregate exposing micro-texture, surfacings subject to uniform trafficking reach an equilibrium value of relatively constant mean summer friction termed the equilibrium SCRIM coefficient (ESC). However, Figure 4.2 demonstrates the point that, not only are friction values seasonal, but traffic volume related and reversible.

Overall, the use of SCRIM on a routine basis, coupled with a policy of initial macro-texture, has led to the UK safety record on its trunk road system which is not only enviable in comparison to other developed

88 Required characteristics

Table 4.4 Investigatory skidding resistance levels for different site categories (HD 28/94, DMRB 7.3.1)

Site Category	Site definition	Investigatory levels MSSC (at 50 km/h)							
		0.30	0.35	0.40	0.45	0.50	0.55	0.60	0.65
		Corresponding risk rating							
		1	2	3	4	5	6	7	8
A	Motorway (Mainline)		■						
B	Dual carriageway (all purpose) – non-event sections		■						
C	Single carriageway – non-event sections			■					
D	Dual carriageway (all purpose) – minor junctions			■					
E	Single carriageway – minor junctions				■				
F	Approaches to and across major junctions (all limbs)				■				
G1	Gradient 5% to 10% longer than 50 m; dual (downhill only), single (uphill and downhill)				■				
G2	Gradient steeper than 10% longer than 50 m; dual (downhill only); single (uphill and downhill)					■			
H1	Bend (not subject to 40 mph or lower speed limit) radius < 250 m					■			
J	Approach to roundabout						■		
K	Approach to traffic signals, pedestrian crossings, railway level crossings or similar						■		

Site Category	Site definition	Investigatory levels MSSC (at 20 km/h)							
		0.40	0.45	0.50	0.55	0.60	0.65	0.70	0.75
		Corresponding risk rating							
		1	2	3	4	5	6	7	8
H2	Bend (not subject to 40 mph or lower speed limit) radius < 100 m			■					
L	Roundabout			■					

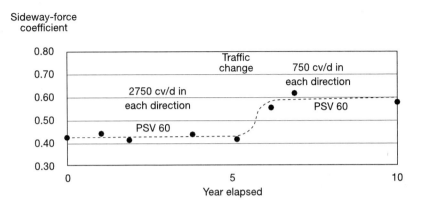

Fig. 4.2 Example of the influence of traffic on SFC

countries in terms of reduction of fatal and serious accidents but is also cost-effective. It has been calculated that each £1 spent on a coherent national policy of maintenance of friction to HD 28/94 (DMRB 7.3.1) saves the national economy £5.50 in the costs of accidents.

It is perhaps useful at this stage to consider wavelength and frequency (section 4.2) because, in the future as greater general understanding of the concepts grows and the ability to measure them accurately grows, they will be specified. All of the surface characteristics identified as desirable can be described by wavelength spectra. Figure 4.3, which was produced by the Permanent International Association on Road Congress (PIARC) technical committee, harmonizes in one table the wavelength spectra for each property. Therefore, it provides a method of describing these characteristics which is independent of the many test procedures currently used in national standards in an attempt to specify these characteristics by surrogate values. However, a method needs to be available to measure these wavelengths rapidly and cost effectively. With the advent of the laser and increasing computer power and resolution, the HRM and other variants in other countries are capable of analysing the road surface in some detail in wavelength spectra (and its reciprocal, spatial frequency). If the complementary data for wavelength and frequency (in cycles per second, Hertz) can be generated, this data can further describe surface characteristics, both desirable and undesirable.

A method of evaluating wavelength spectra, independent of how the data is generated, is power spectral density which is a mathematical process to calculate the vertical displacement from a longitudinal geometric profile. Therefore, given values that lie within the required desirable

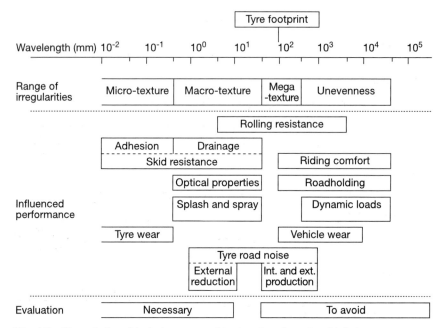

Fig. 4.3 The relationship between road texture levels and vehicle tyre performance

surface characteristics, it should be theoretically possible to design a surface by these end performance criteria only (Figure 4.4). It would then be necessary to match construction techniques and mixture design to produce the required wavelength and frequency criteria. This complete reversal of the current standard, where the surface is laid and then values are generated, lies within the potential of the road engineer but, of course, it needs the addition of the tyre and vehicle designer to optimize their designs accordingly. Whether sufficient agreement could ever be reached on a 'standard' surface in, say, the European Union is unlikely, especially given the reliance on naturally occurring materials, aggregates. Nevertheless, there may well be a convergence in the future on, for example, a desirable texture pattern that would provide safety, comfort, low noise, low spray and the least rolling-resistance. This convergence, in turn, may lead to a type of approval of a road surface. Research is being carried out to attempt to give predictive safety values to texture type and much work has been carried out in the European Union in noise reduction surfacings.

There is a view that an ideal surfacing type already exists, porous asphalt (Chapter 7). It is the policy of the Dutch roads authority that all trunk roads in Holland will be surfaced with porous asphalt by the year 2000. However, there are many variants of the porous asphalt concept.

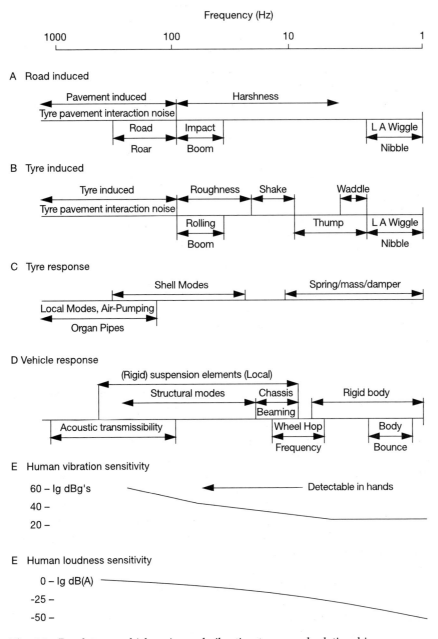

Fig. 4.4 Road, tyre, vehicle noise and vibration terms and relationships

Porous Friction Course (PFC) has existed in the Property Services Agency specification for military airfields for many years and has also been used in commercial airports, including Heathrow, with examples of outstanding longevity by road standards, in excess of twenty years. The drawbacks with porous mixtures are:

- durability and cost compared to dense mixtures, because of the embrittlement, oxidation and hardening of the binder resulting from the greater surface area in contact with air, water and de-icing fluids; and
- the need to employ high-quality aggregate throughout the layer depth in the United Kingdom.

4.3.3 Durability

The retention of the desired characteristics and the rate of deterioration over time are of prime importance in terms of cost and maintenance. In the past, the underestimation of traffic growth in the late 1950s and 1960s and the increasing weight of HGVs caused roads to be designed with inadequate structural thickness, causing the rash of motorway and trunk road strengthening schemes in the 1980s; surface characteristics, and their retention, are inextricably linked with traffic volume and density. Therefore, there are two possible general modes of failure in surfacing materials:

- deformation of profile, rutting, pushing, etc.; and
- the loss of friction.

Both modes require maintenance intervention if allowed to progress beyond certain limits.

In terms of repeated loading causing permanent deformation in the surfacing, cars have a negligible contribution to damage. For comparison purposes, following the European Union directive on harmonization of heavy goods vehicles weights and dimensions, the passage of one 100 kN axle is the same damage factor in structural terms as 160 000 cars and, in terms of permanent deformation (rutting in the surfacing), one 100 kN axle causes the same amount of deformation rutting as 1 300 000 000 cars (EAPA, 1995)! Furthermore, permanent deformation only occurs when the temperature of the surfacing exceeds the softening point of the binder so that rutting occurs only on a few days of the years (although days enough in years such as 1995!). This last fact has, in part, dictated the evolution of surfacing materials in the United Kingdom and overseas. Rolled asphalt surface course mixtures (Chapter 5) performance is defined by the binder, fine aggregate and filler portion of its content while the coarse aggregate content has been likened to plums in a pudding; asphalt concrete surface course mixtures (Chapter 6) depend on inter-locking aggregate particles achieved by a more or less continuous grading envelope. In rolled asphalt the hardness of the binder and its characteristics are very important in resistance to permanent deformation at high temperature whereas the binder grade and type is of lesser importance in asphalt concrete due to the inherent stability of its aggregate skeleton. Therefore, hot countries do not use rolled asphalt but, because rolled asphalt, as it has evolved in the United Kingdom, provides a long-lasting, durable surface which does have good texture with coated chippings to maintain these desirable characteristics, it is widely used in the United Kingdom although, with increasing density of loading and with (perhaps) more frequent spells of warm weather, there are increasing needs to enhance the performance of the binder.

The enhancement of the binder performance can be achieved by modifying the bitumen with polymers (section 3.5) such as ethylene vinyl acetate (EVA), styrene-butadiene rubber (SBR) and styrene-butadiene-styrene (SBS). Generally, they increase the softening point of the binder, improving the resistance to permanent deformation and thus the high-temperature performance. Some polymers also have beneficial effects at low temperature, enhancing elasticity and thus resistance to cracking and therefore low-temperature performance. With these polymer-modified binders (PMBs), the performance of rolled asphalt to cope with increasing traffic is greatly extended, provided macro- and micro-texture values can be achieved and maintained. Asphalt concrete mixtures, porous asphalt and thin surfacings can also be mixed with PMBs, again enhancing performance and durability.

As demonstrated at the beginning of this section, the durability of a surfacing is a direct function of the HGV traffic, topography and climate and therefore difficult to predict. The dry wheel-tracking test (BSI, 1996) is finding increasing use as a predictive test to assess the likelihood of permanent deformation, especially when carried out at enhanced temperatures such as 60 °C. If the engineer is faced with the need to provide a surfacing for a steep gradient, especially south-facing, there will certainly be a case to use a PMB to give the extra performance, durability required of a surfacing. Because, during the last twenty years, the United Kingdom has experienced the two hottest summers of the last two hundred years as well as some near records in between, a more general case could be made for PMBs for greater durability. As an illustration of this, a number of cases have been reported in summer 1995 of surfacings on major schemes rutting severely within a few days of opening to traffic. The causes of these failures are often bound up with the need to maintain traffic flows on major maintenance schemes, which inevitably leads to lower speeds at best and stationary traffic at worst. All asphalt surfacings are vulnerable in early life because as they age, oxidation of the binder occurs (Chapter 3) which hardens the binder, increases the softening point and, therefore, increases the resistance to deformation. The slow-moving or stationary traffic, resulting from contra flows and traffic restrictions on maintenance schemes, causes the severest loading conditions the surfacing is likely to experience in its service life. This, coupled with high temperature, causes a level of permanent deformation that would not be experienced in ten years of service life; hence, the need to recognize in advance that these conditions may exist when planning major maintenance schemes and plan accordingly by specifying higher performance materials.

4.3.4 Restrictions imposed by the site

The prime restriction imposed by site conditions is continuity. One of the main attributes of asphalt is 'buildability' as opposed to white, concrete construction. Buildability is the ability to construct sections of roadway piecemeal in all weathers in accordance with traffic and programme requirements capitalizing on the continuity that is necessary for the use of a concrete train or slip-form paver. To be cost-effective, the majority or all of the pavement to be placed has to be ready at the same time to ensure continuous use of such concrete paving plant. Continuity equates to consistency, and consistency to quality, and thus the ability of asphalt to be placed sectionally leads directly to a reduction in the overall quality that could be achievable if the same approach as is necessary for concrete were employed. The piecemeal approach affects the quality of the profile, the evenness of the texture, the compaction and the overall quality of the finish; too often, the dictates of traffic flows leads to piecemeal programming based on the ability of asphalt to meet such demands. While not dwelling overlong on this topic, planners, engineers and the public alike in this crowded country take for granted the ability of the asphalt industry to respond to such an approach; however, there is a price to pay in reduction of the overall quality that could be achieved. Nevertheless, as 94% of passenger kilometres travelled and 90 out of every 100 tonnes of inland freight moves on roads, this buildability factor is of great importance.

Asphalt surfacings are not designed for static loading because they are thermoplastic in performance. That is not to say that they cannot be but, in general terms, either specialist composite materials (section 4.3.1) or concrete are better suited for heavy static loads. In car and lorry parks, bay markings for parking impose special criteria for performance as they create repetitive static loadings. In these instances, surface course materials such as asphalt concrete or rolled asphalt are unnecessary and that binder course materials have higher stiffness values and are cheaper so that the provision of surfacing in these instances is an aesthetic choice, not a performance choice. Furthermore, these parking areas need to be machine-laid to provide optimum compaction and designers should design layouts with construction in mind and not, as often occurs, place obstructions (such as 'architectural details') that can only be appreciated fully by pigeons!

Dedicated bus lanes and lorry lanes also pose the same problem of canalized, heavy, slow-moving loading. In particular, buses, with their softer suspension, can cause severe deformation in surfacings with, in addition, copious fuel and oil contamination. As in the previous paragraph, binder course materials provide stiffer materials and better texture while the necessary micro-texture can be obtained by specifying higher quality aggregate.

Roundabouts also create a particular set of problems due to the high stresses imposed by the rear tyres of heavy articulated goods vehicles scuffing the surface while the tractor unit swings around the roundabout. Using high-textured rolled asphalt with pre-coated chippings (section 5.4.1) in these circumstances will lead to premature failure and, therefore, the texture needs to be reduced. A high stone content rolled asphalt (section 5.4.2) or a stone mastic asphalt (Chapter 9) would be a more appropriate choice.

Similarly, if a surfacing is to receive further surface treatment, such as the provision of a high-friction surfacing (section 11.3), there is little point in applying coated chippings to a rolled asphalt that is to be covered with such a treatment.

Space does not permit comprehensive discussion of all such cases but in each situation the end use has to be identified and the surfacing requirements tailored to that end use so that an informed choice can be made.

4.4 USES

4.4.1 New works

In this section, a somewhat artificial distinction has been made between new works and maintenance. As the number of new highway schemes decline, such a distinction becomes more and more arbitrary because the criteria for surfacing materials in new works and major maintenance are the same – namely, the provision of safety characteristics, durability and cost effectiveness for the proposed or actual traffic loading. A new surfacing to the relevant specification is a new surfacing however and on whatever it is laid.

For new highway works in the United Kingdom, the predominant surfacing material used up to the early 1990s has been rolled asphalt and pre-coated chippings (section 5.4.1). This material has undergone a number of minor detail changes in the last twenty years from 40 mm of 30% stone content to 45 mm (50 mm if laid in cold weather conditions) of 35% stone content currently. These changes in thickness were initially often options in tenders to enable contractors to include thicker options to allow for winter programming. Policy and budget timings conspire to make laying in the summer months the most suitable time zone, and autumn and winter working the norm. The nominal amount of time available for compaction to the contractor in cold, windy conditions has been calculated (Nicholls and Daines, 1993) and it is largely in response to this data that has made the increase from 40 mm to 45 mm and 50 mm inevitable in such conditions. Rolled asphalt is widely available, durable and a tolerant material to lay and compact. These features apart, its use has been dictated by one surface characteristic alone, namely an initial minimum macro-texture requirement of 1.5 mm by sand patch resulting from the application of coated chippings prior to compaction. Rolled asphalt provides this texture in one sense economically in that approximately 70 m^2 per tonne of 20 mm nominal size high polished stone value (PSV) aggregate gives the texture but uneconomically in that the rest of the thickness (50 mm – 20 mm = 30 mm) does little.

Porous asphalt (Chapter 7) is the only surfacing material in either BS 4987 (BSI, 1992) or BS 594 (BSI, 1993) to provide the initial macro-texture requirement 'naturally'; that is, as a function of its grading, which also gives 20–25% air voids. These voids act as a water reservoir, giving run off of surface water within its nominal 50 mm thickness. As well as water being absorbed, so is sound energy and porous asphalt gives a virtually spray-free surface and reduces noise considerably, even though the UK specification for porous asphalt is optimized for spray-reduction not noise. For trunk road use, its introduction was delayed while trials on the A38 in 1984 and 1987 were evaluated for durability; however, with a minimum binder content of 4.5% for the 20 mm nominal size, it now has approval for use on high-speed trunk roads. The only factor inhibiting wider usage is cost because the whole of the aggregate used in the 50 mm layer requires the same quality of PSV as would be specified for chippings in the same contract.

With the design, build, finance and operate (DBFO) initiatives and the maintenance of networks moving towards the private sector, in theory the choice of materials will be dependent on the road operators' assessment of the whole life cost profile (section 4.5). However, the required surface characteristics will remain a mandatory requirement that the road operator will have to satisfy.

There is an increasing consensus that the preferred solution, in both new works and major maintenance, is to provide a long life (indeed permanent) high stiffness base and binder layers such that only the surface course will need renewal. It follows that providing the required characteristics in the most economical manner (which generally implies in the thinnest layer, especially when high-quality aggregates are necessary) is the most cost-effective way of working.

Thin surfacings (Chapter 10) from 15–25 mm thick provide the required macro-texture as well as the friction necessary for safety together with noise- and spray-reduction and, since their introduction in the United Kingdom in the early 1990s from France (where they have been in use since the 1980s), they are gaining increasing use, predominantly in maintenance but they can be equally effective in new works. Up to mid 1997, only two systems had unrestricted approval for trunk roads; however, there are several more ssystems which have undergone or are undergoing the relevant assessment procedure.

4.4.2 Maintenance

As stated at the beginning of this section, all the surfacing types mentioned are just as appropriate in maintenance. However, one of the benefits of the harmonization of standards in Europe (section 1.5.2) has been that engineers in the United Kingdom are increasingly prepared to look at alternatives to the ubiquitous rolled asphalt. This does not necessarily mean 'new' materials such as stone mastic asphalt (Chapter 9) and thin surfacings (Chapter 10); dense bitumen macadam (DBM) (or, more correctly, asphalt concrete) surface courses are having something of a revival in urban situations. RR 296 (Roe *et al.*, 1991), probably one of the most important reports produced by the Transport Research Laboratory in the last 20 years, showed that, as long as macro-textures of 0.8 mm to 1.0 mm were achieved together with a good micro-texture, asphalt concretes in speed-limited urban situations give a safe, quiet and durable surface.

The transfer of responsibility from the government to the private sector of the road network is in its infancy in the United Kingdom; however, there are implications in maintenance that arise from this change. As will be shown in section 4.5, user delay costs and traffic disruption are the largest cost element in maintenance. This has been graphically illustrated by the punitive charges levied against the road operator on the initial tranche of DBFO schemes if the road is taken out of use for maintenance – a sum approaching £1 million a week on one of the larger schemes. This charge can be effectively eliminated if maintenance takes place during the night, at lowest traffic flow time. These kinds of penalties will affect maintenance choices and surfacing types because a premium will be placed on the speed of one process relative to another.

There is a parallel with other situations where there is a need to eliminate, as far as possible, maintenance altogether. For many years, the City of London used hand laid mastic asphalt and coated chippings which, although very costly compared to rolled asphalt, gave great durability without maintenance. It was cost effective because road works were deemed not to be acceptable due to the congestion, and it must also be borne in mind that the traffic is near stationary for most of the day.

The maintenance of rural roads presents different problems. The pavement depths of such roads are often inadequate and the priority is to keep water out of the foundation. These roads account for by far the largest percentage of roads in the United Kingdom and elsewhere. Asphalt concrete binder course materials give as good or better surface characteristics as asphalt concrete surface course materials in terms of texture, skid resistance and deformation resistance (with the same aggregates), offer better value for money and provide an excellent platform for surface dressing at a later date.

4.4.3 Replacement

All surfacings have a finite life span under given traffic conditions and will therefore need renewal. If there are no constraints of level then an overlay is probably the cheapest option. If levels are a problem, then planing out the existing surfacing and relaying is a very effective way of restoring the required surface characteristics while maintaining the original levels. Indeed, with the accuracy achievable now with cold planing machines, planing can restore longitudinal profile very effectively if working off a datum or wire so that the 'new' surfacing will be better as a result. Often rolled asphalt outlasts its skid-resistance properties in terms of its structural integrity; however, this structural integrity does enable the material to provide a very sound base for a new surfacing if overlaid. It is impermeable and hard, having 'cured' (that is, hardened over time) and consequently can be overlaid with any of the surfacing types previously discussed.

There is a further option available, and that is recycling the old surfacing (Chapter 13). This can be done either in situ using the repave/remix process or when the old surfacing is planed off the quality stone and bitumen can be taken away and reused, either in a new mixture or as lower grade material. Although ecologically sound, recycling has never achieved any prominence in the United Kingdom. This is largely because, in a small densely populated island, no site is far from a quarry or coating plant. Therefore, it is cheaper to supply new material than to recycle existing and, until this balance of costs is shifted, recycling will not take place on any great scale. The present requirement in the *Specification for Highway Works* (MCHW 1) allows for up to 10% of recycled material to be used in the new surfacing material and up to 30% in base and binder courses and it has to comply with the relevant specification as for new material.

The repave/remix process is fully explained in HD 31/94 (DMRB 7.4.1) and has been included in the specification for over 10 years, but is very little used. Providing the penetration of the bitumen in the existing surfacing is no harder than 20 dmm and it is basically structurally sound but worn, it can be repaved or remixed. Repave inlay saves 50% of the material, generally between 15 and 20 mm in thickness of new material, and restores the surface characteristics as new. Remix incorporates part of the existing surfacing with new material to produce a combined new surfacing of the required type and grading. Both processes are very effective but sadly neglected.

Porous asphalt and thin surfacings have not been extant long enough to look at renewal strategies; however, from experience on the Continent, the likely failure mode should be a reducing ability to act as water reservoirs by filling of the voids with detritus. In Austria, porous asphalt is often treated during its life-cycle by cleaning machines that flood the surface with water and then vacuum the water and detritus back into the machine for disposal. This process is claimed to prolong the efficiency of porous asphalt but, because the failure mode is densification of the voided mixture in both porous and thin surfacings, both systems could be easily overlaid, especially the latter.

It can be seen from the foregoing that the replacement of surfacings poses few problems and the engineer has to consider the same parameters of performance when renewing as for new, tailoring the site to the end performance required. The engineer has more choice in replacement as the options for recycling do not exist when new and he can also consider surface dressing a sound existing surfacing if the only requirement is to restore skid resistance.

4.5 WHOLE-LIFE COSTING

Cost assessments of major road schemes have always formed a natural part of successive governments' budgeting considerations and provision. It has become evident over time that, with the ever-increasing density of traffic, traffic disruption and user delay costs represent a greater part of the costs of an individual scheme over its lifetime than its initial construction cost. These costs increase over time so that similar maintenance procedures in, say, year 25 of a 40-year design life will cost more than the same procedure in year 10 because of the growth of traffic. In 1981, the COBA system was introduced by the Department of Transport to assess the long-term benefits of new road schemes by comparing existing journey times and accident rates against the improvements to be gained in both those values from the proposed new road. This system uses a simplified model for maintenance requirements and does not distinguish the merits of alternative designs at construction stage.

The Transport Research Laboratory was commissioned by the Department of Transport to produce a more discriminating comprehensive model for the evaluation of costs over the chosen whole life of a pavement, including initial construction types, maintenance requirement types and timing and consequent user delay costs. This computer model is known as COMPARE (Figure 4.5) and was introduced in 1993/94. The COMPARE program can be run for both new schemes and existing pavements and has been designed for individual schemes not networks. The road and construction type are the most important parameters in the model. The model allows all six pavement options currently in the *Specification for Highway Works* (MCHW 1) to be evaluated because each type will require different maintenance profiles. The road type, say, whether a dual three or four lane carriageway will incur different user delay costs. For its use on existing roads, the condition at the start of the evaluation period needs to be established as accurately as possible by means of various condition indicators. Whether for new or existing roads, the COMPARE model considers costs and maintenance requirements annually over the entire evaluation period and discounts future costs back to year one.

The model allows maintenance treatments to be considered in terms of cost of the treatment, the time taken and the traffic management type and cost, including traffic diverting to other adjacent roads during these periods. As traffic flows vary through the day, the COMPARE program enables user delay costs to be assessed for roadworks at certain times of the day. For flexible pavements, deflection is used to assess structural strength and rut depth and skid-resistance are used to evaluate surface characteristics, the same parameters used by network managers currently.

As pavements deteriorate over time, increasing vehicle costs, such as fuel consumption, are included in the annual costing. Finally, so that alternative strategies can be assessed properly, estimates are made of the residual life of the pavement at the end of the evaluation period. The outputs from the model include maintenance timing and associated costs, user delay costs and an annual assessment of pavement condition. It is clear that, as user delay costs predominate, the duration of maintenance interventions has a large impact in the overall whole life cost which, in turn, has a significant impact on the choice of surfacing type. Following on from the logic of providing a 40-year life structural element of the pavement in either asphalt or concrete material so that only the surfacing requires maintenance, the speed of the surfacing process and the ability to lay in low traffic flow times (such as at night) will be important factors governing choice.

COMPARE should provide a useful tool in assessing the overall costs of individual road schemes, whether they are new works or maintenance, but the quality of the inputs will need to be kept up to date and as accurate as possible to give meaningful costings. The perception to date by contractors of the use of the model by the Highways Agency (HA) is that it is a cost figure added to the tender value to reflect the program's calculation of the cost of a particular choice of construction type. The intent is that COMPARE will interface with a UK pavement management system (PMS) currently being developed by the HA and Local Authority Associations; however, when that becomes operational, the client may have changed to road operators.

4.6 QUALITY SYSTEMS

The origin of standards in asphalt materials dates back to the early thirties in the United Kingdom (section 1.1) to rationalize the many propri-

Whole-life costing

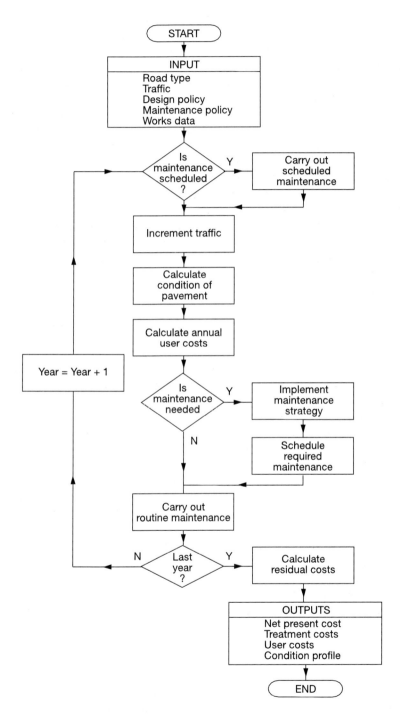

Fig. 4.5 The COMPARE whole life cost model

etary products and processes on offer at that time and to enable the client to assess to a prescriptive standard, compliance to that standard. Definitions of quality in the English language range from a plain description of the nature of a product or service to an implied level of excellence when used as an adjective. This lack of precision, unusual in English which usually has the use of two or three words to describe the same thing, has given rise to some confusion in the industry. Asphalt materials are made from naturally occurring materials processed to particular sizes and then coated; not uniform definitively sized particles to plus or minus tolerances, before coating. The current evolution of standards is rightly starting to address the required performance of materials instead of, as has been the case previously, the somewhat sterile debates about strict compliance to standards. Nevertheless, there is clearly a need to employ quality controls in the production of asphalt materials to give the client assurance that what is ordered to a standard is delivered. There are now also standards that address quality systems, specifying the form, content and degree of such systems.

4.6.1 Quality control

Asphalt surfacings are subject to quality controls at all stages in manufacture, whether mixed at quarry based coating plants or satellite coating plants at railheads or docks. The stone, sand, fillers and bitumen used in manufacture are governed by individual standards because stone, for example, has numerous end uses apart from incorporation in asphalt materials. Quality controls are used in quarrying to encompass the blasting, crushing, washing, screening, stockpiling and sampling of product as well as the maintenance of the very large plant involved, much of which is subject to a high degree of wear, especially when processing high PSV aggregate which, because of its resistance to polishing, is highly abrasive to metals.

Quality controls in surfacing are also fundamental, even if less tangible than in manufacturing process, because laying surfacing has end-performance criteria to satisfy. In order to achieve the requirements of level, evenness, compaction, texture depth and uniformity of appearance, the laying contractor has to be knowledgeable about the materials used, the plant employed and its maintenance and have properly trained personnel. The latter need a lot of craft-skills and judgement despite the machine capabilities which are ever improving. The final aspect to add to this list is the weather, the single largest variable in quality of surface characteristics. Both ends of the weather envelope affect quality directly, causing:

- poor compaction, level control and non-uniformity of texture in cold and windy conditions; and
- over compaction, level control problems, loss of texture and, if trafficked too early, rutting in very hot conditions.

The weather cannot be controlled, but the response to it can be.

The experienced contractor (a phrase often invoked in the standards but difficult to prescribe) can, by timely and effective quality controls, overcome the many problems involved in surfacing operations. This involves planning from the onset of a contract so that the proper resources are in the right place at the right time with safety of the operatives and public alike as a foremost consideration. The machine runs, with the specified staggering of joints if multi-lift construction, have to be planned as well as the avoidance of trafficking freshly laid material by construction plant, which can be difficult on some sites. If laying rolled asphalt surfacing, the stockpiling of chippings, adjustment of chip spreader to the required density of chippings to achieve texture, checking the adjustment and functioning of the paver, checking the frequency and amplitude of vibration of the compaction equipment, monitoring compaction equipment with nuclear density meters (NDMs) and many more activities form the quality controls necessary to achieve the end result looked for by the client. It is a but a short step from quality controls to quality management or, as more commonly expressed, quality assurance, that is placing the quality procedures detailed in the foregoing within a documented formal procedure as the basis of a management system.

4.6.2 Quality assurance

Quality assurance is a management system that requires the suppliers of goods and services to adopt quality of output as a fundamental concept in that supply. The history of quality assurances dates back to the 1930s, originating in the US when some literature was published, outlining a system to improve the quality of manufactured goods which, if adopted, would give the enterprise concerned:

- enhanced reputation and, thereby, sales; and
- by reduction of rejection rates of sub-standard goods, reduce costs and, hence, enhance profitability.

These publications had little impact until the post war period in Japan when quality assurance systems were adopted by Japanese industry to overcome the perception of Japanese goods as inferior quality copies of western goods. The success of this policy has been evident, especially in the car and consumer electronics industries, both of which production line industries being particularly suitable to quality assurance systems and have led to Japanese dominance in world trade in these fields.

In the US, partially in response to the perceived success in Japan, the defence and space procurement agencies started to specify quality assurance procedures from their suppliers to reduce waste, delays and costs. Similarly but later, the UK Ministry of Defence in the 1960s started to require that their suppliers adopt quality assurance systems. In the early

1980s, the government under Margaret Thatcher promoted quality assurance to industry as a whole to improve the quality and competitiveness of British exports and quality assurance gained momentum in the asphalt industry from the mid to late 1980s, mainly in response to client-driven initiatives. The great majority of coated stone suppliers now operate quality assurance systems and contractors also are increasingly adopting quality assurance.

Quality assurance requires that the supplier of goods and services operates to a written policy in respect of quality and has documented procedures covering all aspects of the business that have a bearing on the quality of those goods or services which in practical terms is the entire business. The original BS has been expanded and (adopted) as the BS EN ISO 9000 (BSI, 1994) series international standard and has achieved widespread acceptance and implementation around the world with the stated aim of harmonizing and facilitating international trade.

Any individual system has to address the scope of ISO 9000 and have documented procedures covering that scope. The intention is not that, in complying with that scope, all businesses in a sector operate in precisely the same manner but that each business develops procedures that reflect that particular business and how it operates, but with quality of the end product or service addressed at all stages of the business. The original British Standard edition was drafted in terminology that reflected its manufacturing origins but the current standard has improved, although the definition of product ('activities or processes') requires three footnotes to clarify that it can apply to almost anything offered for sale.

There is not the space in this chapter to go into exhaustive detail which can be derived from the standards themselves and, as stated previously and shown in Figure 4.6, embraces the whole of the product.

There are some common misconceptions about quality assurance. BS EN ISO 9001 covers design, development, production, installation and servicing, whereas 9002 covers systems which do not involve design and development and 9003 covers final inspection and testing. It is not the case that 9001 is better than 9002, it is necessary for an organization that wishes to use the quality standards to decide which standard is appropriate for their business objectives. If the organization is involved in design and development, it must use 9001 to adequately cover in quality terms its business. The asphalt industry does not design its products in complying with British Standards, even though some binder and surfacing materials are called 'design mixtures', and consequently all Coated Material Suppliers and Contractors quality assured to date operate to BS EN ISO 9002. But this compliance may change to BS EN ISO 9001 with the implementation of end performance standards and specifications where the responsibilities of the supplier will include the design procedure.

The other misconception is about auditing. Quality systems operate as first-, second- and third-party schemes. A fundamental part of the scope

Quality systems

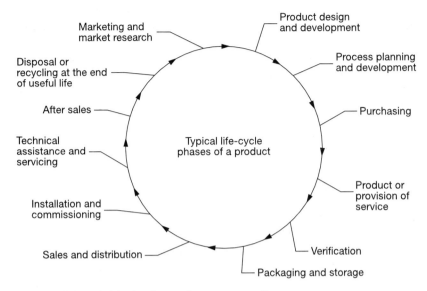

Fig. 4.6 Main activities having an impact on quality

of any quality assurance scheme is the auditing procedure, which is the internal verification of the working of the quality assurance scheme by qualified assessors. Any organization with a quality scheme has a quality assurance manager whose responsibility it is to plan a continuous monitoring of the system. This situation for first-party schemes is that it is the organization's own assessment. In a large organization which may have a number of diverse businesses, it is not uncommon for one business to audit another. Second-party schemes are audited by the client. Many examples of this type of arrangement exist currently in our industry, groupings of counties such as Mercia, South West counties and individual counties such as Kent and Surrey, auditing the coated material suppliers supplying into their respective counties. Third party schemes have independent assessing bodies. These bodies, such as British Standards Institution, Lloyd's and Cares, must belong to the National Accreditation Council Certification Bodies (NACCB) and will register the organization if, after assessment, its quality assurance system is satisfactory; thereafter, it will monitor the organization's scheme on a regular basis for a fee.

The misconceptions arise around the audit level. The organization has to audit and monitor itself, as no amount of 'papering' over deficiencies should escape the notice of a qualified assessor, either internally, second party or third party. Sometimes, much is made of the need for assessors to be knowledgeable about the business they are assessing but it is not the business they are assessing, it is the quality system of the business. While specialist knowledge may be helpful in contextual terms, one of the quickest ways to learn about any business is to have sight of its quality assurance procedures; hence, they are confidential and any experienced assessor knows the potential weak points in systems from experience, these being common to most systems.

Currently, the Highways Agency (HA), County Surveyors' Society (CSS) and Quarry Products Association are finalizing a national scheme for the supply of coated materials, which is to be third party accredited and due to be introduced in 1997. How long it will be before such a similar scheme will be developed for contractors is a matter for conjecture but again, individual counties such as Kent and Surrey are requiring their surfacing contractors to have QA schemes to BS EN ISO 9002 to be included on their tender lists. Quality assurance is self-promoting, because schemes require the organization to have its own suppliers quality assured or places on the organization the need to audit its suppliers to assure quality. As all suppliers of coated materials have quality assurance and increasing numbers of contractors, the need for monitoring outputs from these suppliers of goods and services by the client on site diminishes, with increasing emphasis on statistical records of quality. Quality assurance and the private finance initiative are all part of the transfer of responsibility for the end product, be it the road, airport or other paved area, from the client to the contractor which, in the author's opinion, is as it should be.

4.7 REFERENCES

Manual of Contract Documents for Highway Works, Her Majesty's Stationery Office, London.
    Volume 1: Specification for Highway Works (MCHW 1).
    Volume 2: Notes for Guidance on the Specification for Highway Works (MCHW 2).
Design Manual for Roads and Bridges, Her Majesty's Stationery Office, London.
    HD 26/94 Pavement Design (DMRB 7.2.3).
    HD 28/94 Skidding Resistance (DMRB 7.3.1).
    HD 31/94 Maintenance of Bituminous Roads (DMRB 7.4.1).
Abell, R. (1994) Whole life costing of road pavements, *TRL Annual Review 1994*, Transport Research Laboratory, Crowthorne.
British Standards Institution (1989) *Testing Aggregates; Part 114, Method for Determination of the Polished-Stone Value. BS 812: Part 114: 1989*, British Standards Institution, London.
British Standards Institution (1990) *Sampling and Examination of Bituminous Mixtures for Roads and Other Paved Areas; Part 105, Methods of Test for the Determination of Texture Depth. BS 598: Part 105: 1990*, British Standards Institution, London.
British Standards Institution (1992) *Hot Rolled Asphalts for Roads and Other Paved Areas; Part 1, Specification for Constituent Materials and Asphalt Mixtures; Part 2, Specification for Transport, Laying and Compaction of Rolled Asphalt. BS 594: Part 1: 1992, BS 594: Part 2: 1992*, British Standards Institution, London.
British Standards Institution (1993) *Coated Macadam for Roads and Other Paved Areas; Part 1, Specification for Constituent Materials and for Mixtures; Part 2,*

References

Specification for Transport, Laying and Compaction. BS 4987: Part 1: 1993, BS 4987: Part 2: 1993, British Standards Institution, London.

British Standards Institution (1994) *Quality Management and Quality Assurance Standards. BS/EN/ISO 9000: 1994*, British Standards Institution, London.

British Standards Institution (1996) *Sampling and Examination of Bituminous Mixtures for Roads and Other Paved Areas; Part 110, Methods of Test for the Determination of the Wheel-Tracking Rate of Cores of Bituminous Wearing Courses. BS 598: Part 110: 1996*, British Standards Institution, London.

European Asphalt Pavements Association (1995) *Heavy Duty Pavements – The Arguments for Asphalt*, European Asphalt Pavements Association, Breukelen, The Netherlands.

Hosking, R. (1993) *Highways: Practice to Principles. Seminar Notes PA 3011/93*, Transport Research Laboratory, Crowthorne, 35–47.

Nicholls, J.C. and M.E. Daines (1993) *Acceptable Weather Conditions for Laying Bituminous Materials*, Department of Transport TRL Project Report 13, Transport Research Laboratory, Crowthorne.

Roe, P.G., D.C. Webster and G. West (1991) *The Relation between the Surface Texture of Roads and Accidents*, Department of Transport TRL Research Report 296, Transport Research Laboratory, Crowthorne.

Sabey, B.E. (1966) *Wet Road Skidding Resistance at High Speeds on Variety of Surfaces on A1*, Department of Transport TRRL Laboratory Report 131, Transport and Road Research Laboratory, Crowthorne.

Williams, A.R. (1994) Are roads worthy of today's vehicles and tyres?, *Tyretech '94*, Paper 7, Munich.

CHAPTER 5

Rolled asphalt surface courses

S.M. Child, Surrey County Council

5.1 CONCEPT AND HISTORY

5.1.1 Concept

Rolled asphalt (or hot rolled asphalt, HRA) surface course is a dense mixture of mineral filler, sand and bitumen into which a coarse aggregate is added. It is a gap-graded material, the mechanical properties of which are dominated by the mortar. The material is extensively used for surfacing major roads in the United Kingdom as it provides a dense, impervious layer resulting in a weather resistant durable surface able to withstand the demands of today's traffic loads providing good resistance to fatigue cracking but some susceptibility to deformation. Pre-coated chippings are normally rolled into the surface to provide the required skid resistant properties of the surface course.

5.1.2 History

Powdered natural rock asphalt (Spielmann and Hughes, 1936) was the type of asphalt first used for surfacing city streets and was laid warm and compacted by hand. Developments in America around 1870 resulted in the use of synthetic asphalts incorporating fine mineral matter giving mixtures easily spread with rakes and compacted with a steam roller. An American engineer, Clifford Richardson, undertook a survey in 1894 of different asphalt mixtures that had been laid up to that time. He compared composition of the mixture with actual behaviour and durability of the asphalt (Richardson, 1905) which, at that time, contained very little, if any, coarse aggregate. Having reviewed his findings, Richardson produced comprehensive specifications that gave an asphalt with the best workability qualities combined with durability in use.

Rolled asphalt was first used in the United Kingdom in 1895 on the King's Road, Chelsea, and Pelham Street, Kensington (Attwooll, 1955). This came about following the visit of T.G. Marriott of the Trinidad Lake Asphalt Paving Company Ltd to America. The importance of sand grading and filler content was, unfortunately, not appreciated at that time and the material failed within a matter of months in the damp English climate. Richardson visited the United Kingdom in 1896 and, based on his experience, he recommended the use of an asphalt mixture which was similar in all respects to sand carpet and which performed well for many years.

Extensive practical research undertaken by those companies involved with the development of rolled asphalt, together with the experience and data they collected, made it possible for the British Engineering Standards Association, in conjunction with the Asphalt Roads Association and the Ministry of Transport, to issue British Standards specifications in 1928 (BSI, 1928a, 1928b, 1928c). In 1935, these standards were combined and the number changed to BS 594 (BSI, 1935). Since that time, BS 594 has been revised as a single standard five times (BSI, 1945a, 1945b, 1950, 1958, 1961, 1973). A sixth revision in 1985 resulted in a two part standard; Part 1 covering constituent materials and asphalt mixtures (BSI, 1985a) and Part 2 covering transport, laying and compaction of rolled asphalt (BSI, 1985b). The revision of 1992 (BSI, 1992a, 1992b) maintained the two part standard as each revision throughout its development has reflected the current thinking of research, specifiers and practitioners. End performance, the outcome of harmonized European specifications and new developments based on worldwide research applied to the UK industry, as highlighted in the Introduction (section 1.5), are the next steps to be taken.

5.2 PROPERTIES AND APPLICATIONS

5.2.1 Properties

A typical rolled asphalt surface as shown in Figure 5.1, together with a cross-section in Figure 5.2, will provide the Highway Authority with a material that exhibits the following characteristic properties:

- a dense, stiff layer that resists compaction under traffic loading resulting in no significant deformation and spreads the traffic load through the layer;
- a near impervious layer that prevents ingress of water into the lower pavement layers and sub-grade below thus preserving the stability and life of the lower layers;
- a durable layer that can withstand the climatic influences of rain, sun and temperature range; and
- a durable layer that can withstand man-made influences such as salt and oil deposits.

With regard to the road user, the 'customer', the Highway Authority must satisfy highway law and here safety is paramount. A rolled asphalt surface can be produced to provide the following surface characteristics:

110   *Rolled asphalt surface courses*

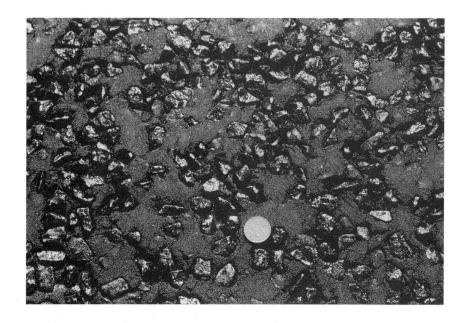

Fig. 5.1   Rolled asphalt surface course

Fig. 5.2   Cross-section of rolled asphalt layer

Properties and applications 111

- a skid-resistant surface for safety by using pre-coated chipping of adequate characteristics and an asphalt mortar having weathering properties;
- an even surface for user comfort in respect of riding quality;
- a deformation resistant surface that keeps its shape and sheds water hence no risk of ponding water; and
- resistance to tractive forces both braking and accelerating.

If a rolled asphalt surface course is correctly designed, produced, laid and compacted as a finished layer, it is one of the best, most proven materials for achieving the required characteristics highlighted in Chapter 4 with a longevity that has been tried and tested over the years. The advantages of rolled asphalt as compared to other surfacings are shown in Table 5.1 (Hunter, 1994) which is based on performance after six years' service.

Today's traffic is taking this material to its limit and designs are being re-evaluated and re-directed in order to provide a rolled asphalt surface course that can sustain many of the above characteristics and give a satisfactory life. Section 5.3 highlights the influence of the component materials on the above characteristic properties together with those of the mixed and laid material.

Table 5.1 Relative properties of asphalt mixtures

Characteristic	Effectiveness ratio (1 = poor, 6 = very good)				
	Open-textured macadam	Porous asphalt	Dense macadam	Hot-rolled asphalt	Surface dressing
Structural strength	2	3	3	6	1
Rut resistance	2	3	4	4	1
Crack resistance	3	3	3	4	1
Skid resistance, low speed	5	5	5	4	5
Skid resistance, high speed	4	5	4	4	5
Texture depth	2	2	3	4	6
Spray suppression	3	6	1	2	3
Low noise	4	4	3	2	1
Ease of application	3	3	3	2	5
Area treated/ unit cost	2.5	1.5	2	1.5	6

5.2.2 Applications

In the mid to late 1940s, it was recognized that rolled asphalt had a large part to play in the surfacing of the highway network. Table 5.2 is an

Table 5.2 Types of asphalt used for various purposes

Type of surface	City streets	Dock traffic	Main and trunk roads	Suburban secondary roads	Housing estate roads	Renovation of existing surfaces
Compressed natural rock Asphalt (hot process)	*	*				
Mastic asphalt	*	*	*			*
Rolled asphalt (two-course process)	*	*	*	*	*	
Rolled asphalt (single-course process)			*	*	*	*
Thin application mixtures (carpet coats)			*	*	*	*

NOTE: This table should be regarded as no more than a rough and ready guide, and does not imply that the various types of asphalt can only be employed for the purposes shown.

extract from the Highway Engineers' Reference Book (Anon, 1947) which shows the advice given on use of rolled asphalt at that time.

In theory, rolled asphalt could be used in all surfacing types from footway to motorway; however, economics and practicalities ensure that it is used in the most appropriate situations. Any situation that demands a surface layer of excellent structural strength, resistance to deformation and cracking, skid resistance and water resistance can use rolled asphalt to the most appropriate design.

The current UK standards for pavement design can be found in HD 26/94 Pavement Design in Volume 7 of the Design Manual for Roads and Bridges (DMRB 7.2.3). HD 26/94 states that, for rolled asphalt surface course, it shall be 45 mm or 50 mm thick to Clause 911 of the Specification for Highway Works (MCHW 1) and to Table B1 of Annex B from BS 594: Part 1 (BSI, 1992) for references to stability and flow values related to traffic loading. Design values for Scotland are given in the special requirements of Clause NG 911SO (MCHW 2). This standard design applies to rolled asphalt surface course for flexible pavements, flexible composite pavements or continuous rigid composite pavements.

Rolled asphalt can be used under heavy traffic loading as its ability to spread load means that it can be considered as a 'structural element' of the pavement. On heavily trafficked roads, channelization of traffic, particularly on uphill gradients, on exposed sites and where tractive/braking forces are greatest, means that rolled asphalt is suscepti-

Table 5.3 Table B1 from Annex B of BS 594: Part 1: 1992

Traffic (in commercial vehicles per lane per day)	Stability of complete mixture (kN)
Less than 1500	3 to 8*
1500 to 6000	4 to 8
Over 6000	6 to 10

*It may be necessary to restrict the upper limit where difficulties in the compaction of materials might occur. Type R enriched mixtures conforming to Table 6 are intended for use with this traffic category.
NOTE 1: For stabilities up to 8.0 kN the maximum flow value should be 5 mm. For stabilities in excess of 8.0 kN a maximum flow of 7 mm is permissible.
NOTE 2: The stability values referred to should be obtained on laboratory mixtures.
NOTE 3: The stability and flow values are those pertaining to the target binder content.

ble to deformation in the wheel tracks and this must be allowed for by selection of the appropriate mixture properties. By choosing the appropriate stability and flow, the rate of deformation can be kept to an acceptable level even though the stability is not perfect at predicting permanent deformation (Szatkowski and Jacobs, 1977; Daines, 1992). Table 5.3 highlights the current BS 594 criteria for the stability of laboratory design asphalt as referenced in the Specification for Highway Works (MCHW 1) for design purposes.

Current practice in the United Kingdom is to use rolled asphalt surfacing, with pre-coated chippings applied to give surface texture and achieve a satisfactory level of skidding resistance, on the motorway trunk road and principal county road networks. Beyond this, it is used on all other types of road down to unclassified and housing estate roads to varying degrees. The choice may depend on the highway designers' preference or the national thinking at that time.

Fifty-five per cent coarse aggregate content rolled asphalt surfacing, with no chippings applied but with its own inherent texture, is often used on lighter trafficked roads where the material gives the benefits of durability and skid resistance appropriate for the site while deformation is not a problem due to reduced loading.

In areas of high traffic stress, such as approaches to roundabouts, junctions, pedestrian crossings and traffic lights, the tractive and braking forces are such that the surface stresses can result in severe loss of chippings particularly on a heavily chipped, high-texture depth, rolled asphalt. If the appropriate level of skid resistance cannot be achieved or maintained at this type of site, the rolled asphalt surface may be laid and a high friction surface (section 11.3) applied.

It is unusual for rolled asphalt to be used in car parks or footways due to practicalities of laying and economics. Rolled asphalt is not a material for extreme point loads and situations such as distribution areas and lorry parks should be surfaced in an alternative material. General parking areas associated with the highway may use a 40% coarse aggregate content rolled asphalt to give a layer with increased load-spreading ability.

Tracked vehicles, such as tanks, are very destructive to road surfaces on bends. Experiments and trials have shown that a hard, smooth-textured material containing more filler than for normal use on roads, can provide a satisfactory surface for tracked vehicles. Recommendations for asphalt for tank traffic are given in Road Note 13 (RRL, 1952).

Rolled asphalt can be used to provide coloured surfaces and an example of this is The Mall in London, which was first surfaced with red asphalt in 1949. The colour is obtained by the use of a red aggregate and the addition of a red pigment to the bitumen.

5.3 COMPONENT MATERIALS

5.3.1 Aggregates and filler

5.3.1.1 Coarse aggregates

BS 594: Part 1 (BSI, 1992) defines coarse aggregate as that 'substantially retained on a 2.36 mm BS test sieve' and permits the use of:

- crushed rock;
- gravel;
- blast-furnace slag; and
- steel slag.

The size and grading of the coarse aggregate is governed by the proposed layer thickness and its ability to be compacted satisfactorily. The resultant mixture has a relatively single sized coarse aggregate fraction of minimum intermediate size between 2.36 mm and 10 mm and is therefore 'gap-graded'. Figure 5.3 is a typical grading for a 30/14 BS mixture (where 30 is the proportion of coarse aggregate in % and 14 is the maximum nominal size of coarse aggregate in millimetres). The maximum size of the coarse aggregate must be related to the proposed layer thickness and the general rule in BS 594 is that the maximum nominal size should be not greater than one-half and not less than one-third the thickness of the course.

Grading is of secondary importance in BS 594 mixtures as the mortar primarily governs mechanical stability. However, it is essential that the correct proportion of coarse aggregate is used otherwise binder drainage could occur, resulting in a variation in binder content of the finished surface course and hence reduced durability, together with excess binder at the surface.

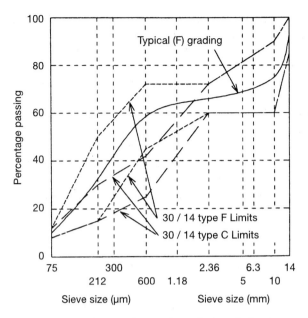

Fig. 5.3 Typical grading and limits for a 30/14 rolled asphalt mixture to BS 594

The BS specifies a maximum flakiness index (section 2.2.1.3) of 45% for crushed rock and crushed gravels and 50% for uncrushed gravels, together with a maximum fines content passing the 75 μm sieve of 6% for crushed rocks and slags and 4% for gravels.

Blast furnace slag is required to comply with BS 1047 (BSI, 1983).

It is important that coarse aggregate used is 'clean, hard and durable' (section 2.3.1). BS 594 states that:

> there is no acceptable method for determining clay content of coarse aggregate but an indication of quantity is given using the sedimentation test described in BS 812: Section 103.2.

It also states that:

> there are no satisfactory tests for determining the amount of deleterious materials in aggregates. Any obviously degraded or dirty stone or other contamination, such as by roots, vegetation or particles of lignite, should be avoided. If an aggregate is known to be prone to stripping then addition of an adhesion agent, Portland cement or hydrated lime might be beneficial.

Coarse aggregate may have a 10% fines requirement to ensure satisfactory strength and also a soundness requirement using the magnesium sulphate test.

Coarse aggregate contents of 0%, 30%, 40% and 55% are permitted by BS 594: Part 1 (BSI, 1992) although Amendment No. 1 replaced the 40/14 mixture with a 35/14 mixture from 15 February 1995. When the higher proportions of coarse aggregate are used, without the application of pre-coated chippings (section 5.4.2), the wearing properties of coarse aggregate must be considered. Polished stone value (section 2.2.4.4), aggregate abrasion value (section 2.2.4.3) and 10% fines value (section 2.2.3.2) should be specified in individual contracts or local specifications.

Some highway authorities may exclude the use of gravel or limestone in surface course mixtures in order to reduce the risk of failure due to polishing at the surface or lack of stability due to aggregate shape (section 2.2.1.3) or bond.

Coarse aggregate is primarily included to extend the mortar in the total mixture, hence reducing the total cost of the product. However, it does contribute to the overall stiffness of the mixture. Mixtures can be defined as 'F' or 'C'. Type F (fine) is a gap graded surface course usually associated with the use of sand fines whereas Type C (coarse) is a coarser grading which normally uses crushed rock or slag fine aggregate. F reflects the finer grading of fine aggregate compared to C as highlighted in the comparison shown in Figure 5.3.

5.3.1.2 Fine aggregate

BS 594 defines fine aggregate as that which 'shall substantially pass a 2.36 mm BS sieve' and permits the use of:

- sand;
- fines produced by crushing material that would be acceptable as a coarse aggregate (generally referred to as crushed rock fines); and
- a mixture of sand and crushed rock fines (generally referred to as a blend).

Fine aggregate provides the major proportion of the mortar and is that material passing a 2.36 mm but retained on a 75 μm BS sieve. Specification details for 600 μm and 212 μm sieves are given in the BS for each mixture type.

The BS requirements for fine aggregate are given in Table 5.4.

The characteristics of the fine aggregate are very important as it forms a large element of the mortar. Different aggregates will require differing amounts of binder to achieve the same mixture properties. This occurs as

Table 5.4 Requirements for fine aggregates

Mixture type	Retained on 2.36 mm BS sieve (by mass)	Passing 75 μm BS sieve (by mass)
F	Not more than 5%	Not more than 8%
C	Not more than 10%	Not more than 17%

a result of differences in grading, shape, surface area and porosity which affect binder demand.

Natural sands or sand blends have been traditionally used as fine aggregate in rolled asphalt surface course. The advantages of using a fine sand consisting of fairly rounded particles are that it enables the rolled asphalt to be readily worked on site and also permits the use of a binder content sufficiently high for the material to be compacted to a dense, impervious surface. The need to increase mixture stability to meet the demands of today's traffic has resulted in wider use of crushed rock fines which tend to be coarser and harsher (more angular and elongated) and hence increase the stability while reducing the workability. Blast furnace slags are also of coarse grading and rough texture giving a similar increase in stability.

5.3.1.3 Filler

The filler is that portion of aggregate which passes a 75 μm BS sieve and will be present in the aggregate mixture from the following sources:

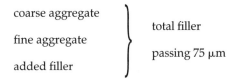

coarse aggregate
fine aggregate } total filler passing 75 μm
added filler

Added filler can be limestone or Portland cement and shall contain not less than 85% by mass of particles passing the 75 μm sieve and have a bulk density in toluene[1] of not less than 0.5 g/ml and not more than 0.9 g/ml.

As with fine aggregate, filler can affect the binder demand. A high filler content will result in a large surface area of aggregate hence increased binder demand. The filler can modify the grading of the fine aggregate to give a denser mixture with greater aggregate contact.

The filler in conjunction with the bitumen forms the 'binder' which lubricates and binds the fine aggregate to form the 'mortar'. It is present to stiffen and strengthen the bitumen so that the voids in the fine aggregate are filled with a filler/bitumen mixture of higher strength and lower temperature susceptibility than could be achieved with a harder binder. Clifford Richardson wrote the following on this matter (Richardson, 1905):

> The only way to keep water out of an asphalt surface is to have the voids in the asphalt mixture as small as possible in size, but not necessarily in volume, to fill them with bitumen of a consistency which will permit of contraction and to stiffen the latter with a proper amount of filler which will alone permit the use of a sufficiently soft cement. If the interstitial spaces are few in numbers but large in size, the asphalt occupying them will be in such large masses, if the voids are entirely filled, that they will easily yield to

stress and cause the surface to mark and push and the pavement to appear soft. If the voids are not filled, water quickly enters and destroys the pavement.

The Energy Efficiency Office has had research undertaken to investigate the feasibility of using pulverized fuel ash (PFA), a by-product of coal-fired power stations, as a replacement for limestone filler in rolled asphalt surface course. Laboratory testing of mixtures indicates that mixtures with PFA filler exhibited better compacting and mixing characteristics than mixtures made with limestone. The conclusion of the project (Energy Efficiency Office, 1990) was that:

> by using PFA as a filler, the temperature of mixing and compaction can be reduced to as low as 88 °C without adversely affecting the performance of the road. The reduction in heating energy, required to prepare the mix is a significant benefit ...

5.3.2 Binders

Traditionally 'binder' in rolled asphalt was known as 'asphalt cement' but this was dropped in BS 594: 1985 to avoid confusion with concrete terms and to coincide with terminology used in coated macadam.

BS 594: Part 1 (BSI, 1992a) permits the use of either penetration grade bitumen or lake asphalt–bitumen mixture. These can be 35, 50, 70 or 100 penetration (bitumen only) complying with BS 3690: Part 1 (BSI, 1989d) and BS 3690: Part 3 (BSI, 1990a), respectively. Table 5.5 details the BS requirements for penetration grade bitumens.

Table 5.5 BS 3690 requirements for bitumen

Property	Test method			Grade		
		35 pen	40 pen HD*	50 pen	70 pen	100 pen
Penetration at 25 °C	BS 2000: Part 49 (BSI, 1993d)	35 ± 7	40 ± 10	50 ± 10	79 ± 10	100 ± 20
Softening point (°C) (min) (max)	BS 2000: Part 58 (BSI, 1993c)	52 64	58 68	47 58	44 54	41 51
Loss on heating for 5 h at 163 °C Max. loss by mass (%)	BS 2000: Part 45 (BSI, 1993b)	0.2	0.2	0.2	0.2	0.5
Max. drop in penetration (%)		20	20	20	20	20
Solubility in trichloroethylene Min. proportion by mass (%)	BS 2000: Part 47 (BSI, 1993c)	99.5	99.5	99.5	99.5	99.5

Editions of BS 594 prior to 1992 included pitch-bitumen binders which had the advantage of giving a degree of weathering to the bitumen which maintained the texture depth, the disadvantage being embrittlement of the pitch-bitumen. It has been shown that changes in bitumen production could produce a similar effect and the use of a pitch-bitumen blend declined.

The hardness of the bitumen, as defined by penetration, is the most important factor in determining the resistance of the rolled asphalt to deformation under the effect of traffic. In hot weather when the pavement surface may exceed 50 °C, the bitumen must not become too soft, while in cold weather it must not be too brittle.

Bitumen is specified according to penetration and softening point. This will ensure that it has suitable flow characteristics. The maximum temperature reached by the surface during the summer cycle must not exceed the softening point value otherwise the bitumen may soften and deformation can occur. The grade of bitumen used will depend on climatic conditions, softer binder in the cooler north, and traffic levels, harder binder in the more densely trafficked areas. Thirty-five penetration is used for severe traffic conditions while 70 penetration is used for light traffic; 50 penetration is the most commonly used grade for pavement surfaces; heavy-duty 40 penetration bitumen has been used in the recent past to provide improved resistance to deformation however it proved very unworkable and has fallen into decline, although it is still an acceptable alternative in the Specification for Highway Works (MCHW 1). 100 pen bitumen is used on minor roads or in secondary applications.

The type of bitumen will affect the nature of the rolled asphalt surface. The surface is exposed to air and will harden with time. Hardening of the bitumen results in abrasion by traffic which exposes 'fresh' aggregate that can be subject to further wear by traffic. Lake-asphalt will harden more quickly than normal bitumen and can enhance the skidding resistance characteristics of the surface.

BS 594: Part 1 (BSI, 1992) defines 'design' and 'target' binder contents as follows:

- design binder content: The binder content of a surface course design mixture determined in accordance with the procedure described in BS 598: Part 107: 1990.
- target binder content: The nominal soluble binder content of a surface course design mixture specified in Clause 3.2 and which is equal to or greater than the design binder content and the specific requirements given in tables 3, 4 or 5.

BS 598: Part 107 (BSI, 1990c) is used to determine a suitable binder content for a chosen aggregate composition. The result is a design binder content which is a compromise between the conflicting needs of stability, durability and workability.

Bitumen properties will influence the performance of a rolled asphalt surface course from manufacture to in-service life. Bitumens are visco-elastic materials and exhibit behaviour from purely viscous to elastic depending on loading time and temperature. During mixing and compacting, as well as at high service temperatures, properties can be considered in terms of viscosity. However for most of the time, bitumen will behave visco-elastically and properties can be considered in terms of stiffness modulus.

Mixing temperature must be high enough to allow rapid distribution of bitumen on to the aggregate but short enough to prevent excessive oxidation of the bitumen exposed on the aggregate in a thin film. Transporting and laying are also governed by viscosity in that, at viscosities lower than 5 Pa.s, the material is too mobile to compact while, at viscosities greater than 30 Pa.s, the material is too stiff to be workable.

In-service, it is necessary to consider the influence of bitumen properties at high and low service temperatures. At high temperatures, say 30 °C to 60 °C, deformation and fatting-up are the problem while at low temperatures, less than 5 °C, it is cracking and fretting.

A surface course can deform plastically under moving or stationary traffic and particularly under the high shearing stresses imposed by braking, accelerating or turning traffic. Deformation is highest at high service temperatures and the cumulative effect of short duration repeated loadings will be governed by viscosity. Fatting-up takes place when bitumen is forced to the surface, this will be exacerbated if bitumen content is high or void content too low. Fatting-up results in a smooth, shiny surface which has low skidding resistance in wet weather. Limiting the softening point or viscosity at 60 °C will limit fatting-up.

Low-temperature failure is associated with cracking of the surface course or fretting of individual particles from the surface. The stiffness modulus of the mixture will determine the resistance to these two problems. High stiffness of the bitumen will lead to cracking being induced while fretting is likely at low temperatures in similar conditions. It is important to determine the design criteria required and ensure that the bitumen proposed will provide a mixture to satisfy them.

5.3.3 Binder modifiers

The increasing volume of traffic, increased axle loads and higher tyre pressures have taken conventional rolled asphalt to its limit. Although it has performed satisfactorily on a wide range of roads in the past, the last decade has seen increased use in a range of modifiers that have been designed to enhance the properties of a mixture by modifying bitumen performance. It is those areas of road that are more highly stressed or more exposed to climate influence that need particular attention.

Component materials

The degree of improvement will depend on the site to be treated and various modifiers have been trialled. Denning and Carswell (1981) stated that, for a modifier to be effective and for it to be practicable and economic, it must:

- be readily available;
- blend with bitumen;
- resist degradation at asphalt mixing temperatures;
- improve resistance to flow at high road temperatures without making the bitumen too viscous at mixing and laying temperatures or too stiff or brittle at low road temperatures; and
- be cost-effective.

One of the main roles of a modifier is to increase the resistance of the asphalt to permanent deformation without adversely affecting its properties at other temperatures.

This is achieved by either:

- stiffening the bitumen, thus reducing the visco-elastic response with a corresponding reduction in permanent strain; or
- increasing the elastic component of the bitumen, thus reducing the viscous component which has a similar effect to a reduction in permanent strain.

The visco-elastic response to static and varying wheel load is shown in Figures 5.4 and 5.5.

Table 5.6 highlights examples of categories of binder modifiers and the characteristic that is modified.

Table 5.6 Binder modifiers and the characteristics modified

Type of modifier	Characteristic modified	Remarks
Reclaimed tyre rubber	Deformation resistance	Difficult to disperse in
Sulphur; organo-manganese; thermoplastic polymers	Deformation resistance. Workability during compaction	Sulphur rarely used for health and safety reasons. Organo-manganese (Chemcrete) no longer used in UK. EVA is the principal thermoplastic polymer
Thermoplastic rubbers	Deformation resistance. Fatigue characteristics	Must be compatible with hot bitumen[a]. SBS is a typical example
Thermosetting binders	Deformation resistance. Flexibility	Blending two components which act like a resin. Good resistance to oil and chemicals

(Increasing benefit and cost ↓)

[a] Bitumen/thermoplastic rubber compatibility is measured using the hot storage stability test. The polymer content can be determined using an infrared spectrophotometer.

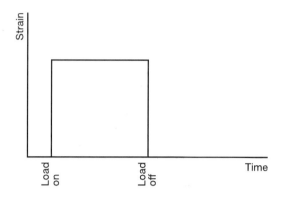

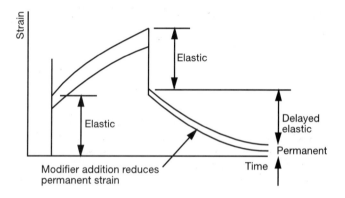

Fig. 5.4 Visco-elastic response to static loading

Table 5.7 Comparison of wheel-tracking rates of rolled asphalt surface course mixtures with various modifiers

Binder	Penetration at 25 °C	Wheel-tracking rate at 45 °C (mm/hour)
50 pen	56	3.2
50/50 TLA + 50 pen bitumen	24	1.6
70 pen + 5% EVA (45/33)	57	1.0
200 pen + 7% SBS	84	0.7
HD40	42	0.7
70 pen + 5% Asphapol 2000	59	0.5
50 pen + EVATECH	56	1.3
Cariphalte DM	93	0.8
Multiphalte 35/50	40	0.9

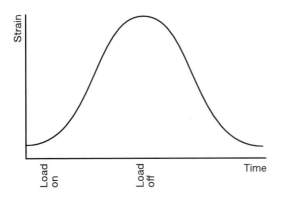

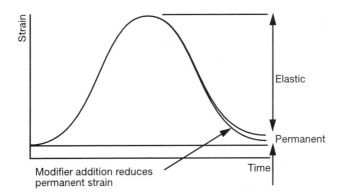

Fig. 5.5 Visco-elastic response to a moving wheel-load

The common characteristic improved by the addition of a modifier is that of deformation resistance, which can be determined using the wheel-tracking test. A summary of various sources of data highlights the benefits of using a modifier to achieve increased deformation resistance as set out in Table 5.7.

Figure 5.6 is a photographic comparison of the wheel-tracking properties of asphalt mixtures with and without modifier to highlight the benefit of adding a modifier to enhance the deformation resistance characteristics.

Polymer modification reduces the temperature susceptibility of bitumen, and hence the low-temperature characteristics, thus reducing the likelihood of rolled asphalt surface course made with modified binder to crack in colder weather.

Nicholls (1994) highlighted that, for surface course, the property changes described in the mixed material are as follows:

Fig. 5.6 Result of wheel-tracking cores with modified (left) and unmodified (right)

- increased workability;
- reduced permanent deformation;
- increased load spreading ability; and
- reduced embrittlement.

The binder in a surface course needs to be stiff, even during hot weather, and reasonably durable, even during cold weather. It is essential that the use of a modifier in a designed rolled asphalt surface course material is carefully considered. The modifier must be selected for a specific purpose and the use of a modifier, which is a costly element, must be shown to be cost-effective over the life of the surface course.

5.3.4 Mixed material

Until the early 1970s, rolled asphalt surface course was supplied to a recipe specification. However, BS 594 (BSI, 1973) introduced the concept of design asphalt using the Marshall Test. This test was used to estimate the bitumen content required to optimize the properties of the fine aggregate and filler, the overall aim being to produce mixtures of adequate resistance to deformation and fatigue and of adequate workability and durability. In 1985, this was further developed and the design was based on an 'all-in' grading including the coarse aggregate. BS 598: Part 107 (BSI, 1990c) gives the design method and states that:

the procedure gives a binder content, called the design binder content, which is a compromise between the conflicting needs for stability, durability and workability.

The method involves the determination of a binder content of the complete mix using Marshall Test equipment.

The design method of 'Marshall Test' involves the manufacture and test of three cylindrical specimens at each bitumen content to be tested over a range of at least nine bitumen contents at 0.5% intervals, a total of 27 specimens. From the tests, graphs of mixture density, compacted aggregate density, stability and flow against bitumen content are plotted. Figure 5.7 is the output from a Marshall design showing typical curves. The three bitumen contents corresponding to maximum mixture density, maximum compacted aggregate density and stability are determined and the mean value calculated. An empirical value is added to the optimum bitumen content to give the 'design binder content'. The engineer can determine the 'target bitumen content' of the mixture by adding extra bitumen, typically 0 to 1.0%, to allow for such properties as workability and durability.

Appendix B of BS 594: Part 1 (BSI, 1992a) gives recommendations for the application of design test criteria. It is essential that the properties of a

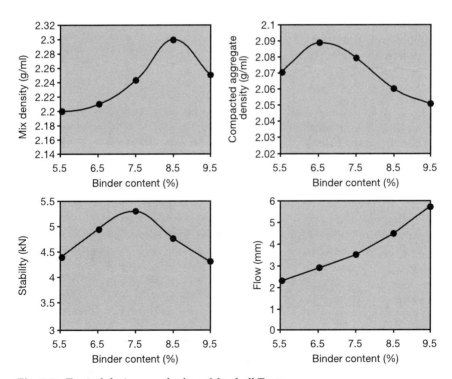

Fig. 5.7 Typical design graphs from Marshall Tests

rolled asphalt surface course are appropriate to the site on which it is to be used. Table 5.3 highlights the criteria given in the British Standard and the values quoted are on laboratory mixtures at the target binder content.

Marshall stability is a measure of the maximum load a specimen can withstand before fatigue while flow is the maximum diametral compressive strain measured at the instance of failure. The Marshall Quotient is the ratio of stability to flow and represents the ratio of load to deformation. The Marshall Quotient may be used to give an indication of mixture stiffness while specifying a minimum flow value may prevent mixtures susceptible to embrittlement being used. Increased stability means increased stiffness of the rolled asphalt which will result in:

- increased resistance to deformation;
- increased likelihood of embrittlement; and
- increased working difficulties in respect of compaction, temperature loss and chip retention.

Increased flow means increased susceptibility to deformation as the material is likely to have a higher binder content and be more 'fluid'. It is, however, less likely to fail by fatigue cracking.

Surrey County Council has reflected the above in its design philosophy for rolled asphalt surface course mixtures specified in Term Tender specification used on county roads (Surrey CC, 1995). The philosophy is shown in Table 5.8 which highlights the use of Marshall Quotient and minimum flow values.

Rolled asphalt surface courses are mixtures of constituent parts, coarse aggregate, fine aggregate, filler and bitumen. It is essential, particularly considering the demands of today's traffic on the pavement, that the 'designed' surface course provides:

- resistance to deformation;
- resistance to fatigue;
- adequate stiffness to withstand traffic loading; and
- adequate flexibility to prevent embrittlement and avoid cracking.

The mixed material has strength that depends on the stiffness of the mortar. A harder bitumen or higher filler content will result in a stiffer, hence stronger, mixture. The fine aggregate and mortar provide the load spreading characteristics of the mixture.

Table 5.8 Surrey County Council rolled asphalt surface course design criteria

Mixture	Stability (kN)	Minimum (mm)	Quotient (kN/mm)
A	2–4	2.0	0.6–1.1
B	4–6.5	2.5	1.1 (minimum)
C	6.5+	3.0	1.1 (minimum)

Air voids can have a significant effect on performance (Hill, 1983) with Daines (1995) concluding that:

> a durable surfacing should result when the air voids content is less than 4% at design stage and properly compacted during construction.

Air void results as a measure of compaction will depend on aggregate shape and grading and bitumen content.

Recent developments have seen an increased use of the wheel-tracking test to measure deformation resistance and this, although normally used at 45 °C, is proving an excellent discriminatory tool at 60 °C. In addition, asphalt testers such as the Nottingham Asphalt Tester (NAT) (Cooper and Brown, 1995) have been developed and can be used to measure stiffness by means of the repeated load indirect tensile test and to assess resistance to deformation using the creep test. A fatigue test has also been developed for use in asphalt testers. The way forward for rolled asphalt surface course has already been demonstrated by the Highways Agency in that end-performance trials have been undertaken on the M56 and M53 (Nunn and Smith, 1994) using asphalt tester results as a design/control tool.

The performance of rolled asphalt has been tested to the limit in the United Kingdom by high temperatures and channelized traffic. The adverse effect of these conditions has been demonstrated by the severe rutting that has occurred on the inside lane of numerous stretches of motorways and trunk roads across the United Kingdom. The rutting is the result of the high summer temperatures, heavier traffic and 'super-single' tyres, which have accelerated the process. Experience in the Middle East using the Marshall stability design parameters for pavements where surface temperatures exceeded 60 °C has demonstrated that rolled asphalt can be designed and laid to withstand such forces. Compaction is vital to the life expectancy, as is the design of lower layers with some of the problems being in the lower layers and 'reflected' through to the surface.

It is imperative that the highway construction industry moves quickly towards end-performance and end-product criteria for all functional properties of the rolled asphalt mixture. A major step forward was taken with the introduction of Clause 943 into the *Specification for Highway Works* (MCHW 1). This clause introduces a requirement whereby the rolled asphalt mixture must provide satisfactory deformation characteristics as measured by the wheel-tracking test (BSI, 1996a) at 45° (moderately heavily trafficked roads) or 60 °C (very heavily trafficked roads). The performance-related design mixture must undergo job mixture approval trials which require details to be submitted regarding:

- quantity of binder and aggregate;
- binder grade;
- generic type of any binder modifier;
- proprietary name and quantity of any binder modifier added at the mixer; and

- additional modified binder and mixture data such as those highlighted in Table 5.9.

The job mixture approval tests to be undertaken during a site trial (which need not necessarily be carried out at the site itself) include:

- composition analysis, including binder content;
- nuclear density gauge readings;
- wheel-tracking rate and rut depth (from 200 mm diameter cores); and

Table 5.9 Binder and mixture data for approval of a binder modifier

	Parameter	*Requirement*
Binder	Penetration before and after hardening in the Rolling Thin Film Oven Test (RTFOT) (ASTM D 2872)	25 °C, 100 g, 5 secs; 5 °C, 200 g, 60 secs.
	Rheology	Curve showing relationships of complex stiffness modulus (G^*) and phase angle (δ, delta) to loading time. At least 5 frequency sweeps, 10°C apart in the range −10 °C to +60 °C of which one must be at 25 °C.
	Storage stability	Difference of not more than 5 °C for three determinations to Clause 941 (MCHW 1) for binder to be considered stable.
	Photomicrography	photomicrograph of modified binder (UV).
	Cohesion	Vialit pendulum curve
	Application temperatures	Max^m and min^m mixing temperatures; Max^m and min^m compaction temperatures; Wind-chill factor graphs.
	Fraass brittle point	IP 80.
	Ageing	Ageing in the Pressure Ageing Vessel (PAV) after RTFOT.
	Rheology after ageing	Rheological properties as above but after RTFOT and PAV.
Mixture	Mixture sensitivity	Wheel-tracking rate at target binder content +0.6%.
	Yield strain	TRL yield strain test (Nunn, 1989).
	Indirect tensile stiffness modulus	DD 213 (BSI, 1993a) before and after water immersion.
	Repeated load axial test	DD 226 (BSI, 1996b).
	Indirect tensile fatigue test	ITFT (procedure under development at time of writing) after ageing.

Table 5.10 Key functional requirements for rolled asphalt binders

Property	Rationality
Fatigue cracking Permanent deformation	Major distress modes
Dynamic stiffness Resistance against stripping Resistance again ageing	Major influencing properties
Resistance against thermal cracking Resistance against wear	Secondary distress modes
Compactability Segregation sensitivity Binder drainage	Mixture design and construction
Permeability Skid resistance	Safety factors

- bulk density, maximum theoretical density and air voids content (from 150 mm diameter cores).

Eventually, end-performance criteria related to:

- skidding resistance;
- surface profile;
- surface texture;
- noise;
- strength; and
- durability

will be demanded by the user, specifier and client in order to ensure value for money and long life pavements. The key functional requirements for a rolled asphalt surface course may be defined as in Table 5.10.

5.3.5 Test requirements

A summary of the test requirements for the constituent materials, the mixtures and the finished product are given in Table 5.11.

5.4 VARIETIES

5.4.1 Rolled asphalt with pre-coated chippings

Rolled asphalt has traditionally contained 30% coarse aggregate and been laid 40 mm thick with an application of pre-coated chippings. This has been found to be satisfactory for a wide range of surfacing applications. Increasing traffic loading has resulted in more attention being given to the mixture and the description 'design mixture' can be truly used for rolled asphalt today. The load spreading, strength characteristics must be designed in at manufacture. In addition, mixtures with low proportion of

Table 5.11 Summary of test requirements

Coarse aggregate	Type (petrographic analyses) grading Flakiness index Passing 75 μm	BS 812: Part 105.1 (BSI, 1989b) BS 812: Part 105.1 (BSI 1989b) BS 812: Part 105.1 (BSI, 1989b)
	Blast-furnace slag Compacted bulk density (steel slag)	BS 1047 (BSI, 1983) BS 1047 (BSI, 1983)
	Polished Stone Value (PSV) Aggregate Abrasion Value (AAV) 10 per cent fines	BS 812: Part 114 (BSI, 1989c) BS 812: Part 113 (BSI, 1990e) BS 812 Part 111 (BSI, 1990f)
Fine aggregate	Clay content Proportion passing 2.36 mm Proportion passing 75 μm	BS 812: Part 103.2 (BSI, 1989a) BS 812: Part 103.1 (BSI, 1985c) BS 812: Part 103.1 (BSI, 1985c)
Filler	Grading Proportion passing 75 μm	BS 812: Part 103.1 (BSI, 1895c) BS 812: Part 103.1 (BSI, 1895c)
Added filler	Bulk density in kerosene	BS 812: Part 2 (BSI, 1995)
Bitumen	Viscosity (cut-backs) Penetration Softening point	BS 2000: Part 72 (BSI, 1993f) BS 2000: Part 49 (BSI, 1993d) BS 2000: Part 58 (BSI, 1993e)
Modifier	Storage stability	SHW Clause 941 (MCHW 1)
Mixed material	Stability and flow Composition Indirect tensile stiffness modulus Repeated load axial test Indirect tensile fatigue test Wheel-tracking Water sensitivity Compactability	BS 5908: Part 107 (BSI, 1990c) BS 598: Part 102 (BSI, 1989e) DD 213 (BSI, 1983a) DD 226 (BSI, 1996b) – BS 598: Part: 110 (BSI, 1996a) – –
Coated chippings	Minimum PSV Grading Specified size AAV Hot sand test	BS 812: Part 114 (BBSI, 1989c) BS 812: Part 103.1 (BSI, 1895c) BS 63: Part 1 (BSI, 1987) BS 812: Part 113 (BSI, 1990e) BS 598: Part 108 (BSI, 1990d)
Laid material	Surface texture Rolling straight edge Transverse straight edge Skidding resistance Noise Chipping loss Strength	BS 598: Part 105 (BSI, 1990b) SHW Clause 701 (MCHW 1) SHW Clause 701 (MCHW 1) – – – –

coarse aggregate, such as 30%, must have chippings, normally 20 mm, pre-coated with bitumen and applied to the surface immediately after laying. An adequate binder film is essential to allow chippings to bond into the surfacing, with a minimum of 1.5% being recommended for 20 mm chippings. The application of chippings is essential to provide resistance to skidding when the surface is wet. This is dependent on three factors as follows:

Varieties 131

Fig. 5.8 Laying rolled asphalt surface course

- the nature of the bitumen – Trinidad lake asphalt binder will give a hardening at the surface that enhances the skid resistance properties of the surface;
- the proportion of bitumen in the mixture – lean mixtures result in higher skid resistance properties, although durability relies on a richer mixture; and
- the nature of the coarse aggregate or coated chippings – the application of chippings results in a rugous surface, known as macrotexture, a measure of which is texture depth. A minimum texture depth by the sand-patch method of 1.5 mm is specified for roads carrying high-speed traffic. Polished stone value (section 2.2.4.4), aggregate abrasion value (section 2.2.4.3) and 10% fines value (section 2.2.3.2) are all detailed in a contract specification. Polished stone value is selected based on a design process in HD 28/94 (DMRB 7.3.1) that relates commercial vehicle flow to category of site.

The application of chippings, as shown in Figure 5.8, brings its own problems with regard to cooling of the asphalt layer. A chipper operates behind the paver to apply the chippings while the roller carries out the compaction operation behind the chipper. Layer thickness and wind speed are the two most detrimental effects on the surface. Specification tolerances allow the surface layer to be as thin as 35 mm, although some highway authorities do specify a minimum 40 mm. Wind chill factors, even on a sunny day, can mean that the time available for compaction is very short (Nicholls and Daines, 1993).

For some time, a 40% coarse aggregate laid 50 mm thick, or alternatively the addition of the correct modifier, was used to provide additional time for compaction, and hence enhancing the workability to allow improved durability and chipping retention, by enhancing the heat retention of the surface course. Problems occurred in the use of stiff 40% mixtures, particularly in the winter, and the British Standard was amended with the 40% mixture being replaced with a 35/14 mixture (35% of 14 mm nominal size coarse aggregate) as the optimum, to be laid 45 mm thick. The 35/14 mixture is now widely used in the United Kingdom for major roads.

5.4.2 High stone-content rolled asphalt

High stone content, or medium temperature, asphalt has proved to be very durable in many situations on lightly trafficked country roads. It has a higher strength than 30/14 rolled asphalt and is easier to lay in what are often difficult situations. The material has achieved lives in excess of 10 years when properly compacted. The standard use of 100 pen bitumen in these mixtures, and hence lower mixing and compaction temperatures than with 50 pen bitumen, gave rise to the alternative name of medium temperature asphalt.

The 55/10 and 55/14 mixtures, as given in BS 594: Part 1 (BSI, 1992a), are intended to be self-texturing and care needs to be taken not to overroll the final surface. In some areas such as Cambridgeshire, this has been addressed by specifying a relatively low minimum texture depth as the 'bottom line' before the expected texture has been developed: this approach has generally proved successful. In other areas such as Warwickshire, grit has been applied during the rolling process to enhance the initial skid-resistance.

5.4.3 0/3 and 15/10 rolled asphalts

These two materials are not used in great quantities. Sand carpet, or 0/3 rolled asphalt, is used as a regulating course to the surface layer when thin application is required. Both materials will provide reasonably impermeable and durable footway or play area surfacings when laid and compacted at the right temperature and correct thickness. The County Surveyors Society has stated in the Pavement Design Manual (CSS, 1994) that a number of authorities have successfully used a non-standard rolled asphalt. This is a 45/10 material with 100 or 200 pen binder laid as a single layer 50 mm thick directly on top of sub-base. A variety of this is 45/6 material laid 20 mm thick used to resurface existing footways. The specifications for these materials are in Table 5.12.

5.5 TACK COATS

Tack coat must be used prior to laying rolled asphalt if the underlying surface will not provide sufficient bond or mechanical key with the new surface. This may occur if the underlying surface is:

Maintenance

Table 5.12 Specification for 45% Type F rolled asphalt

BS sieve	Proportion by mass of total aggregate passing BS test sieve (%)	
	45/6F	45/10F
14 mm	–	100
10 mm	100	90–100
6.3 mm	85–100	45–80
2.36 mm	45–57	45–57
600 μm	34–57	34–57
212 μm	8–35	8–35
75 μm	4.0–8.0	4.0–8.0

A maximum of 11% of the aggregate shall pass the 2.36 mm sieve and be retrained on the 600 μm sieve.
The binder content shall be 6.6% by mass
The coarse aggregate shall be crushed rock with a polished stone value as described in the contract. For 45/6, which is for footway use only, the aggregate shall be limestone.
The binder shall be 200 pen bitumen and the maximum temperature at any time for this binder shall be 160 °C.

- dirty or contaminated;
- concrete;
- an exposed binder course following planing of an inadequate surface course; or
- an old hard, smooth asphalt surface.

BS 594: Part 2 (BSI, 1992a) states that 'where a tack coat is required, it shall be bitumen emulsion complying with class A1-40 or K1-40 of BS 434: Part 1 (BSI, 1984), applied at a uniform rate of spread of 0.35 L/m^2. The emulsion shall not be allowed to collect in hollows and shall be allowed to break before the asphalt is laid.'

5.6 MAINTENANCE

Rolled asphalt surfacing has been known to last 30 years in certain circumstances. In today's traffic, 8–15 years' life is more normal, depending on location and level of traffic. Rolled asphalt has been taken to the limit on its capability and has been used extensively in the United Kingdom. Early life failure can occur, particularly if the rolled asphalt is subject to dense, channelized traffic and does not have adequate deformation resistance.

Rolled asphalt will exhibit signs of failure due to surface cracking, deformation and loss of skid resistance. Typical problems that can occur are as follows:

- Loss of chippings — Plucking out due to lack of embedment or fretting of surface mortar.
- Embedded chippings — Too much embedment at time of laying and movement of mortar around pre-coated chippings results in excess mortar at surface and low texture.

- Fretting — Erosion of fines from the asphalt mortar under the actions of traffic. Harder bitumen and higher void content may increase likelihood of fretting.
- Cracking — This can take the form of crazing, transverse cracking, longitudinal cracking, edge cracking or reflection cracking at joints.
- Deformation — Lost riding quality, wheel-track rutting, ponding of water or surface slippage.
- Lack of skidding resistance — Resulting from a reduction in macro-texture of the surface layer and/or micro-texture of the aggregates used.
- Visual appearance — A multi-patched surfacing, particularly with statutory undertakers' works, may warrant treatment to restore the visual appearance.

A maintenance treatment can be designed, depending upon the mode of failure/problem exhibited. The following are typical treatments, together with the problems they solve:

- surface dressing – seal surface, restore skid resistance
- thin slurry – seal surface, restore skid resistance
- thick slurry – restore profile, restore skid resistance
- thin surface course – restore profile, restore skid resistance
- resurface[2] (plane and replace) – remove brittle surface, restore profile, restore skid resistance

The actual treatment undertaken will be determined based on all the factors involved and the time position on the life span of the asphalt surface.

Rolled asphalt is very tolerant of local patching and minor deficiencies can be easily remedied by a small patch properly executed. This would be done by hand, although anything in excess of 15 m long must be laid by machine.

5.7 ECONOMICS

The Institute of Asphalt Technology, in its video on Hot Rolled Asphalt, says that a properly specified, well-laid rolled asphalt should provide a maintenance free running surface for 15–20 years depending on traffic flows and underlying strength of pavement layers. It is against this background that the economics of using a tried and tested product must be viewed. Rolled asphalt with 35% 14 mm nominal size coarse aggregate with 20 mm pre-coated chippings having a polished stone value of 65 will typically cost £4.40/m^2 at 1996 prices.

5.8 CONCLUSION

At a conference held in London in 1979 entitled *The Performance of Rolled Asphalt Road Surfacing*, the following summary was given:

> The three themes of this Conference can be drawn together in the issues of control and specification. Each aspect demands more rigorous control of manufacture and laying. It is imperative to meet the increasing demands imposed by greater volumes of commercial traffic, in which the mean axle weight in the axle spectrum is increasing, if pavements are to fit their purpose. The Department of Transport has increased the demands of its specification and is anxious to achieve end result drafting. However, improved control supervision is essential.

As the millennium approaches, the above statements currently apply to rolled asphalt.

Notes

1. There is a proposal with the British Standards Institution to change from use of toluene to kerosene for health and safety reasons.
2. The plane and replace solution could utilize the recycling process for surface course 'repave'.

5.9 REFERENCES

Annual Book of ASTM Standards. Volume 04.03, Road and Paving Materials; Pavement Management Technologies, American Society for Testing and Materials, Philadelphia.
  ASTM D 2872-88, Standard Test Method for Effect of Heat and Air on a Moving Film of Asphalt (Rolling Thin-Film Oven Test) (ASTM D 2872).
Standard Methods for Analysis and Testing of Petroleum and Related Products, Institute of Petroleum, London.
  Breaking Point of Bitumen Fraass Method (IP 80)
Manual of Contract Documents for Highway Works, Her Majesty's Stationery Office, London.
  Volume 1: Specification for Highway Works (MCHW 1).
  Volume 2: Notes for Guidance on the Specification for Highway Works (MCHW 2).
Design Manual for Roads and Bridges, Her Majesty's Stationery Office, London.
  HD 26/94 Pavement Design (DMRB 7.2.3).
  HD 28/94 Skidding Resistance (DMRB 7.3.1).
Anon (1947) *Highway Engineers' Reference Book for County and Municipal Engineers, Civil Engineers and Contractors*, George Newnes Ltd, London.
Attwooll, A.W. (1955) Rolled asphalt surfacing, *Journal of the Institution of Highway Engineers* 3(6), 59–75.

British Standards Institution (1928a) *Single Coat Asphalt (Sand and Stone Aggregate). BS 344: 1928*, British Standards Institution, London.
British Standards Institution (1928b) *Single Coat Asphalt (Clinker Aggregate). BS 345: 1928*, British Standards Institution, London.
British Standards Institution (1928c) *Two-Course Asphalt. BS 342: 1928*, British Standards Institution, London.
British Standards Institution (1935) *Specification for Rolled Asphalt, Fluxed Lake Asphalt and Asphaltic Bitumen (Hot Process). BS 594: 1935*, British Standards Institution, London.
British Standards Institution (1945a) *Rolled Asphalt, Asphaltic Bitumen and Fluxed Lake Asphalt (Hot Process); Part 1, Single Course. BS 594: Part 1: 1945*, British Standards Institution, London.
British Standards Institution (1945b) *Rolled Asphalt, Asphaltic Bitumen and Fluxed Lake Asphalt (Hot Process); Part 2, Two Course. BS 594: Part 2: 1945*, British Standards Institution, London.
British Standards Institution (1950) *Rolled Asphalt, Asphaltic Bitumen and Fluxed Lake Asphalt (Hot Process). BS 594: 1950*, British Standards Institution, London.
British Standards Institution (1958) *Rolled Asphalt, Asphaltic Bitumen and Fluxed Lake Asphalt (Hot Process). BS 594: 1958*, British Standards Institution, London.
British Standards Institution (1961) *Rolled Asphalt Hot Process. BS 594: 1961*, British Standards Institution, London.
British Standards Institution (1973) *Specification for Rolled Asphalt (Hot Process) for Roads and Other Paved Areas. BS 594: 1973*, British Standards Institution, London.
British Standards Institution (1983) *Specification for Air-Cooled Blast-Furnace Slag Aggregate for Use in Construction. BS 1047: 1983*, British Standards Institution, London.
British Standards Institution (1984) *Bitumen Road Emulsions (Anionic and Cationic); Part 1, Specification for Bitumen Road Emulsions. BS 434: Part 1: 1984*, British Standards Institution, London.
British Standards Institution (1985a) *Hot Rolled Asphalts for Roads and Other Paved Areas; Part 1, Specification for Constituent Materials and Asphalt Mixtures. BS 594 Part 1: 1985*, British Standards Institution, London.
British Standards Institution (1985b) *Hot Rolled Asphalts for Roads and Other Paved Areas; Part 2, Specification for Transport, Laying and Compaction of Rolled Asphalt. BS 594 Part 2: 1985*, British Standards Institution, London.
British Standards Institution (1985c) *Testing Aggregates; Part 103, Methods for the Determination of Particle Size Distribution; Section 103.1, Sieve Tests. BS 812: Part 103.1: 1985*, British Standards Institution, London.
British Standards Institution (1987) *Road Aggregates; Part 1, Specification for Single-Sized Aggregate for General Purposes. BS 63: Part 1: 1987*, British Standards Institution, London.
British Standards Institution (1989a) *Testing Aggregates; Part 103, Methods for the Determination of Particle Size Distribution; Section 103.2, Sedimentation Test. BS 812: Part 103.2: 1989*, British Standards Institution, London.
British Standards Institution (1989b) *Testing Aggregates; Part 105, Methods for the Determination of Particle Shape; Section 105.1, Flakiness Index. BS 812: Part 105.1: 1989*, British Standards Institution, London.
British Standards Institution (1989c) *Testing Aggregates; Part 114, Method for the Determination of the Polished-Stone Value. BS 812: Part 114: 1989*, British Standards Institution, London.
British Standards Institution (1989d) *Bitumens for Building and Civil Engineering; Part 1, Specification for Bitumens for Roads and Other Paved Areas. BS 3690: Part 1: 1989*, British Standards Institution, London.

British Standards Institution (1989e) *Sampling and Examination of Bituminous Mixtures for Roads and Other Paved Areas; Part 102, Analytical Test Methods. BS 598: Part 102: 1989*, British Standards Institution, London.

British Standards Institution (1990a) *Bitumens for Building and Civil Engineering; Part 3, Specification for Mixtures of Bitumen with Pitch, Tar and Trinidad Lake Asphalt. BS 3690: Part 3: 1990*, British Standards Institution, London.

British Standards Institution (1990b) *Sampling and Examination of Bituminous Mixtures for Roads and Other Paved Areas; Part 105, Methods of Test for the Determination of Texture Depth. BS 598: Part 105: 1990*, British Standards Institution, London.

British Standards Institution (1990c) *Sampling and Examination of Bituminous Mixtures for Roads and Other Paved Areas; Part 107, Method of Test for the Determination of the Composition of Design Wearing Course Rolled Asphalt. BS 598: Part 107: 1990*, British Standards Institution, London.

British Standards Institution (1990d) *Sampling and Examination of Bituminous Mixtures for Roads and Other Paved Areas; Part 108, Methods for Determination of the Condition of the Binder on Coated Chippings and for Measurement of the Rate of Spread of Coated Chippings. BS 598: Part 108: 1990*, British Standards Institution, London.

British Standards Institution (1990e) *Testing Aggregates; Part 113, Method for Determination of Aggregate Abrasion Value (AAV). BS 812: Part 113: 1990*, British Standards Institution, London.

British Standards Institution (1990f) *Testing Aggregates; Part 111, Method for Determination of 10% Fines Value (TFV). BS 812: Part 111: 1990*, British Standards Institution, London.

British Standards Institution (1992a) *Hot Rolled Asphalts for Roads and Other Paved Areas; Part 1, Specification for Constituent Materials and Asphalt Mixtures. BS 594: Part 1: 1992*, British Standards Institution, London.

British Standards Institution (1992b) *Hot Rolled Asphalts for Roads and Other Paved Areas; Part 2, Specification for Transport, Laying and Compaction of Rolled Asphalt. BS 594: Part 2: 1992*, British Standards Institution, London.

British Standards Institution (1993a) *Method for Determination of the Indirect Tensile Stiffness Modulus of Bituminous Mixtures. DD 213: 1993*, British Standards Institution, London.

British Standards Institution (1993b) *Petroleum and its Products; Part 45, Loss on Heating Bitumen and Flux Oil. BS 2000: Part 45: 1993*, British Standards Institution, London.

British Standards Institution (1993c) *Petroleum and its Products; Part 47, Solubility of Bituminous Binders. BS 2000: Part 47: 1993*, British Standards Institution, London.

British Standards Institution (1993d) *Petroleum and its Products; Part 49, Penetration of Bituminous Materials. BS 2000: Part 49: 1993*, British Standards Institution, London.

British Standards Institution (1993e) *Petroleum and its Products; Part 58, Softening Point of Bitumen (Ring and Ball). BS 2000: Part 58: 1993*, British Standards Institution, London.

British Standards Institution (1993f) *Petroleum and its Products; Part 72, Viscosity of Cutback Bitumen. BS 2000: Part 72: 1993*, British Standards Institution, London.

British Standards Institution (1995) *Testing Aggregates. Part 2: Methods for Determination of Density. BS 812: Part 2: 1995*, British Standards Institution, London.

British Standards Institution (1996a) *Sampling and Examination of Bituminous Mixtures for Roads and Other Paved Areas; Part 110, Method of Test for the*

Determination of Wheel-Tracking Rate. BS 598: Part 110: 1996, British Standards Institution, London.

British Standards Institution (1996b) *Method of Test for Determining Resistance to Permanent Deformation of Bituminous Mixtures Subject to Unconfined Dynamic Loading. DD 226: 1996*, British Standards Institution, London.

Cooper, K.E. and S.F. Brown (1995) Assessment of the mechanical properties of asphaltic mixes on a routine basis using simple equipment, *The Asphalt Yearbook 1995*, The Institute of Asphalt Technology, Staines, 35-40.

County Surveyors Society (1994) *Pavement Design Manual, Eng/6-94*, Wiltshire County Council.

Daines, M.E (1992) *The Performance of Hot Rolled Asphalt Containing Crushed Rock Fines, A303 Mere*, Department of Transport TRL Research Report 298, Transport Research Laboratory, Crowthorne.

Daines, M.E. (1995) *Tests for Voids and Compacting in Rolled Asphalt Surfacing*. Department of Transport TRL Project Report 78, Transport Research Laboratory, Crowthorne.

Denning, J.H. and J. Carswell (1981) *Improvements in Rolled Asphalt Surfacings by the Addition of Organic Polymers*, Department of the Environment Department of Transport TRRL Laboratory Report 989, Transport and Road Research Laboratory, Crowthorne.

Energy Efficiency Office (1990) *The Use of Pulverised Fuel Ash as a Bituminous Filler. R & D Profile 52*, Energy Efficiency Enquiries Bureau, Harwell.

Hill, J. (1983) Economic use of bituminous materials. *Journal of the Institution of Highway Engineers,* Jan.

Hunter, R.N. (ed.) (1994) *Bituminous Mixtures in Road Construction*, Thomas Telford, London.

Nicholls, J.C. (1994) *Generic Types of Binder Modifier for Bitumen*, SCI Lecture Paper No. 0035, Society of Chemical Industry, London.

Nicholls, J.C. and M.E. Daines (1993) *Acceptable Weather Conditions for Laying Bituminous Materials*, Department of Transport TRL Project Report 13, Transport Research Laboratory, Crowthorne.

Nunn, M.E. (1989) An investigation of reflection cracking in composite pavements in the United Kingdom. *Proceedings of the Conference on Reflective Cracking in Pavements*, Liege, 146–53.

Nunn, M.E. and T. Smith (1994) *Evaluation of a Performance Specification in Road Construction*, Department of Transport TRL Project Report 55, Transport Research Laboratory, Crowthorne.

Richardson, C. (1905) *The Modern Asphalt Pavement*, John Wiley & Sons Inc., New York.

Road Research Laboratory (1952) *Recommendations for Bituminous Surfacings for Roads Carrying Tracked Vehicles*, Department of Scientific and Industrial Research Road Note 13, Her Majesty's Stationery Office, London.

Spielmann, P.E. and A.C. Hughes (1936) *Asphalt Roads*, Edward Arnold & Co. Ltd, London.

Surrey County Council (1995) *Term Tender 1*, Surrey County Council, Guildford.

Szatkowski, W.S. and F.A. Jacobs (1977) Dense wearing courses in Britain with high resistance to deformation, *Colloquium 77, Plastic Deformability of Bituminous Mixes*, Zurich, 65–7.

CHAPTER 6

Asphalt concrete (and macadam) surface courses

J.T. Laitinen, Associated Asphalt Ltd

6.1 CONCEPT AND HISTORY

Currently, 36 million tonnes of coated materials are produced in the United Kingdom, of which approximately 7 to 9 million is rolled asphalt. There are no precise figures for the amount of macadam surface course produced, as the amounts produced vary from location to location with plants producing between 40% and 80% of their surface course tonnage as close graded macadams. These will with European harmonization be termed asphalt concretes.

The draft European Standard (CEN, 1995a) for asphalt concrete defines asphalt concrete as:

> Asphalt concrete is a continuously graded mixture of mineral aggregate, filler and bituminous binder which form an interlocking structure. This interlocking aggregate structure is the major contributor to the strength and performance of the laid material. The surface course will have good resistance to deformation as measured by the Wheel Tracking test.

The draft classifies mixtures into three categories based on grading and voids as follows:

- 'Dense-graded'   target value passing 2 mm > 25% (expected voids < 5%);
- 'Medium graded'   target value passing 2 mm 15 to 25% (expected voids 5–10%); and
- 'Open-graded'   target value passing 2 mm < 15% (expected voids > 10%).

These classifications cover most asphalt mixtures, although not rolled asphalt (Chapter 5) to BS 594 (BSI, 1992) which does not have an interlocking structure nor continuous grading, the key to asphalt concrete type classification.

Asphalt concrete mixtures are those that have a particle size distribution that in general follows the Fuller Curve (Fuller and Thompson, 1907)

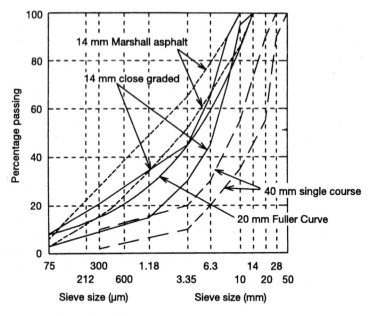

Fig. 6.1 Comparison of grading curves

(section 1.3). There are numerous types of asphalt concrete mixtures produced worldwide with Fuller Curve indices between 0.35 and 0.6, some of which are shown in Figure 6.1. It is interesting to note that the UK BS 4987: Part 1 (BSI, 1993a) close-graded macadam follows most closely the classic 0.45 curve index which has been found to have an aggregate grading with the least voids.

The UK version of asphalt concrete is a development from tar macadams laid in the 1800s culminating in an extensive trial when 700 different compositions were laid and monitored by the Road Research Laboratory in 1939 (RRL, 1962). The best of these trials formed the basis of BS 1241 for tar macadam and BS 1621 for bitumen macadam, issued in 1945 and 1950 respectively. BS 1621 was updated and revised in the 1960s and 1970s when it became BS 4987 *Coated Macadams for Roads and Paved Areas*, incorporating both tar and asphalt binders. The most recent version is BS 4987: Parts 1 and 2 (BSI, 1993a), which incorporated heavy duty macadam bases developed by TRRL (Nunn *et al.*, 1987) and a recognition that surface course mixtures previously termed dense were not truly dense and are now designated close-graded. However, because the proportion of aggregate passing 2 mm is greater than 25% with an interlocking grading, close-graded macadams fall into the **dense** category of asphalt concrete mixtures along with 6 mm medium, dense and 3 mm fine-graded surface course macadams.

Properties and applications 141

All the surface course mixtures in BS 4987 are essentially recipe mixtures with relatively wide grading tolerances in contrast to the designed asphalt concrete mixtures which are commonly produced for airfield surfacing in the United Kingdom and for general highway surfacing in the United States of America and most other parts of the world. Such asphalt concretes are designed and produced to very tight specification tolerances which cover gradation, voids and stability. The specifications in the United Kingdom which cover the composition and design for these materials in detail are those for airfield surfacing issued by the British Airports Authority Group Technical Services (BAA, 1993) and the Defence Works Services, formerly Property Services Agency (DWS, 1995a). Other specifications are issued by national standard organizations; for example, in the US ASTM D 3515-89 (ASTM D 3515) covers their composition requirements and the American Asphalt Institute Manual Series No. 2 (Asphalt Institute, 1993) describes the design procedures. All the standards cover surfacing materials ranging from 10 mm to 20 mm nominal size mixtures, although the UK airfield specifications have a significantly finer grading than the other asphalt concretes (see Figure 6.1).

6.2 PROPERTIES AND APPLICATIONS

Asphalt concrete covers a wide range of materials, properties and applications ranging from a private driveway surfaced with 6 mm medium-graded macadam or a non-trunk road surfaced with 40 mm single course macadam to BS 4987 (BSI, 1993a) to 14 mm or 20 mm Marshall asphalt laid on an international airport, requiring sufficient stiffness to cope with 400 t passenger airlines exerting up to 180 t per gear-leg on the surfacing without fracture of the aggregate or fretting to produce foreign object damage (FOD), which can cause catastrophic damage if ingested into the jets.

This section examines the mixture properties and application of the following commonly used materials:

- **UK highways**
  - 40 mm Single Course BS 4987: Part 1: Group 6: Table 7, 9
  - 14 mm Close Graded Macadam BS 4987: Part 1: Group 7 T 23, 425
  - 10 mm Close Graded Macadam BS 4987: Part 1: Group 7 T 26, 27, 28
  - 6 mm Dense Graded Macadam BS 4987 Part 1: Group 7 T 29, 30, 31
  - 3 mm Fine Graded Surface Course BS 4987: Part 1: Group 7 T34, 35, 36
- **Highways and airfields: UK and overseas**
  - British Airports Authority (BAA)
  - 14 mm and 20 mm Marshall Surface Course British Airport Authorities Plc GTS Section 14
  - Marshall Surface Course Defence Works Services (DWS) Functional Standard 13

- 14 mm and 20 mm asphalt concrete various worldwide specifications
- 10/14 mm Airfield Macadam BAA and DWS.

6.2.1 Asphalt concrete for UK highways and byways

BS 4987, *Coated Macadam for Roads and Other Paved Areas*, covers the asphalt concrete materials currently used for UK highways with the materials listed above. Due to the all encompassing nature of the specification, the range of properties and uses is varied and, depending where and with what raw materials the mixture is produced, the end result will vary from a highly rut resistant material with excellent load-spreading properties to one that may rut or fret with the action of traffic and/or weather. This problem is further exacerbated by the possible lack of knowledge by the user on its expected properties, especially on work not controlled by a highway authority. This is not surprising because the specification does not state any performance criteria, it only dictates the type of aggregate, binder and composition which is permitted with limited guidance as to when and where a specific type should be chosen.

The *Specification for Highway Works* (MCHW 1) recognizes the 14 mm close graded macadam as being suitable for some trunk roads if it is produced to Category A specifications, that is with higher viscosity (lower penetration grade) binders such as 100 to 300 pen. However, there is a confusing overlap in that a Category B specification could be mixed with 200 pen bitumen, which could also be acceptable for Category A. The performance criteria for Categories A and B is that Category A is suitable for roads carrying more than 250 commercial vehicles a day, which equates to 2.5 million standard axles (MSA) for a 20-year design life (MCHW 1). However, there is no guarantee that the user will achieve this goal with these materials without the skill and past experience with these materials of both the producer and Client. It is a credit to the industry and Client that (macadam) asphalt concrete has been successful as they have been, although there is a trend away from these materials due to early life failures.

The early failures can be ascribed to:

- increased traffic loading that is occurring throughout the network;
- the wide tolerances permitted;
- the limited emphasis on design to allow for aggregate characteristics such as absorption, micro-texture, shape and type;
- the limited appreciation of the need to adjust the binder content; and
- the difficulty of achieving adequate compaction.

Perhaps the most important comment on binder content in BS 4987: Part 1 (BSI, 1993a) is Note 2 of Clause 4.5.1, which states that the producer and supplier should agree an alternative target binder if evidence is available. For example, some county specifications call for an increased binder content when local granites are used, due to microtexture and absorbency, allowing all suppliers to quote on a level playing field and supply a suitable mixture.

However, if the County (Client) Specification does not allow for such factors, there is little or no incentive to change the composition providing the material will survive the first year of maintenance. It is not possible in this chapter to cover all the performance characteristics produced by the variations of mixture composition permitted in BS 4978, but it may be of interest to the reader to see the effect of changes in grading composition within specification and changes in fines type. Tables 6.1–6.4 show properties on laboratory prepared specimens of the following mixture types and aggregate compositions tested. However, the properties are only examples,[1] and the properties will vary with the composition and grading of each fraction of aggregate and the binder properties. Nevertheless, they do provide a basis for comparison.

6.2.1.1 14 mm close-graded surface course

- binder grade — 100 pen
- coarse aggregate — granite
- fine aggregate — granite or sand
- filler — limestone
- compositions — fine, mid and coarse graded were prepared with mid point binder

The results are shown in Table 6.1 with grading curves shown in Figure 6.2.

Table 6.1 14 mm close-grade surface course

Grading Fine aggregates	Coarse		Mid		Fine	
	Granite	Sand	Granite	Sand	Granite	Sand
Stability (kN)	9.9	10.4	8.6	11.1	10.2	9.5
Flow (mm)	3.1	2.4	2.7	2.1	2.4	1.9
Max. density (kg/m^3)	2.435	2.430	2.437	2.426	2.438	2.426
Voids (per cent)	5.4	1.6	6.4	1.5	4.8	1.8
Voids in mineral aggregate	17.4	14.2	17.7	14.4	16.8	14.6
Voids filled with binder	65.2	83.4	64.1	82.2	71.1	80.5
ISTM @ 20 °C (MPa)	2038	2567	1645	2451	2294	2263
Repeated load axial test @ 30 °C (microstrain)	2970	1270	5109	1323	3345	2253
Wheel-tracking rate @ 45 °C (mm/h)	0.4	n/a	0.7	1.6	0.6	0.9

n/a = not available

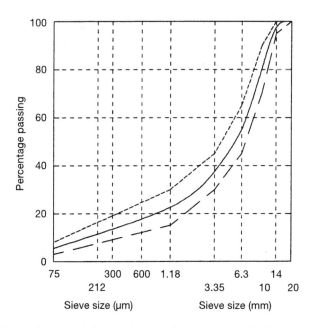

Fig. 6.2 14 mm close-graded macadam grading curves to BS 4987

6.2.1.2 10 mm close-graded surface course

- binder grade — 100 pen
- coarse aggregate — granite
- fine aggregate — granite or sand
- filler — limestone
- compositions — mid-point grading on both sand fines and granite fines mixtures with mid-point binder content.

The results are shown in Table 6.2 and grading curve in Figure 6.3.

Table 6.2 10 mm close-graded surface course

Fine aggregate	Granite	Sand
Stability (kN)	7.3	9.3
Flow (mm)	2.9	2.2
Max. density (kg/m^3)	2.421	2.411
Voids (%)	6.4	1.3
Voids in mineral aggregate	18.6	13.8
Voids filled with bitumen	62.9	89.1
ITSM @ 20 °C (MPa)	1235	2076
Wheel-tracking rate @ 45 °C (mm/h)	0.37	1.7

Properties and applications

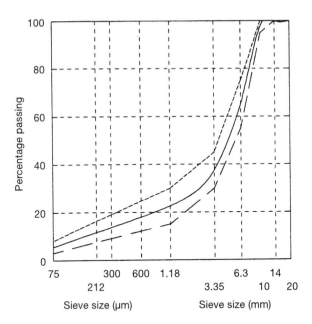

Fig. 6.3 10 mm close-graded macadam grading curves

6.2.1.3 6 mm dense surface course

- binder grade — 100 pen and 200 pen
- coarse aggregate — granite
- fine aggregate — granite or sand
- filler — limestone
- compositions — mid-point grading with mid-point binder for both 100 and 200 pen mixtures using sand and granite fines.

The results are shown in Table 6.3 and grading curve in Figure 6.4.

Table 6.3 6 mm dense surface course

Fine aggregate	Granite		Sand	
Grade of bitumen	100 pen	200 pen	100 pen	200 pen
Stability (kN)	11.8	10.9	3.7	2.9
Flow (mm)	2.4	2.7	2.1	1.7
Max density (kg/m3	2.386	2.391	2.397	2.397
Voids (%)	3.3	3.2	4.1	4.3
Voids in mineral aggregate	17.4	17.3	18.1	18.7
Voids filled with bitumen	81.3	81.9	77.8	75.3
ITSM @ 20 °C (MPa)	1645	719	995	552
Wheel-tracking rate @ 45 °C (mm/h)	1.5	3.0	8.9	10.4

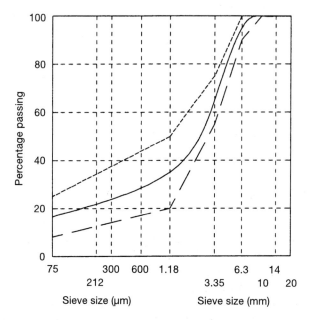

Fig. 6.4 6 mm close-graded macadam grading curves

6.2.1.4 40 mm single course 200 pen granite coarse and fine

Although this mixture is categorized as a binder course material in BS 4987, it is effectively a surface course because it can be trafficked for 1–2 years prior to being surface dressed, hence its inclusion as a surface course.

- binder grade — 200 pen
- coarse aggregate — granite
- fine aggregate — granite
- filler — limestone
- compositions — mid-point of specification for binder and grading

The results are shown in Table 6.4 and grading curve in Figure 6.5.

Tables 6.1–6.4 give an indication of the properties that may be obtained from standard mixtures with grading curves as shown in Figures 6.1–6.5. It is interesting to note that, in the case of 14 mm and 10 mm close graded macadam asphalt concrete, the change from granite fines to sand fines did not affect stability or rutting resistance but it did have a marked effect on voids in the mixture, with far fewer voids in the sand fines mixture than the granite fines mixture, which did not compact as well due to its harsher texture and coarser grading.

Table 6.5 gives the properties of a 14 mm close graded macadam asphalt concrete prepared in the laboratory to the mid-point grading of

Properties and applications

Table 6.4 40 mm single course

Grading	Mid value
Mixture density (kg/m^3)	2.024
Maximum density (kg/m^3)	2.482
Voids (%)	18.5
Voids in mineral aggregate	26.2
Voids filled with bitumen	29.2
ITSM @ 20 °C (MPa)	900
Repeated load axial test @ 30 °C (microstrain)	24 070
Relative hydraulic conductivity (s^{-1})	0.05

BS 4987 (BSI, 1993a) with an all limestone aggregate blend mixed with 100, 50 and 15 penetration grade bitumen. The results demonstrate the increase in elastic stiffness as harder grade binders are used, although laying 50 pen or 15 pen 14 mm close-graded macadam may cause some problems. Additionally, as seen from Table 6.1, the 'compactability' on limestone mixtures is similar to that of a granite and sand blend, producing fewer voids than an all granite mixture.

This limited comparison has also shown that the mid-point grading with granite fines had a higher void content and, had the mixture been laid on site, it could be reasonable to expect the void content to be at least 2% higher, bringing into question its durability and highlighting the need for mixture design.

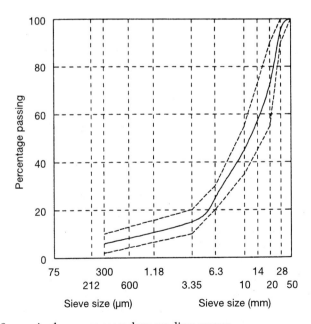

Fig. 6.5 40 mm single course macadam grading curves

148   *Asphalt concrete (and macadam) surface courses*

Table 6.5 14 mm close-graded surface course limestone aggregate

Bitumen grade (penetration) @ 25 °C	*100 pen*	*50 pen*	*15 pen*
Max. density (kg/m^3)	2.460	2.460	2.460
Voids (%)	3.1	2.9	3.5
Voids in mineral aggregate	14.9	14.7	15.3
Voids filled with bitumen	78.8	80.6	76.8
ITSM @ 20 °C (MPa)	2586	5546	10 758

The 6 mm dense macadam asphalt concrete mixture comparisons demonstrated that, by changing from 100 penetration grade to 200 penetration grade, there is approximately 1 kN decrease in stability and 50% reduction in load spreading properties. But the greatest change was when the granite fines were substituted for sand, with the stability dropping by approximately 8 kN and the wheel-tracking rate increasing by 7 mm/h. It is not surprising that, when 6 mm dense asphalt concrete with sand fines is used for driveways and car parks, it is prone to deformation (Figure 6.6), especially in its early life.

Fig. 6.6 6 mm dense macadam car park surface course

6.2.2 Asphalt concrete airfield surfacing

There are two main specifiers of airfield pavements in the United Kingdom, the British Airports Authority (BAA) and the Defence Works Services (DWS), which cover civil and military airfields respectively. Both authorities have similar, and essentially evolving, specifications which incorporate experiences and knowledge gained on preceding contracts. The evolutionary aspect requires the contractor to ensure that he or she is fully aware of the current requirements which tend to revolve around the use of certain aggregate fillers, binders and additives: the current requirements are covered in sections 6.3 and 6.4.

However, the basic mixture properties requirements have remained constant with regard to the voids in the total mixture, the voids filled with binder, the flow and the stability, although the minimum stability requirement for DWS contracts was 2 kN lower than BAA contract up to 1995. In the current DWS specification (DWS, 1995a), the minimum stability requirement ranges from 6 kN to 10 kN depending on the traffic frequency and tyre pressures, whereas BAA has a minimum of 10 kN for all airfields. This 6 to 10 kN range of stabilities has a significant effect on mixture design, production, placement and surface characteristics even if the voids requirement does not alter.

In section 6.2.1 on close-graded macadam type asphalt concrete, the range of properties obtained on nominally the same mixture was shown. In contrast, due to the strict design procedures covering grading and binder type, the range of properties are more limited for Marshall design asphalt concrete. Tables 6.6 and 6.7 give the mixture properties requirements for BAA and DWS Marshall and airfield macadam (DWS, 1995b) and Figure 6.7 shows the grading curve of 20 mm Marshall surface course with 14 mm airfield macadam (14 mm Marshall surface course was shown on Figure 6.2).

The requirements are similar, although there are differences in the aggregate requirements which will be covered in section 6.3. The optimum binder requirements appear to have a difference of 0.3%, but they are the same because the minimum binder content permitted for surface course is 5.0% and both effectively require that the minimum design target is 5.3% with tolerances of ± 0.3%. Typically, the materials when laid will have the properties shown in Table 6.8 with the grading curve shown in Figure 6.7.

The object of the properties in Table 6.8 is to provide the following characteristics (DWS, 1995a):

Table 6.6 Design requirements airfield surfacing

	Marshall asphalt		Airfield macadam	
	BAA	DWS	BAA	DWS
Optimum binder (% of total mixture)	5–7	5.3–7.3	5–7	>5
Stability (kN)	>10	Table 6.7		
Flow (mm)	<4	<4		
Voids in total mixture (%)	3–4	3–4	Not specified	
Voids filled with bitumen (%)	76–82	76–83		

Table 6.7 Minimum stability requirements for Defence Works Services

Tyre pressure	Traffic[a]		
	Low	Medium	High
Up to 1.4 MPa (200 psi)	6	8 [6]	10 [8]
More than 1.4 MPa (200 psi)	8	10	10

[ ] Values for cooler regions of UK
[a] Traffic categories: low frequency — Maximum of 50 movements per week by aircraft in the critical type pressure range.
Medium frequency — Maximum of 500 movements per week by aircraft in the critical type pressure range

Table 6.8 14 mm UK airfield asphalt concrete

	Marshall asphalt	Airfield macadam
Stability (kN)	9–13	6–11
Flow (mm)	2–4	1–5
Voids in mixture (%)	3.5–4.5	2–6
Voids filled with binder (%)	71–85	60–90
Voids in the mineral aggregate (%)	15–18	15–19
ISTM @ 20°C (MPa)	1500–3000	n/a
Repeated load axial test @ 30 °C (microstrain)	10–20 000	n/a
Wheel-tracking rate @ 45 °C (mm/h)	0.5–3.5	n/a

n/a not available

- good rideability;
- good friction characteristics;
- high enough strengths and stability to withstand the sheer stresses induced by heavy wheel loads and high tyre pressures;
- a durable and hard-wearing weatherproof surface free from loose material and sharp edges which might endanger aircraft;
- resistance to fuel spillage and jet blast (although this requirement may not be fully met with asphalt surfacings); and
- facilitate economic maintenance.

Good friction characteristics cannot be completely met by mixture design alone and requires some form of surface treatment or additional surfacing such as porous friction course (Chapter 7). The alternative is grooving, which is generally carried out on new surfacing after it has been placed for a minimum of 24 hours for DWS works and 72 hours for BAA works. The grooves are cut transversely across the runway 4 mm wide by 4 mm deep at 25 mm centres to produce a texture depth not less than 1.0 mm. In hot climates, the spacing of the grooves may be increased

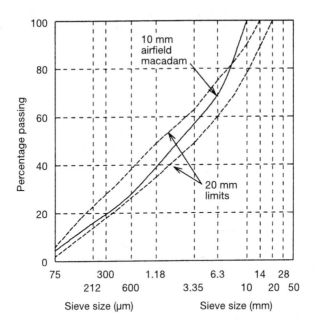

Fig. 6.7 Comparison of 20 mm Marshall asphalt and 10 mm airfield macadam

to 40 mm centres and the grooves 5 mm x 5 mm (BAA, 1993). Great care is required in this operation, with specialist contractors undertaking it and ensuring all the arising are vacuum swept to produce a FOD free surface for aeroplanes and so that runway marking can be painted permanently.

An ungrooved airfield asphalt concrete will have a texture depth in the range 0.15 mm to 0.5 mm, as measured by the sand patch method in Appendix L of the BAA specification (BAA, 1993), with 0.25 mm as the minimum requirement. As with highways, texture depth is critical in terms of skid-resistance and it has been found that a new surface with a texture of less than 0.4 mm will not give a greater grip value than 0.5 as measured on a Findley Irvine GripTester or 0.5 on the Mu Meter when tested at 40 mph. The current CAA Specification calls for a grip value of 0.63 on completion of construction and 0.80 at the end of the maintenance period; these values are only obtainable with grooving or a porous friction course surfacing.

6.2.3 International asphalt concrete

Tables 6.9 and 6.10 illustrate some of the gradings and properties required for highway and airfields in Europe, United States of America, Middle and Far East in contrast to the UK airfield and highway asphalt concretes.

Table 6.9 Comparison of asphalt concretes in Europe

Parameter	Austria	Germany	France	Italy	Netherlands
Design method	Marshall	Marshall	Gyratory	Marshall Duriez Wheel-tracking Fatigue Stiffness	Marshall
Stability	–	–	–	>9 kN	>7.5 kN
Flow (mm)				2–4	2–4
Voids (%)	3–5	2–5	–	–	2–6
Voids filled with binder	–	–	–	–	<82%
Grading: >2 mm	50–60	40–60	–	–	60
<63 mm	5–8	5–8	–	50–60	7
Binder content (%)	5.9–7.2	5.9–7.5	5.5–6.2	4.5–6.5	6.0–6.4

Table 6.9 gives a comparison of asphalt concretes in Europe (EAPA, 1995) and Table 6.10 asphalt concretes in some other parts of the world.

Tables 6.9 and 6.10 show that the requirements are similar with allowances for expressed use, such as low or high traffic loadings.

All asphalt concretes require a design procedure, usually Marshall with additional requirements such as those shown for the United States of America and Malaysia becoming more common as the equipment and technology for measuring performance criteria becomes economically more accessible to the client and contractor. The Strategic Highways Research Project (Anderson, 1994) carried out in the United States of America has generated a vast amount of information on asphalt materials with new or revised testing methods being developed, such as pressure ageing vessel to give better data on possible long-term performance. Much emphasis has been placed on bitumen performance quality, as measured on a dynamic shear rheometer (Figure 6.8), which allows elastic testing to be carried out over a wide temperature and loading range. These data have been developed into tables giving minimum binder performance requirements for any given climatic conditions; an example is reproduced in Table 6.10.

6.3 COMPONENT MATERIALS

Asphalt concrete is a combination of the following basic components:

- coarse aggregate – material > 3.35 mm;
- fine aggregate – material < 3.35 mm and > 75 μm;
- filler – material < 75 μm; and
- binder – usually petroleum bitumen.

Table 6.10 Asphalt concrete worldwide

Parameters		USA[a] highway			Middle East (Yemen) highway	Far East (Malaysia) airfield
Grading – % passing:						
20 mm	(19 mm)[b]	100			100	100
14 mm	(12.5 mm)[b]	90–100			80–95	76–100
10 mm	(9.5 mm)[b]	–			–	64–89
6.3 mm	(4.75 mm)[b]	44–74			48–62	56–81
3.35 mm	–	–			–	41–55
2.36 mm	(2 mm)[b]	28–58			32–45	–
1.18 mm	–	–			–	16–31
425 μm	(425 m)[b]	–			16–25	12–16
300 μm	(300 m)[b]	5–21			–	–
150 μm	(150 m)[b]	–			8–18	6–10
75 μm	(75 m)[b]	2–10			4–8	3–7
Binder		4–11			5–7	min 5
		Light	Med	Heavy		
Min. stability (kN)		3.3	5.3	8.0	9.0	10
Flow (mm)		3–7	3–6	3–5	3–5	
Voids (%)		–	3–5	–	3–5	–
Voids in mineral		–	13–15	–	16–19	–
Aggregate		70–80	65–78	65–75	70–80	–
Voids filled with bitumen						
Bitumen penetration grade		120/150	–	40/50	60/70	80/100
Binder rheology:						
Unaged			✓			✓
Aged			✓			✓
DSR			✓			✓
						✓
Wheel tracking						✓
Elastic stiffness						✓

[a] ASTM D 3515 (12.5 mm) gradation
[b] size for the United States of America and Yemen

The purpose is to provide a stable mixture by means of a well-graded aggregate structure with a good mechanical interlock held together with a binder. The binder needs to be able to maintain its cohesiveness while, at the same time, providing sufficient lubrication and workability during placement and compaction to ensure long-term durability. The mixture also needs load-spreading properties, which is a function of the binder properties, aggregate type and gradation.

Good asphalt concrete can only be produced with careful design and correct choice of components for the function it is required to fulfil (i.e. the traffic it will carry). Each component will have an effect on the

Fig. 6.8 Dynamic shear rheometer

mixture properties and its durability, some of these effects have been described already in section 6.2, such as changes in aggregate type and/or grading. This section will cover the main requirements for aggregates, filler and binders when intended for use in asphalt concrete; additionally, the effect of binder-modifiers will be covered.

6.3.1 Coarse aggregates

The function of the coarse aggregate (that retained on the 3.35 mm sieve) is to provide stability to the mixture by means of aggregate particle interlock ensured by a well-proportioned grading and frictional resistance to creep. The latter is a function of the aggregate's texture and shape.

Most asphalt concrete surface course specifications call for aggregates that are clean, hard and durable, although 'numbers' are not always put to this requirement. The specifications **may** allow the use of the following types of aggregate:

- crushed rock, such as basalt, gabbro, granite, gritstone hornfels, limestone, porphyry or quartzite, as in BS 4987 (BSI, 1993a) and the airfield specifications (BAA, 1993; DWS, 1995a);
- gravel, as in BS 4987 (BSI, 1993a);
- blast-furnace slag complying with BS 1047 (BSI, 1983a), as in BS 4987 (BSI, 1993a);
- steel slag, as in BS 4987 (BSI, 1993a).

Component materials

Transport Research Laboratory (Bullas and West, 1991) investigated various tests for establishing the suitability of an aggregate for roadbase material in terms of being clean, hard and durable (section 2.3.1). The proposed criteria for acceptability are based on tests and values that are reproduced in Table 6.11. Although these are for roadbase materials, the tests themselves are also valid for surface course materials.

6.3.1.1 Adequate skid resistance – for either runway or highway

Guidance on skid resistance for (trunk) roads in terms of the polished-stone value of the aggregate is given in HD 28/94 (DMRB 7.3.1), which also advises on aggregate abrasion values. For runways, similar requirements are specified in the current standards (BAA, 1993; DWS, 1995a) and are reproduced in Table 6.12.

Table 6.11 Criteria for defining clean, hard and durable aggregates for bitumen macadam roadbase

Property	Test		Criterion for acceptability	
			By recipient	By supplier
Cleanness	Content of adherent silt and clay size particles	Proportion passing 63 μm Appendix B, based on BS 812: Part 103	1.2 % or less	0.8% or less
Hardness	Resistance to crushing	Ten per cent fines value BS 812: Part III: 1990	132 kN or more	148 kN or more
	Resistance to impact	Aggregate impact value BS 812: Part 112: 1990	24 or less	22 or less
Durability	Soundness	Magnesium sulphate soundness value BS 812: Part 121: 1989	66 or more	84 or more

6.3.1.2 Adequate strength

To ensure that the coarse aggregate does not crush, either during placement or with trafficking, it has to have sufficient strength; the strength may be measured by either the 10% fines test (BSI, 1990c) or aggregate crushing value (BSI, 1990b) (section 2.2.3.2).

6.3.1.3 Durability

Durability is the aggregate's resistance to traffic and/or weathering, and there is no uniformity in the tests used to assess it. BS 4987 does not have any test specifically for this property, but both the Los Angeles abrasion test (ASTM C 131 or ASTM C 535) and micro-Deval (section 2.2.4.3) have been used elsewhere for their usefulness in this respect. Airfield runway specifications specify a minimum magnesium sulphate value (BSI, 1989) (section 2.2.4.2) of 82%, which has proven satisfactory for BAA and DWS specifiers. This test will highlight weaknesses and impurities, such as shale, in an aggregate which may cause premature failure. Finally, a simple water absorption test can be used as a 'vetting' test when examining aggregates for suitability from unknown sources with results over 2% indicating a possible weakness.

6.3.1.4 Shape

The shape is important, both for achieving a mixture with minimum voids and good particle inter-lock. A flaky or elongated shape (section 2.2.1.3) will produce a mixture with excess voids, even when compacted to refusal, as well as being more susceptible to breaking down during laying and subsequent trafficking. Rounded aggregates, such as dredged gravels, will produce less stable mixtures, although they may be suitable for footpaths and domestic parking areas.

6.3.1.5 Cleanliness

Aggregates should not contain any deleterious matter such as clay, shale, organic impurities or excessive dust, all of which may cause weakness within the mixtures or cause coating problems. The magnesium sulphate soundness test (BSI, 1989) (section 2.2.4.2) or methylene blue test (CEN, 1995b) (section 2.2.1.2) can be used as indicators of the above impurities. Extensive work on aggregate bitumen adhesives has been carried out to develop the net absorption test (section 2.2.5), which is proving useful in ranking good and susceptible aggregates.

The simplest test for cleanliness, in terms of possible clay or excess fines, is the determination of passing 75 µm or 63 µm by means of wet sieving to BS 812: Part 103.1 (BSI, 1985) (section 2.2.1.1). In coarse aggregate, it is not desirable to have more than 1.5% because this may cause coating problems when the fine material is 'baked-on' to the aggregate particles during the drying process, leading to subsequent stripping problems.

Another test, although not necessarily related to cleanliness, is the binder stripping test which is an indicator of the aggregate's affinity, or otherwise, to bitumen. The test involves immersing samples of 10 mm aggregate coated with bitumen for 24 hours, then counting the number of any stripped particles. If more than 6 out of 150 show signs of stripping, the aggregate should not be used. The test does not always fail poor aggregates and possibly the aggregate binder adhesion tests may prove more reliable.

6.3.2 Fine aggregate

The types of fine aggregate (that passing the 3.35 mm sieve but retained on the 75 μm sieve) permitted for close-graded macadams to BS 4987 (BSI, 1993a) and airfield asphalt concretes (BAA, 1993; DWS, 1995a) are similar, being either;

- natural bank, river dune or pit sand;
- crushed rock fines; or
- blends of sand and crushed rock fines.

The requirements for cleanliness are similar to coarse aggregates, with the main controlling factor the amount passing the 75 μm sieve. The permitted maximum is 8% for close graded macadam asphalt concrete and 5% for airfield asphalt concretes. However, the choice of fines aggregate combination for the various asphalt concretes does not differ significantly depending which country or specifier has set the standard. The requirements for specific standards include:

- BS 4987 allows any of the above mentioned fines including sea dredged sand, hence there is a possible wide variation on nominally the same mixture;
- US and other overseas specifiers do not prescribe any combination of the above, only that the grading envelope should be met leaving the choice to the designer who will be governed by the design requirements for grading, voids and stability, all of which are affected by aggregate choice;
- Defence Works Services specification for airfield surface courses allows any combination; however, it specifies that all crushed rock fines should be washed;
- British Airports Authorities specification allows the use of all the above fines, although it is very specific about the actual blend and the amount passing the 75 μm sieve with the following requirements:
  - the fines shall contain at least two types of natural sand, the proportions are not specified;
  - the fines shall contain at least 30% of crushed rock fines having the same PSV requirement as the coarse aggregate (a minimum of 60 or 55 for taxiways and low category use areas); and
  - the crush rock fines shall not have more than 5% passing 75 μm sieve and should be washed unless otherwise approved.

These requirements will be discussed further in section 6.4. Nevertheless, these requirements impose real logistical and financial considerations and should be treated with due care and attention at the tender and design stage.

6.3.3 Fillers

As with the fines aggregate requirements, the permitted types of filler (the material passing the 75 μm sieve) in close graded macadam asphalt concrete are wide and allow the use of any of the fines from the permitted aggregates, Portland cement or limestone as long as a minimum of 75% passes the 75 μm sieve. Hence, the whole of the required content of the mixture passing the 75 μm sieve can be the filler collected from the plant. In practice, this does not produce any problems with performance providing the amount added is well controlled.

For airfield surface course mixtures, the filler type and content is more restrictive, generally the permitted proportion of the added filler that passes the 75 μm sieve is a minimum of 85% and must account for at least 60% of the material passing 75 μm in the total mixture. The type of filler is limited to either crushed limestone or Portland cement, complying with BS 12 (BSI, 1991). The effect of filler will be discussed in section 6.4.

6.3.4 Hydrated lime

UK airfield runway surfacing specifications call for the addition of 1.5% for civil work (BAA, 1993) and 2% high calcium hydrated lime for military work (DWS, 1995a). The BAA specification may call for an addition of anti-strip agent as well.

The addition of hydrated lime improves the adhesion between aggregate and bitumen and reduces the chance of failure caused by stripping of the binder from the stone, which is most likely during its early life especially with certain acidic type aggregates. This type of failure is less likely once the binder has hardened/oxidized. However, in general these types of failure are also related to other factors, such as wet weather during production period, poor drying-mixing and compaction problems. It is relatively rare to actually see stone stripping, that is the binder migrating off the stone.

Hydrated lime can also improve mixture stiffness and stability because it is very effective at extending and increasing the binder viscosity and

Table 6.12 Aggregate requirements for asphalt concrete

Test	BS 4987	BAA	DWS
Magnesium sulphate soundness	—	82%	82%
Maximum flakiness	45%	30%	30%
Maximum elongation	—	—	30
Minimum aggregate crushing value	—	30	30
Maximum water absorption	—	1.5	2
Maximum clay silt[a]	1/8	5	1/5
Minimum polished stone value	—	60	55
Minimum aggregate abrasion	—	10	—

[a] Based on the amount passing the 75 μm sieve.

Component materials

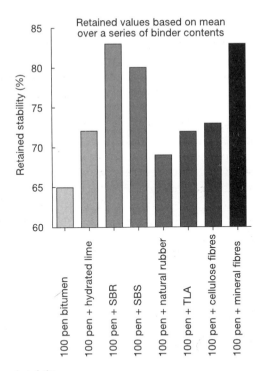

Fig. 6.9 Retained stability

softening point; it can have far more effect on the mixture than a cohesion aid. Figure 6.9, although not based on asphalt concrete, demonstrates the effect of various additives on retained stability in a 20 mm porous asphalt, an ideal mixture to demonstrate the improvement or otherwise of cohesion improving agents.

A comparison of the type and content of the filler and the resulting mixture properties (Al-Suhaibani and Al-Mudanheem, 1992) found that:

- the binder softening point was increased by over 50% when 3% of the mixture filler content was replaced by hydrated lime; and
- the optimum binder content was increased by up to 20% when 3% of filler is replaced with hydrated lime because both the air voids and the voids in the mineral aggregate are increased.

The reasons for the increase in voids are the greater surface area of hydrated lime relative to limestone or other fillers and the resulting stiffening effect on the mixture which requires higher compaction temperatures. Other work has confirmed the stiffening effect (Shahrour and Saloukeh, 1992), but achieved similar voids both with and without hydrated lime although with more stable mixtures – presumably the compaction temperatures were adjusted for the increases in filler binder

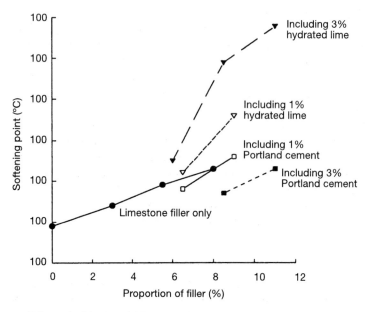

Fig. 6.10 Effect of addition of filler on softening point of bitumen

viscosity. Figure 6.10 (Al-Suhaibani and Al-Mudanheem, 1992) shows the change in softening point with different filler combinations. This indicates Portland cement is not an effective substitute, although BS 4987 recommends either Portland cement or hydrated lime as an adhesion agent.

6.3.5 Binders

The range of types and grades of binders are broadly similar for all asphalt concrete surface course materials. They cover 60 to 450 penetration grade of petroleum bitumen, with most materials produced within the range 70 to 200 penetration for highways and airfield runways, with harder grades utilized where high ambient and/or intense traffic is expected and the softer grades in cooler or less intense traffic locations. Softer grades tend to be used in lightly trafficked areas such as driveways, car parks or in colder climates. However, the performance of the asphalt concrete can be maintained by ensuring a well graded and compacted mixture utilizing aggregates with good micro-texture such as crushed rock rather than gravel or rounded sands (Table 6.13 shows the effect of binder grade and aggregate type).

Tar-based binders to BS 76 (BSI, 1974) (section 3.2) are less frequently used but have certain advantages over bitumen in locations where fuel spillage may occur due to the greater resistance of tars to fuel damage. Tars with equi-viscous temperatures (EVT) of 34 to 42 are permitted in

BS 4987 macadam asphalt concretes, but only for category B traffic. The reason for this limitation is their greater temperature susceptibility, with a greater change in stiffness than bitumen between being brittle in the cold and having low viscosity at higher temperatures; comparing, say, 100 pen bitumen with C42 tar, the bitumen would have a softening point of about 45 °C, at which temperature the tar would flow freely whereas the bitumen would need a load, such as a ball in the softening point test to BS 2000: Part 58 (BSI, 1993b), to achieve measured flow.

However, as with softer grade bitumens it is possible to design a suitable tar-bound mixture but due to the health risks associated with tar, it is preferable to use alternative materials to achieve fuel resistance. The alternatives include thin epoxy sprays or cementitious grouts, especially where resistance to point loadings is combined with chemical or fuel spillages, as on dock or warehouse pavements.

6.3.6 Binder modifiers

Chapter 3 covers many of the modifiers commonly used in some detail. The effects of these modifiers are equally valid in the binders used to coat asphalt concretes, although their effect may be less dramatic than in, say, a gap-graded rolled asphalt surface course (Chapter 5) to BS 594 (BSI, 1992) where the bitumen–filler–fines mortar largely dictates its performance.

6.3.6.1 Macadam asphalt concrete

BS 4987 macadam asphalt concretes are rarely modified, the reason being that macadams are precluded from high-speed surfacing contracts as it is not possible to achieve a texture depth in excess of 1.5 mm and is less likely to be used in high stress areas where binder modification would be cost-effective. However, some modifiers are used.

Table 6.13 Typical binder properties

Grade		70 pen	100 pen	200 pen
Penetration @ 25 °C	(dmm)	65	98	197
Softening point	(°C)	45	49	38
Viscosity @ 150 °C	(cP)	275	206	140
Dynamic viscosity @ 60 °C & 1.5 Hz	(Pa.s)	3.23×10^2	12.15×10^2	4.67×10^1
Ductility @ 5 °C	(cm)	1–3	1–5	15–25
Typical use			Asphalt concrete Highways and airfields	Asphalt concrete Cat B (BS 4987) Drives, parking areas

Anti-strip agents

The only modifiers commonly used in BS 4987 macadams are anti-strip agents, which are added at a rate of 0.3 to 0.5% of the binder content to improve binder adhesion to certain aggregates, typically granites, and hence minimize fretting (Figure 6.11). The use of anti-strip agents is at the discretion of the producer and can improve the durability, especially when mixed and laid during the period from November to April in the United Kingdom. Although they do not ensure against failure, only an adequate binder content and compaction will.

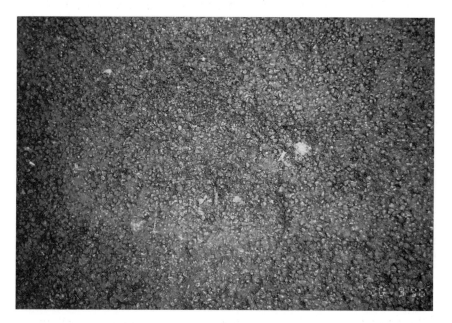

Fig. 6.11 Surface fretting

Pigments

Pigments are modifiers that are used extensively to produce asphalt concretes with decorative finishes (see also section 12.8). The most commonly used colour is red (from iron oxide), which is possibly the only pigment colour that is powerful enough to colour black bitumen mixtures. The amount of red oxide required is generally about 5% of the total mixture and replaces part or all of the fraction passing the 75 μm sieve. Alternatively, there are various proprietary binders produced from resins or synthetic bitumens, such as Shell's Colouradd binder, which require only 1 to 2% of pigments, such as titanium oxide (white), chromium oxide (green), to produce very bright colours.

With conventional 'black' bitumen and a pigment, the resultant mixture properties are similar to the unpigmented version. However, with synthetic or low asphalt 'light' binders, the resultant performance will be inferior unless attention is paid to the mixture design to ensure sufficient mechanical inter-lock. This is particularly the case in areas where there is likely to be standing, turning traffic, as in a typical shopping centre car park which is the most common use of coloured surfacing.

With the need for more aesthetically pleasing surfacing or ones designating safety zones, the need to further modify and improve these mixtures is needed. Generally, these light binders have lower viscosity compared to conventional binders and, as such, are more likely to deform. Their performance can be improved by the addition of polymers such as styrene-butadiene-styrene (SBS) or styrene-butadiene rubber (SBR), which can increase their resistance to rutting and fretting, especially in hot weather. With pigmented mixtures, it is very important to ensure that the aggregate colour is complementary to the mixture colour because, when the surfacing is trafficked and starts to wear, the aggregate colour will tend to predominate even though the surfacing looked, say, bright orange (or whatever exotic colour the client wanted) when it was first laid.

6.3.6.2 Asphalt concretes

Asphalt concretes used for surfacing runways and highways often require modification to enhance the properties of conventional binder or to compensate for deficiencies in locally available materials; the aggregate type or grading may not impart the required properties to the mixture for it to resist either the climate, the traffic loading or a combination of both. This recognition of minimum performance requirement has been highlighted by the United States SHRP programme and SUPERPAVE procedures (section 1.5.4). Table 6.14 reproduces some of the binder properties necessary to meet climatic condition.

Although the programme has only recently been completed, its methodology has already been adopted on contracts outside the United States of America from Mongolia to Malaysia, and in many cases it has indicated that modification of the base binder was necessary. In the United Kingdom, both BAA and DWS have been investigating and experimenting with modification of runway surfacing mixtures to increase the performance of current surfacing, which may not always be able to give the required life and performance. The potential need is increasing as airplane size and flight frequencies increase, making it ever more costly and difficult to carry out resurfacing work; hence, the need to design materials with modifiers such as SBS and SBR which increase resistance to rutting and cracking or ethylene vinyl acetate (EVA) which will increase rut resistance. All of these modifiers can be supplied pre-

Table 6.14 SHRP binder specification

	\multicolumn{10}{c}{Aged asphalt binder grades}

	AB 1-			AB 2-			AB 3-			AB 4-
	1	2	3	1	2	3	1	2	3	1
Highest mean monthly temperature °F	70 to 80			80 to 90			90 to 100			100 to 110
Lowest mean monthly temperature °F	−30 to −20	−20 to −10	>−10	−30 to −20	−20 to −10	>−10	−30 to −20	−20 to −10	>−10	>−10
Stiffness temperature dependancy	◣	◣	◣	◣	◣	◣	◣	◣	◣	◣
Low temperature cracking (PAV residue) Stiffness at 2 hours at designated temperature, * max, ksi	30	30	30	30	30	30	30	30	30	30
Failure tensile strain at designated temperature, (Direct Tension Test, SHRP B-003) max, %	5	5	5	5	5	5	5	5	5	5
Designated temperature, °F	−30	−20	−10	−30	−20	−10	−30	−20	−10	−10
Permanent deformation (RTFOT**residue) Various components of stiffness at designated temperature, * min, psi	15	15	15	15	15	15	15	15	15	15
Designated temperature, °F	80	80	80	90	90	90	100	100	100	110
Fatigue cracking (PAV residue) Minimum allowable m-value at designated cross-over time and temperature, *	0.5	0.5	0.5	0.5	0.5	0.5	0.5	0.5	0.5	0.5
Designated temperature, °F	25	30	35	30	35	40	35	40	45	50
Constructability *** Mixing EVT for viscosity of 1.70±0.20 P, °F	250 to 350			250 to 350			250 to 350			250–350
Compaction EVT for viscosity of 2.80±0.30 P, °F	200 to 300			200 to 300			200 to 300			200–300
Safety Flash point (COC flash point, ASTM D92), min. °F	450			450			450			450

* Calculated from master stiffness curve using the Bending Beam Shear Rheometer (SHRP B-001) at 5 °F and the Dynamic Rheometer Test (SHRP B-002) at 104 °F.
** TFOT can possibly be utilized at the discretion of the specifying agency.
*** EVT – Equi-Viscous Temperature per ASTM D4402.

blended into bitumen for ease of production but SBR has an additional advantage in that it can, if required due to production requirements, be added directly into the pug-mill or drum mixer without the need to have an extra bitumen storage tank.

The immediate benefits of modification are discussed in Chapter 3, such as the increased flexibility of SBR and SBS and the rut resistance of EVA, but, in addition, it is their resistance to ageing that is important. Therefore, prior to using a modifier, it is advisable to test the modified-binder both before and after accelerated ageing in the thin film oven test (ASTM D 2872) and/or the pressure ageing vessel (AASHTO PP1-93).

Component materials

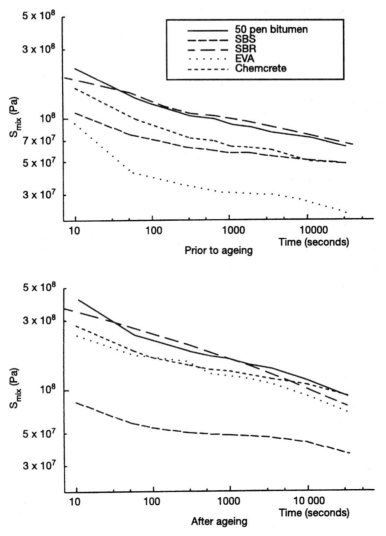

Fig. 6.12 Creep master curves (test temperature 40 °C) for mixtures containing various binders before and after ageing

Figure 6.12 compares the master curves profile of various binders before and after ageing (Davies and Laitinen, 1995), as measured on a dynamic shear rheometer; as may be seen, some binders lose performance more rapidly than others with ageing.

Asphalt concretes are based on an aggregate grading which is designed to produce materials with minimum voids, but this can create problems of mixture performance due to practical production limitations on grading and binder content because the materials will tend to have a far narrower band in which the optimum properties are present. Too

Fig. 6.13 Fatigue cracking

Fig. 6.14 Excessive binder and deformation

little binder may produce a material prone to fatigue cracking (Figure 6.13) and too much binder may result in deformation (Figure 6.14) and bleeding (Figure 6.15), which are real problems.

Component materials

Fig. 6.15 Bleeding and lack of voids

Because of this limited acceptable range, the allowable production tolerances for binder and grading are generally half of that permitted for macadam asphalt concretes, as shown in Table 6.15.

However, even with these tighter limits, it is not always possible to avoid performance problems and this is where certain modifiers can assist in increasing the mixture tolerance to variation in production, load and climatic conditions. Figure 6.16 shows the effect of SBR in reducing the temperature susceptibility of an asphalt concrete laid on all sections of a highway project where the incline was greater then 10%. The project was in an area of high day-time temperatures and low night-time temperatures.

Table 6.15 Comparison of production limits for 14 mm surface course

	14 mm close-graded macadam (BS 4987)	14 mm Marshall asphalt (BAA or DWS)
Aggregate passing:		
20 mm	−5	−5
14 mm	±5	±5
10 mm	±10	±5
6.3 mm	±10	±5
3.35 mm	±7.5	±4
1.18 mm	±7.5	±4
425 μm	−	±4
150 μm	−	±4
75 μm	±2.5	±1.5
Binder	±0.5	±0.3

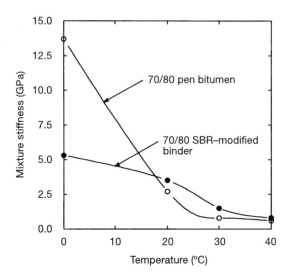

Fig. 6.16 Comparison of mixture stiffness with change of temperature

Although it is not always essential to use a modifier, there are heavy-duty applications where it will be necessary, such as the runway and taxiway surfacing of the new Kuala Lumpur International Airport where over 1 million tonnes of asphalt concrete modified with various polymers has been laid.

6.4 DESIGN

6.4.1 Need for design

BS 4987 macadam asphalt concretes are all recipe mixtures with little allowance made for aggregate variations to the binder content, although there is a generalization for aggregate type; 0.2% less binder is required for limestone aggregates than for other crushed rock and there is an allowance for blast furnace slag. Table 24, Binder content for 14 mm close graded surface course is reproduced as Table 6.16. The note at the bottom of the table regarding gravel aggregates should to be treated with caution because it recommends higher binder contents for gravel while, in practice, they are more likely to drain and strip with more binder.

As shown in section 6.2, the properties of macadam asphalt concrete can vary and a design procedure, similar to other asphalt concretes, can prove useful in establishing a durable mixture, especially where new aggregate sources or mixtures with poor performance on the road are being evaluated. All design procedures involve evaluating a series of asphalt mixtures where one component, usually the bitumen content, is varied over a controlled set of mixtures. Compacted samples are tested

Table 6.16 Binder content for 14 mm close graded macadam

Aggregate	Proportion of total mixture (% by mass ± 0.5%)		
	Bitumen	High-temperature tar	Low-temperature tar
Crushed rock (excluding limestone)	5.1	6.1	5.5
Limestone	4.9	6.1	5.5
Blastfurnace slag of bulk density			
1440 kg/m³	5.5	6.6	6.6
1360 kg/m³	6.0	7.0	7.0
1280 kg/m³	6.6	7.5	7.5
1200 kg/m³	7.0	8.0	8.0
1120 kg/m³	7.5	8.4	8.4
Steel slag	4.8	6.1	5.5
Gravel[a]	–	–	–

[a] The information on the binder contents required for these mixtures made with gravel is insufficient for a single target value to be specified. The binder content to be used should be chosen within the range 5.5% to 6.5% and should be approved by the purchaser. As a guide, lower values should be selected for pavements designed to carry category B traffic and higher values for footway work. The tolerance of ± 0.5% should apply to the selected and approved binder content.

against the specified requirements and the mixture that most closely meets or surpasses these requirements is chosen either as the final mixture or, most commonly, for a further series of specialist tests. These tests very from country to country, or even county to county, depending on the contract requirements.

The object is to design material that is:

- durable;
- stable;
- flexible;
- fatigue resistant; and
- workable.

The above is affected by:

- binder content;
- void content;
- type of aggregate;
- grading; and
- viscosity of binder.

Some of the specifications described earlier ensure that these requirements are met by specifying:

- minimum / maximum binder content;
- minimum / maximum void contents;
- aggregate physical properties;
- grading curve; and/or
- binder properties before and after ageing.

The above are not intended to be a complete list of parameters that need to be addressed but general requirements/criteria.

6.4.2 Marshall design

To meet these criteria, a number of methods of design have evolved and been refined. Perhaps the most well-known method of design asphalt concrete is the Marshall method developed by Bruce Marshall, an engineer with Mississippi State Highway Department in the 1940s (Asphalt Institute, 1993), and standardized in ASTM D 1559-89 (ASTM D 1559). This is principally the same as the design method for rolled asphalt described in BS 598: Part 107 (BSI, 1990a). Both specifications give full details of method, which involves making a series of specimens by means of an impact hammer under controlled conditions. These specimens are made with a range of binder contents and are then tested for voids, stability, and flow. The results are plotted against binder content and the optimum mixture is chosen.

The method allows for a very precise measurement of voids and stability, although it does not necessarily guarantee good performance as regards durability and resistance to rutting. However, it does give the basis for a mixture that can be further developed with other tests, such as wheel-tracking (BSI, 1996a).

6.4.3 Hveem method of design

The Hveem method of design is perhaps less well known in the United Kingdom despite being well established in the United States of America, especially in the state of California where it was developed by Francis Hveem in the 1930s, and has been standardized in ASTM D 1560-92 (ASTM 1560) and 1561-92 (ASTM D 1561). The method differs from the Marshall method in that the object is to determine the maximum binder content in a mixture that will give sufficient resistance to deformation and be sufficiently impervious not to swell during an immersion test while maintaining a minimum voids content of 4%.

Essentially, the principle behind this method is to produce a durable mixture by means of maximizing binder content while maintaining minimum stability requirements. The method has some unique features which are summarized below.

6.4.3.1 Aggregates

These should meet all physical requirements of contract and produce a grading blend complying with specification (as is common to all design methods).

6.4.3.2 Estimation of binder content

There are a number of stages to this which begin with the following techniques:

Fine aggregate
Binder demand is estimated using the centrifuge kerosene equivalent (CKE), which involves soaking the passing 4.75 mm fraction in oil then centrifuging the sample and determining the amount of oil retained.

Coarse aggregate
The binder is estimated from the amount of oil retained after the aggregate has been soaked and allowed to drain in a funnel.

These estimated amounts are then combined from a series of standard charts which take into account factors for surface area and specific gravity. The final chart corrects for binder grade.

6.4.3.3 Specimen preparation

Once the CKE procedure has been carried out, asphalt mixtures are prepared at 0.5% increments, 2 above and 1 below the CKE value.

The mixtures are compacted in a gyratory compactor with specimens made for specific tests:

- density/voids
- swell test
- stabilometer test

The size of the specimen is similar to standard Marshall, 62 mm high by 102 mm diameter.

6.4.3.4 Tests

Stabilometer test
A specimen is tested in a cell under axial and lateral pressure at a temperature of 60 °C and the displacement is recorded and relative stability calculated.

Swell test
The amount of swell in a contained specimen is measured after 24 hours of 'soaking' with a 50 mL head of water.

Table 6.17 Hveem mixture design criteria

Traffic category	Heavy		Medium		Light	
Test property	Min.	Max.	Min.	Max.	Min.	Max.
Stabilometer value	37	–	35	–	30	–
Swell	Less than 0.762 mm					

6.4.3.5 Interpretation of results

The result should meet the requirements of Table 6.17, reproduced from AIM Series 2 (Asphalt Institute, 1993).

The final choice of binder content is made by using the Hveem design pyramid, which is a series of eliminations to determine the highest binder content meeting the requirements in Figure 6.17.

This description is only an outline of a method which may be valid for non-trunk roads where durability is of more importance than stability.

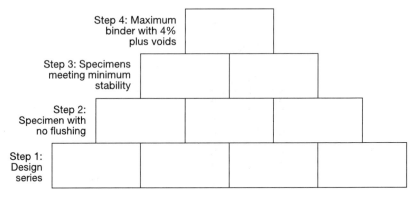

Fig. 6.17 Design binder

6.4.4 Mixture approval and verification

Sections 6.4.2 and 6.4.3 have briefly covered two laboratory design procedures to determine an optimum binder content for a particular aggregate blend. However, this should only be treated as a starting point. From the test data obtained, it is possible to gain some insight into the sensitivity of a mixture to variations in the binder content from studying the plotted values for mixture density, voids and stability. Ideally, the mixture should comply with specification requirements not only at the optimum binder content, but at binder contents at least 0.5% either side of it. This may appear obvious, but a mixture where the stability changes by more than 30% up or down within a 1% range should be treated with caution.

Design   173

It could be that the aggregates available are limited and their shape, texture and grading dictate the mixture properties, but it may be possible to adjust the aggregate proportions and improve the grading by increasing or decreasing the voids.

6.4.5 Voids

The amount of, and type of, voids in a mixture is not only a function of the amount of compaction or quality of binder but also the grading of the aggregate, all of which will affect performance. There are several classifications of voids in asphalt and the relationships between them which can be measured; these are categorized in Figure 6.18 (Asphalt Institute, 1993).

Figure 6.18 includes effective binder and density, which is the amount of binder less that which is absorbed into the aggregate and is commonly used in the United States of America, together with other locations where US engineers have prepared the specification. In calculating mineral

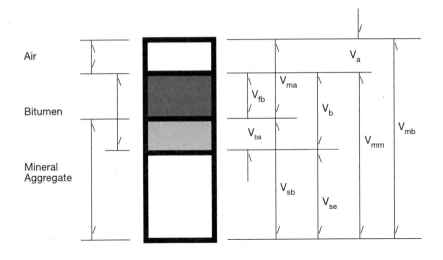

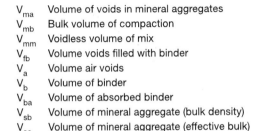

V_{ma} Volume of voids in mineral aggregates
V_{mb} Bulk volume of compaction
V_{mm} Voidless volume of mix
V_{fb} Volume voids filled with binder
V_a Volume air voids
V_b Volume of binder
V_{ba} Volume of absorbed binder
V_{sb} Volume of mineral aggregate (bulk density)
V_{se} Volume of mineral aggregate (effective bulk)

Fig. 6.18 Volumes in a compacted asphalt mixture

aggregate and filled voids, UK and US specifications will quote different values for an identical mixture due to the difference in definitions. The difference can be shown using the following data:

Bulk density (oven dry) of aggregate	= 2.703 kg/m³
Bulk density of compacted mixture	= 2.442 kg/m³
Voids in mixture	= 3.7%
Binder in mixture	= 5.3%
Density of binder	= 1.03 kg/m³

Using Asphalt Institute Manual Series No 2:

$$V_{ma} = 100 - \frac{\text{Mixture bulk density} \times \% \text{ total aggregate}}{\text{Bulk density of total aggregate}}$$

$$= 100 - \frac{2.442 \times 9.47}{1.03} = 14.4\%$$

Using BAA/DWS Method:

$$V_{ma} = \%\text{voids in mixture} + \frac{\%\text{binders} \times \text{Bulk density of mixture}}{\text{Density of binder}}$$

$$= 3.7 + \frac{5.3 \times 2.442}{1.03} = 16.3\%$$

Had the apparent density been used in the MS2 calculation, the results would have been similar. This applies to the voids filled with binder as well. However, as long as the calculations are carried out consistently and compared to the appropriate specification the curves produced will give the same binder contents.

The V_{ma} curve will help to judge if the mixture is likely to be overfilled and a binder content at, or just below, the minimum V_{ma} content is likely to have the greatest resistance to deformation. Typically, dense asphalt concrete has a V_{ma} of 15 to 19% (13 to 16% in the United States of America). This must always be related to a voids filled with binder of 71–87%.

The above examples are only to illustrate the importance of being aware of which density figures are used and reference should be made to the appropriate specification (BAA, 1993; DWS, 1995a) for full details on formulae, especially as they can change. For example, the Defence Works Services specification now incorporates maximum mixture density as determined by the method commonly known as the 'Rice' method and given in ASTM D 2041-91 (ASTM D 2041) and DD 228 (BSI, 1996b). The Rice method may give a slightly different maximum to that determined by calculating the individual components but it will prove more accurate and quicker for a specific mixture.

6.4.6 Grading

In section 6.3, aggregate properties and requirements relating to durability and performance (such as strength and polished stone value) were covered. However, the grading of the mixed aggregates will determine the performance characteristics of the asphalt concrete as much as, if not more than, the type of aggregate used. For example, a grading power factor of 0.7 would produce a mixture with voids in the range of 12–20% while a power factor of 0.45 would give maximum packing properties, minimum voids and possible poor performance due to lack of voids in the mineral aggregate (VMA).

The grading has to take into account the thickness the material is to be laid which will determine the maximum size of aggregate. It is generally accepted that the thickness should be approximately 2.5 times the maximum size of the aggregate. From the maximum size, it is possible to calculate the individual fractions (or proportion passing the various sieve sizes) using the power factor appropriate for the type of mixture which can then be used to compare grading within a particular grading specification band because small changes to the grading, especially deviations from a smooth curve below 3.35 mm, can affect the VMA, the workability and the performance of the mixture.

These deviations in grading curve can affect some mixtures more than others. Examples of this variation in effect are asphalt concretes produced to the airfield specifications (BAA, 1993; DWS, 1995) which tend to have a curve with a factor of 0.3 (Figure 6.1), that is a relatively high proportion passing the 3.35 mm and finer sieves. These fines can produce unstable mixtures, especially if the amount of crushed rock fines is less than 25 to 30% of the fines content (passing 3.35 mm). The reason for this particular grading is to produce a mixture which, when compacted, will be dense and have a minimal texture depth (0.2 to 0.4 mm) to eliminate the possibility of fretting and FOD (foreign object damage) to jet aircraft.

This type of mixture is more difficult to produce because the amount of crushed rock fines needs to be at least 30% to maintain stability (section 6.3.2). However, the material is also more difficult to compact due to the coarser micro-texture of crushed rock fines and, with the additional requirement regarding sand blends (BAA, 1993), tends to increase the amounts passing the 150 μm to 425 μm sieves and producing a slight 'hump' in the grading curve. This 'hump' has the effect of making the mixture less stable during compaction, especially if caused by natural sand fraction, due to the reduced VMA. This manifests itself in the material pushing excessively under the compaction roller, forming a bow wave and possibly leading to transverse surface crazing due to the inherent stability of the mixture. The crazing is not necessarily detrimental to long-term performance but can cause concern to the client and contractor.

This type of problem is less likely in a mixture with little or no natural sand and the VMA are maintained at the correct level, such as 16% VMA at 3% voids in the mixture. The voids in the mixture are essentially affected by the amount of bitumen and compactive effort whereas VMA is relatively constant and mainly affected by the grading curve; variations of 3% from the curve will affect the VMA (Asphalt Institute, 1993).

6.5 PRODUCTION AND PAVING

Asphalt concrete could be produced on any coating plant with the capability of heating aggregate and binder and blending the two. However, the quality and performance of the mixture will be dependent on the design of that mixture and the proficiency of the coating plant to reproduce that mixture consistently. This proficiency is a function of the type of coating plant and the quality assurance and controls measures in place to produce a mixture with the correct mixture properties and temperature which can be transported, laid and compacted to reproduce the laboratory designed properties. This, in turn, will be determined by the specification and contract requirements.

In this chapter, two types of asphalt concrete which are widely used for different applications in the United Kingdom have been discussed; those covered by BS 4987 (BSI, 1993a) for highways and those covered by specifications for airfields (BAA, 1993; DWS, 1995). Table 6.18 gives an indication of the difference in requirements to produce a material with double the grading tolerances and no voids requirements (BS 4987 surface course materials) to one that has tight grading tolerances and an air voids requirement.

Considerably more quality control effort is put into Marshall-type asphalt concrete than close-graded macadam type asphalt concrete, although this does not mean close-graded macadams are necessarily inferior for their intended purpose. Nevertheless, they can, in some instances, be improved upon by being aware of the effect of aggregate and binder type on the mixture properties (section 6.2). It is essentially a question of economics and performance requirements, highway and general paving mixtures need to be mixed and laid to a standard which is attainable by as many producers/contractors as possible to ensure that the client gets a material which is economical and available to most locations. Provided the contractor and client are aware of the material's possible performance limitation and each of their responsibilities with regard to quality control/assurance and expected performance criteria, then close-graded macadam asphalt concrete will continue to provide a surfacing material for most areas where there is no requirement for texture depth.

Marshall-type asphalt concrete may not necessarily always be a higher performance material in terms of stiffness and resistance to rutting when compared to close-graded asphalt concrete, but what the specifications ensure is that the surfacing will be consistently produced (Figures 6.20 and 6.21) and laid to meet defined specification criteria and performance

Production and paving

Table 6.18 Comparison of criteria for highways and airfields

Criteria	Highways	Airfields
Coating plant type	Batch or drum	Batch or drum*
Number of cold feeds	Not specified	1 for each aggregate type (normally 6 to 8)
Number of screened hot bins	Not specified	Minimum 3 (normally 5 or 6)
Number of filter silos	Not specified	Minimum 2 (hyd. lime & L/s)
Mixing time	Not specified	Minimum 1 minute
Insulated lorries	Yes	Yes
Mixing to paving time	Not specified	Maximum 3 hours
Ambient temperatures:		
Surface	Minimum −3°C	Minimum 5°C
Air	Minimum −1°C	Minimum 0°C
Tack coat:		
Onto new basecourse	No	0.25 to 0.35 L/m^2
Onto old basecourse	Yes	0.35 to 0.55 L/m^2
Paver	Yes	Yes, with insulated hopper
Mixing and paving trials	Not specified	All materials
Rollers	One 6 to 12 t dead-weight or equivalent vibrating plus one when over 100 t	One 10 to 12 t (25 to 70 N/mm) One 10 to 12 t (52.5 to 75 N/mm^2)† plus back-up as required
Compaction requirement	Number of rollers	Coring at 1 per 500m2 for voids
Layer bond	Not specified	Coring check‡
Testing		
Aggregates	For approval	For approval plus density and stockpile gradings
Binder and filler	For approval	Approval plus density
Hot bins aggregates	Not specified	Daily#
Mixed material	Specified in SHW (MCHW 1), not BS	Every 100 t
Stability	Not specified	Every 100 t
Mixture of water content	Not specified	0.5% maximum weekly

* Drum mixers may be more difficult to control the grading, requiring consistent supplies of graded material.
† Generally a combination of vibrating rollers, one drum rubber coated, is required to achieve the required finish.
‡ Bond is checked using cores but care is needed because newly laid material may not have developed the bond (see Figure 6.19, a ground radar plot showing delamination caused by coring from a runway surfacing trial area). Coring can also affect bond around the core hole over a radius of about a metre.
\# Hot bin samples are taken to verify the properties of the materials in the hot bins; however, in practice they are difficult and disruptive to take on a regular basis. Mixed material samples will achieve the same result and the samples rate can be increased if necessary to check a mixture that may need adjusting.

while a close-graded macadam asphalt concrete (Figure 6.22) will vary depending on the aggregate type even when mixed to the same clause.

To ensure this consistency, strict quality assurance and quality control procedures are required to closely monitor all stockpiling, processing, mixing, paving and compaction by highly skilled personnel at all levels. This also requires that the production equipment is accurately calibrated more frequently and that more of the normally required plant and machinery to process and pave the material are used, for example extra

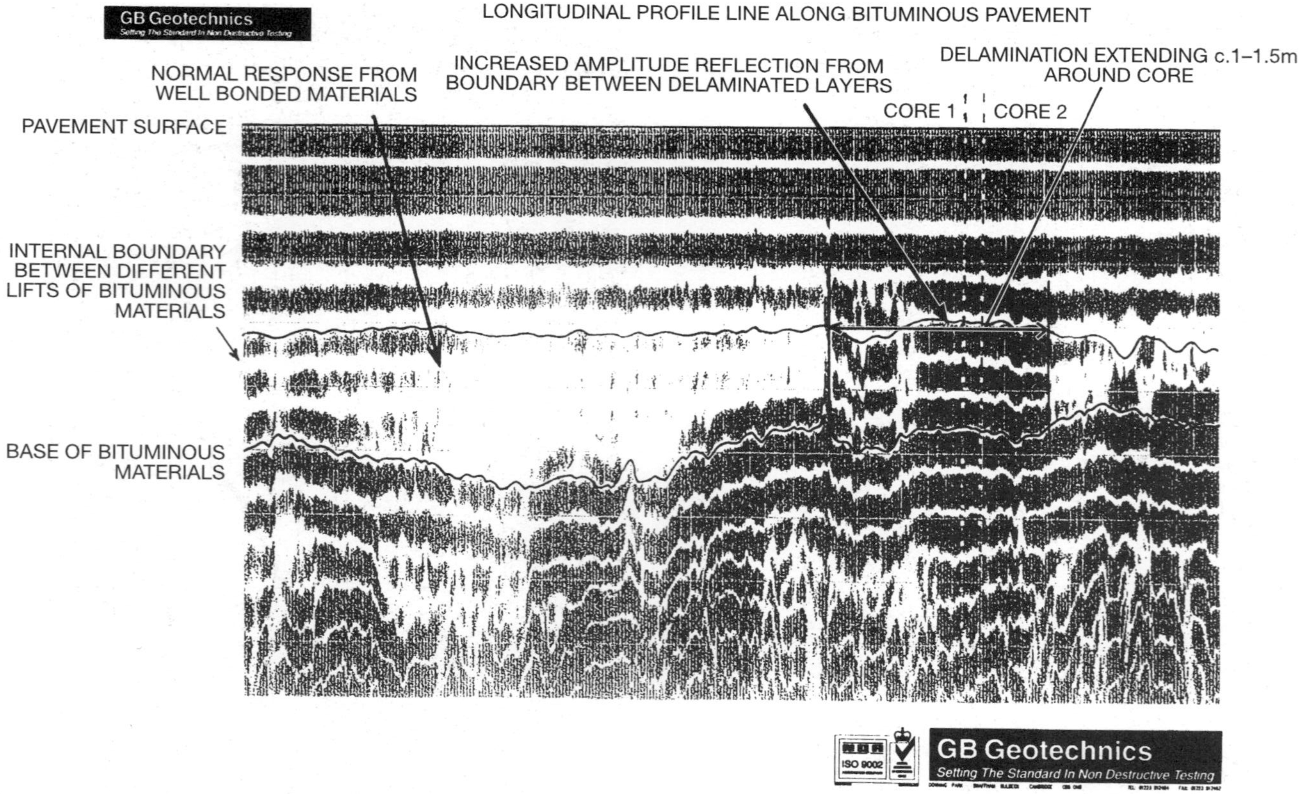

Fig. 6.19 Ground radar plot showing delamination caused by coring

Fig. 6.20 14 mm Marshall asphalt surface course

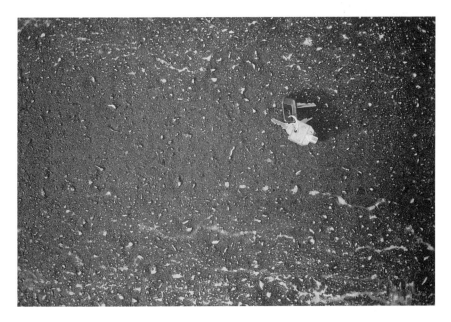

Fig. 6.21 Typical surface finish with 14 mm Marshall asphalt surface course

180 *Asphalt concrete (and macadam) surface courses*

Fig. 6.22 Typical surface finish with 14 mm close-graded macadam surface course

cold feeds at the coating plant, extra compaction plant and, quite often, specialist compaction rollers and techniques to achieve the required level of compaction and surface finish.

The compaction of any surface course can be difficult simply because it is generally the thinnest layer and cools most rapidly while ultimately it has to be resistant to traffic and the elements. It needs to be compacted to its optimum void content before it has cooled to a level where its stiffness (viscosity) is greater than the compactive effort available. With asphalt concretes, there is generally little time to spare due to their basic interlocking grading structure and hence the specification requirement for airfield work is to have double the number of rollers and of certain type compared to highway work.

If, having designed a material to produce the best possible properties and optimum void contents, it is still necessary to vary the paving and compaction techniques to suit that particular material in order to achieve the required compaction and finish, the following are **some** of the factors which need to be addressed.

Paver

- General condition – should be well maintained with no worn parts.
- Heated, vibrating tampers/screed.
- Angle of attack.
- Speed of paving.

Rollers

- **Three-point static**
  - Used as breakdown roller.
  - May be too heavy/light.
  - Speed.
  - Ideally will close tearing/crazing caused by paver/material – not always.
- **Tandem**
  - Used to provide finish and compaction.
  - Vibration to be used with caution as on thin lifts could cause cracking.
- **Pneumatic Type Roller (PTR)**
  - Used as secondary roller because it will produce kneading action to close and compact the material.
  - May leave tyre marks when the tyre pressure may need to be adjusted to suit the material; a lower pressure will increase the surface contact area and, in some instances, help to eliminate surface crazing.
  - Can also be used as a 'breakdown' roller on some mixtures, although PTR are most commonly used between three-point roller and tandem rollers.
- **Combination rollers**
  - Tandem rollers with one drum covered with rubber.
  - Can be effectively used as a finishing roller to seal the surface and remove other roller marks.

The above is not intended as a definitive guide to rollers but as an indication to the wide range of techniques required and available. If a particular mixture is proving to be difficult to compact, consideration should be given to the design, in particular the VMA.

6.6 CONCLUSION

It is hoped that this chapter has given the reader some further appreciation of the range of asphalt concrete materials and their properties although the data presented here should be treated as typical rather then absolute and as an encouragement to determine the properties.

Notes

1. The author wishes to thank his colleagues within Associated Asphalt Pavement Technology and CAMAS Aggregates for assisting in preparing and testing the samples used for these illustrations.

6.7 REFERENCES

Manual of Contract Documents for Highway Works, Her Majesty's Stationery Office, London.
  Volume 1: Specification for Highway Works (MCHW 1).
Design Manual for Roads and Bridges, Her Majesty's Stationery Office, London
  HD 28/94 – *Skidding Resistance* (DMRB 7.3.1)
Annual Book of ASTM Standards. Volume 04.02, Concrete and Aggregates, American Society for Testing and Materials, Philadelphia.
  ASTM C 131-89, *Test Method for Resistance and Degradation of Small-Size Coarse Aggregate by Abrasion and Impact in the Los Angeles Machine* (ASTM C 131).
  ASTM C 535-89, *Test Method for Resistance and Degradation of Large-Size Coarse Aggregate by Abrasion and Impact in the Los Angeles Machine* (ASTM C 535).
Annual Book of ASTM Standards. Volume 04.03, Road and Paving Materials; Pavement Management Technologies, American Society for Testing and Materials, Philadelphia.
  ASTM D 1559-89, *Test Method for Resistance to Plastic Flow of Bituminous Mixtures Using Marshall Apparatus* (ASTM D 1559).
  ASTM D 1560-92, *Test Methods for Resistance to Deformation and Cohesion of Bituminous Mixtures by Means of Hveem Apparatus* (ASTM D 1560).
  ASTM D 1561-92, *Practice and Preparation of Bituminous Mixture Test Specimens by Means of California Kneading Compactor* (ASTM D 1561). ASTM D 2041-91, *Standard Test Method for Theoretical Maximum Specific Gravity and Density of Bituminous Paving Mixtures* (ASTM D 2041).
  ASTM D 2872-88, *Standard Test Method for Effect of Heat and Air on a Moving Film of Asphalt (Rolling Thin-Film Oven Test)* (ASTM D 2872).
  ASTM D 3515-89, *Specification for Hot-Mixed, Hot-Laid Bituminous Paving Mixtures* (ASTM D 3515).
AASHTO Provisional Standards, March 1995 edn, American Association of State Highway and Transportation, Washington DC.
  AASHTO PP1-93, Edition 1A, *Standard Practice for Accelerated Aging of Asphalt Binder Using a Pressurized Aging Vessel* (PAV) (AASHTO PP1-93)
Al-Suhaibani, A.R. and J. Al-Mudanheem (1992) *Effect of Filler Type and Content on Properties of Asphaltic Concrete*, ASTM STP 1147, American Society for Testing and Materials, Philadelphia.
Asphalt Institute (1993) *Manual Series No. 2*, 6th edn, Asphalt Institute, Maryland, USA.
Anderson, D. (1994) The SHRP Binder Test methods and specification, *Revue Generale des Routes et des Aerodromes*.
British Airports Authority (1993) *Civil Engineering Section 14: Bituminous Materials – Aircraft Pavements. Quality System Reference Specification SP-CE-014-04-0A*, British Airports Authority Technical Service Division.
British Standards Institution (1974) *Specifications for Tars for Road Purposes. BS 76: 1974*, British Standards Institution, London.
British Standards Institution (1983) *Air-cooled Blastfurnace Slag Aggregate for Use in Construction. BS 1047: 1983*, British Standards Institution, London.
British Standards Institution (1985) *Testing Aggregates; Part 103, Methods for the Determination of Particle Size Distribution; Section 103.1, Sieve Tests. BS 812: Part 103.1: 1985*, British Standards Institution, London.
British Standards Institution (1989) *Testing Aggregates; Part 121, Methods for the Determination of Soundness. BS 812: Part 121: 1989*, British Standards Institution, London.
British Standards Institution (1990a) *Sampling and Examination of Bituminous Mixtures for Roads and Other Paved Areas; Part 107, Method of Test for the*

References

Determination of the Composition of Design Wearing Course Rolled Asphalt. BS 598: Part 107: 1990, British Standards Institution, London.

British Standards Institution (1990b) *Testing Aggregates; Part 110, Methods for the Determination of Aggregate Crushing Value (ACV). BS 812: Part 110: 1990*, British Standards Institution, London.

British Standards Institution (1990c). *Testing Aggregates; Part 111, Methods for the Determination of 10% Fines Value (TPF). BS 812: Part 111: 1990*. British Standards Institution, London.

British Standards Institution (1990d) *Testing Aggregates; Part 112, Methods for the Determination of Aggregate Impact Value (AIV). BS 812: Part 112: 1990*, British Standards Institution, London.

British Standards Institution (1990e) *Testing Aggregates; Part 113, Methods for the Determination of Aggregate Abrasion Value (AAV). BS 812: Part 113: 1990*, British Standards Institution, London.

British Standards Institution (1991) *Specification for Portland Cement. BS 12: 1991*, British Standards Institution, London.

British Standards Institution (1992) *Hot Rolled Asphalts for Roads and Other Paved Areas; Part 1, Specification for Constituent Materials and Asphalt Mixtures; Part 2, Specification for Transport, Laying and Compaction of Rolled Asphalt. BS 594: Part 1: 1992, BS 594: Part 2: 1992*, British Standards Institution, London.

British Standards Institution (1993a) *Coated Macadam for Roads and Other Paved Areas; Part 1, Specification for Constituent Materials and for Mixtures; Part 2, Specification for Transport, Laying and Compaction. BS 4987: Part 1: 1993, BS 4987: Part 2: 1993*, British Standards Institution, London.

British Standards Institution (1993b) *Petroleum and its Products; Part 58, Softening Point of Bitumen (Ring and Ball). BS 2000: Part 58: 1993*, British Standards Institution, London.

British Standards Institution (1996a) *Sampling and Examination of Bituminous Mixtures for Roads and Other Paved Areas; Part 110, Methods of Test for the Determination of the Wheel-Tracking Rate of Cores of Bituminous Wearing Courses. BS 598: Part 110: 1996*, British Standards Institution, London.

British Standards Institution (1996b) *Methods for Determination of Maximum Density of Bituminous Mixtures. DD 228: 1996*, British Standards Institution, London.

Bullas, J.C. and G. West (1991) *Specifying Clean, Hard and Durable Aggregate for Bitumen Macadam Roadstone*, Department of Transport TRL Research Report 284. Transport Research Laboratory, Crowthorne.

Comité Européen de Normalisation (1995a) *Asphaltic Concrete Mixtures for Roads and Other Paved Areas*, Draft European Standard Work Item 227100, Comité Européen de Normalisation TC 227/WG1/TG3.

Comité Européen de Normalisation (1995b) *Tests for Geometric Properties of Aggregates, Part 12: Assessment of Fines – Methylene Blue Test*, Draft European Standard prEN 933-12, Comité Européen de Normalisation TC 154/WG.

Davies, E. and J.T. Laitinen (1995) Effects of Ageing on Durability of Modified Bitumen and Bitumen Aggregate Mixtures, *Second ICPT '95*, Singapore.

Defence Works Services (1995a) Marshall asphalt for airfield pavement works, *Defence Works Functional Standards*, Her Majesty's Stationery Office, London.

Defence Works Services (1995b) Hot rolled asphalt and coated macadam for airfield pavement works. *Defence Works Functional Standards*, Her Majesty's Stationery Office, London.

European Asphalt Paving Association (1995) *Heavy Duty Pavements: The Argument for Asphalt*, European Asphalt Paving Association.

Fuller, W.B. and S.E. Thompson (1907) The laws of proportioning concrete. *Trans. American Society of Civil Engineers* **59**, 67–172.

King, M. (1992) *Survey of Asphalt Concrete Airfield Trials*, GB Geotechnics.

Nunn, M.E., C.J. Rant and B. Schoepe (1987) *Improved Roadbase Macadams; Road Trials and Design Considerations*, Department of Transport TRL Research Report 132, Transport Research Laboratory, Crowthorne.

Road Research Laboratory (1962) *Bituminous Materials in Road Construction*, Her Majesty's Stationery Office.

Shahrour, M.A. and B.G. Saloukeh (1992) *Quality and Quantity of Locally Produced Filler. ASTM STP 1147*, American Society for Testing and Materials, Philadelphia.

CHAPTER 7

Porous asphalt surface courses

T.R.J. Fabb, Refined Bitumen Association[1]

7.1 TERMINOLOGY

Porous asphalt was originally known as 'Friction Course' material. It was later termed 'Pervious Macadam' in the United Kingdom and in 1992 the term 'Porous Asphalt' was adopted, following agreement to include this term in the draft CEN terminology for highways materials. It has had other names in different countries, including 'Drainasphalt' in France, 'Flüsterasphalt' (whispering asphalt) in Germany and, if rather colloquially, 'Popcorn Mix' in the United States of America.

7.2 CONCEPT AND HISTORY

The original concept was to use porous asphalt to reduce greatly the amount of rain water lying on airfield pavement surfaces and thereby to reduce aquaplaning and skidding of aircraft. To achieve this aim, the material must have a very high content of interconnected air voids. A concomitant of its use is that the binder course must be impermeable in order to protect the remainder of the pavement and its foundation from the damaging effect of the ingress of water.

It was developed by the UK Air Ministry in the mid-1950s. The formulation adopted at that time was a very open-graded 10 mm macadam, having about 20% air voids when laid. It proved to be very successful and, since the early 1960s, many civil and military airfield runways have been surfaced with it. When used for this purpose, the material is still known as 'Airfield Friction Course'.

Subsequently, attempts were made by the (then) Road Research Laboratory to adapt this material for use on highways. It was considered that clogging with detritus would be a greater problem on highways than on airfields and, consequently, a major change was the adoption of a 0 to 20 mm grading on the assumption that it would have more and larger voids than the 10 mm variety and thereby remain unclogged for longer. One of the earliest road trials, on the High Wycombe bypass (Please *et al.*,

1972; Jacobs 1983), was not entirely successful in that the porous asphalt failed to be effective after only a few years, although remaining structurally intact for 15 years. The reasons were not clear, but it may have been connected with the relatively high fines content tending to destabilize the coarse aggregate skeleton and with the very low traffic volume on the road at that time.

Further trials followed (Nicholls, 1997). Although there were early indications that the use of modified binders helped to offset compaction by traffic, no serious attempts were made for some time to introduce the material on to UK highways because of strong concern by the Department of Transport (DoT) about the durability of the material. By the early 1980s, awareness of the potential benefits of the material had greatly increased, leading to further road trials in Staffordshire in 1983, 1984 and 1987. The results of the latter two (Daines, 1986, 1992; Colwill *et al.*, 1993) were very influential in the preparation of an Advice Note and Specification for porous asphalt surface course, published by the Government Overseeing Departments early in 1993 and subsequently incorporated into *Design Manual for Roads and Bridges* (DMRB 7.2.4) and *Specification for Highway Works* (MCHW 1) with associated Notes for Guidance (MCHW 2).

The current policy of the UK Highways Agency appears to be to support the use of porous asphalt in circumstances where the financial advantages of doing so outweigh the disadvantages. However, porous asphalt is now accepted as the quietest road surfacing material available. In the United Kingdom, this is seen as its greatest advantage and it is the main reason for its use to be considered. It tends to be used only when the additional cost of using it is calculated to be less than the total cost of mitigating the effects of road noise on nearby properties (by providing sound barriers and/or double glazing) and paying compensation.

Although porous asphalt was developed in the United Kingdom, in several continental European countries it has been adopted to a much greater extent and more quickly than in the United Kingdom, due to different assessments of the balance between advantages and disadvantages. Table 7.1 (section 7.4.5) indicates the usage in several countries in 1994. The United Kingdom is not the only major country to have used porous asphalt sparingly; Germany, with its very smooth asphalt concrete surface courses and many high-speed roads (and corresponding high accident rates), has so far used porous asphalt only in trials.

7.3 PROPERTIES

7.3.1 General

The main characteristic properties of porous asphalt:

- It contains a high volume of interconnecting air voids, typically at least 20% for new material. These voids enable a surface course of porous asphalt to absorb a great deal of rain. For example, a course 50 mm thick containing 20% air voids could absorb 10 mm of rain if it all fell instantaneously. In practice, drainage across the fall starts as soon as sufficient water has been absorbed to establish continuous water films at the bottom of the layer. One effect of this water-absorbing property is to greatly reduce the amount of water lying on the running surface, and thereby to greatly reduce both the amount of spray generated by traffic and the risk of aquaplaning by high-speed traffic.
- During compaction on the road, most of the coarse aggregate particles at the surface are orientated so as to present a large face parallel to the road surface, thus producing a relatively flat, planar running surface containing surface voids. This type of surface is said to have a 'negative texture' (section 10.4.8) which:
  - does not excite tyre treads to the same extent as conventional positively textured road surfaces and, therefore, produces low levels of tyre noise; and
  - has excellent high-speed skid-resistance (provided, of course, that high quality aggregate is used).

7.3.2 Aquaplaning and skidding accidents

As illustrated in Figure 7.1, aquaplaning can occur in very wet conditions when vehicles are travelling at relatively high speeds. It was the need to reduce the risk of aquaplaning that led to the original development of porous asphalt as 'Airfield Friction Course'.

Research carried out at the TRL in the 1950s and 1960s established relationships between skid-resistance, speed and macro-texture (Roe *et al.*, 1988), later expanded into the influence of macro- and micro-texture on accidents (Roe *et al.*, 1991). Figure 7.2 demonstrates these relationships and the importance of adequate macro-texture.

On the basis of these relationships, highways authorities in the United Kingdom and some other countries specify minimum values for macro-texture in order to reduce the loss of skid resistance as speed increases, and to reduce the risk of skidding accidents and aquaplaning.

A variety of techniques has been used to impart the necessary macro-texture to a road surface:

- Concrete surfaces are generally textured by dragging wire brushes across the concrete as it is setting. As shown in Figure 7.3, this produces a series of closely and regularly spaced ridges across the direction of the road.
- Different asphalts are textured in different ways. In the United Kingdom, the commonest technique is to roll large chippings into

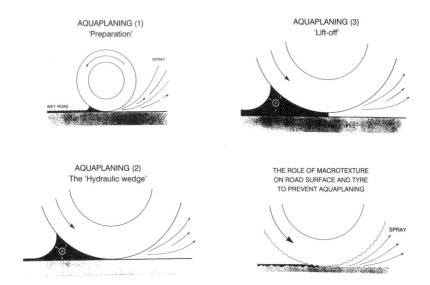

Fig. 7.1 Schematic cause of aquaplaning

newly-laid rolled asphalt (Chapter 5). Gussasphalt (Chapter 8) is textured with a roller which imprints a pattern of square depressions into the cooling surface.
- Surface dressing (section 11.1) is widely used to impart a high level of texture – generally as a remedial measure for surfaces which have lost macro-texture.

Achieving the desired macro-texture, either by the use of chippings rolled into asphalt or by surface dressing, produces a random pattern of small bumps, as illustrated in Figure 7.4.

This macro-texture reduces loss of skid-resistance and delays the onset of aquaplaning by reducing the film thickness of water in contact with the tyre and by providing gaps to dissipate pressure in the hydraulic wedge. However, the provision of macro-texture by such methods was found not to prevent aquaplaning by aircraft on very wet surfaces. This inability appears to be due to a combination of:

- the very high speed of aircraft when landing compared with the speed of road traffic; and
- the slow-draining character of runways.

These factors can lead to very wet, or even flooded, conditions. The replacement of conventional airfield surfacings by porous asphalt produced an enormous improvement. This improvement can be ascribed to:

Properties 189

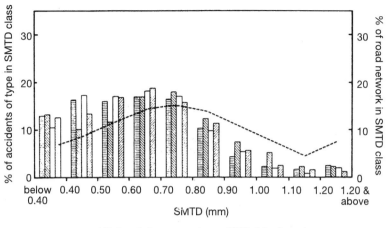

a) Network A, averaged over 1982–84, all roads

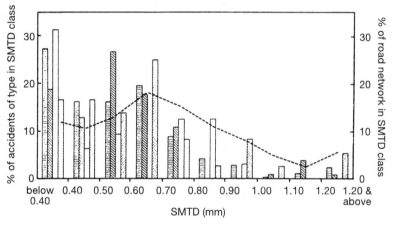

b) Network B, 1984 all roads

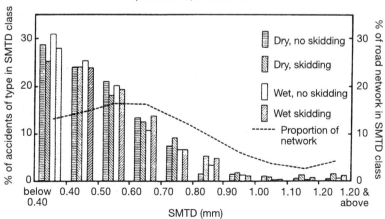

c) Network C, 1984 all roads

Fig. 7.2 Proportion of accidents at different texture levels

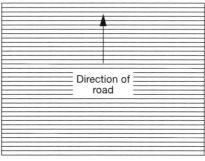

Fig. 7.3 Surface texture on concrete

- porous asphalt greatly reducing the amount of water lying on the surface of runways; and
- even when saturated, porous asphalt dissipates hydraulic pressure through its pores.

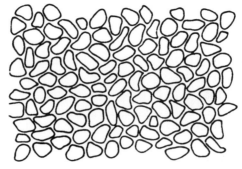

Fig. 7.4 Surface texture on traditional asphalt surfacings

The UK criteria for macro-texture for highways were developed with conventional 'positively textured' surfaces, as were the criteria for the polished stone values of aggregates. There are indications that, with equal PSV, porous asphalt performs better in terms of skid-resistance than conventional asphalt surfacings, although other investigations show similar performance. Consequently, the texture and PSV criteria for negatively textured asphalts are under review. Further, the standard measurement of skid-resistance is not carried out at the speeds and extremely wet conditions that pertain when aquaplaning may occur. Under such conditions, porous asphalt is markedly superior to conventional surfacings (which is why it was first used on airfield runways). Because the worst type of multiple-vehicle 'pile ups' occur in very wet weather on high-speed roads, the comparisons of the skid-resistance of alternative surfacing materials provided by the normal measurements under non-critical conditions do not adequately indicate the accident-reducing potential of porous asphalt.

7.3.3 Tyre noise

The proportion of the UK population seriously affected by traffic noise has increased significantly in recent years, mainly due to increased traffic volumes and the construction of new trunk roads, including bypasses. This has led to increased levels of complaints concerning existing roads and to greater objections to proposals for new roads or the widening of existing ones. This has been happening at a time when many people have an increased awareness of environmental matters and increased concern for 'quality of life'. Advances in vehicle design and construction have greatly reduced noise emissions from engines and transmissions, leaving tyre noise as the main source of vehicle noise. The public has also become better informed about the differences between various types of road surfaces in relation to noise.

The factors that affect the 'noisiness' of a road can be divided into those related to noise generation and those which affect the noise after generation. The principal factors include the following.

Noise generation

- vehicle speed;
- vehicle type and size;
- engine type and output;
- design of the vehicle;
- level of vehicle maintenance; and
- the tyre/surface interaction noise.

Noise after generation

- noise attenuation by the road surface and adjoining ground surfaces (commonly referred to as the ground effect);

- the distance from the noise source; and
- the influence of barriers and other objects which screen the noise.

The factors on which the road surface have an effect are the tyre/surface interaction noise and the noise attenuation by the road surface.

Porous asphalt is accepted as the least-noisy road surfacing currently available. The tyre/surface interaction noise is generally affected by the smoothness of the road surface, as is demonstrated by smooth road surfaces being so quiet, as anyone who has driven on the smooth asphalt concrete surfaces which are common in Germany can confirm. Unfortunately, smooth surfaces (that is, surfaces deficient in macro-texture) have very poor skid-resistance, especially at high speeds. To obtain high skid-resistance and hence reduce the probability of skidding accidents or aquaplaning (section 7.3.2), high texture depths are required.

For most surfacings, rough textures are obtained at some cost in terms of noise. A simplified illustration of the mechanism of noise generation at the tyre/road surface is given in Figure 7.5. As any part of a rolling tyre meets a texture-giving asperity on the road surface, it is slightly compressed. As the point of contact moves forward off the asperity, it relaxes and another following point of contact is compressed on the same asperity. It is this alternating compression and relaxation (that is, vibration) of the tyre which generates the noise, as well as causing the tyre to heat up. The noise generation mechanism also includes the compression and release of air in the tyre tread.

The quietness of porous asphalt is also partly due to generated noise being absorbed by the voids in the surface. Surface voids do absorb sound by the Gauss–Helmholz effect, but the contribution made by this effect would appear to be relatively small, because:

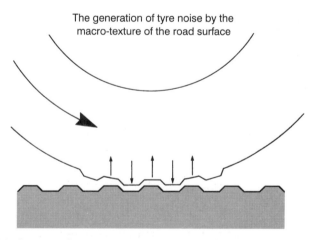

Fig. 7.5 Mechanism of noise generation at the tyre/road surface

- porous asphalt which has become heavily clogged up with detritus loses little of its noise reducing property (Nelson and Abbott, 1990); and
- smooth surfaces without surface voids are also very quiet.

Therefore, the major benefit is often assumed to be due to the reduced generation of noise rather than to its absorption. Experience in Holland and Austria has demonstrated an inverse relationship between noise reduction and maximum size of the aggregate used. Thus, from their experience (PIARC, 1993), a 0 to 10 mm grading is significantly quieter than a 0 to 14 mm grading, and a 0 to 8 mm grading gives even greater benefit. Therefore, such small nominal aggregate mixtures are now used in some of the most noise-sensitive areas of these countries. However, a trial on the M1 in the United Kingdom of porous asphalts with different sized nominal aggregate shows that the largest sized aggregate is, marginally, the quietest (Nicholls, 1997). The reason for this inconsistency is not known; it may be due to the use of different equipment, techniques or criteria for measuring noise.

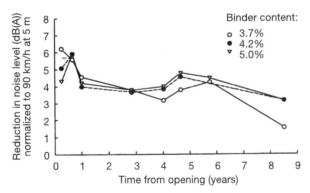

Fig. 7.6 Reduction in maximum noise for light vehicles on porous asphalt relative to rolled asphalt

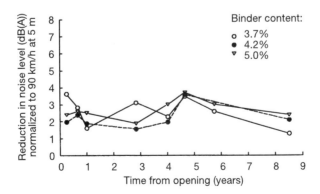

Fig. 7.7 Reduction in maximum noise for heavy vehicles on porous asphalt relative to rolled asphalt

Figures 7.6 and 7.7 (Nicholls, 1997) show the reduction in noise achieved by the typical UK mixture for both light and heavy vehicles and demonstrates that this benefit is maintained over time. The 'noise bonus' shown by these figures is very significant. Compared with rolled asphalt surfacings with pre-coated chippings, porous asphalt reduces tyre noise by between 3 and 4 dB(A). Compared with brushed concrete surfacings the benefit is even more dramatic – about a further 2 to 3 dB(A). Because the decibel scale is logarithmic, a reduction of 4 dB(A) is very significant. It is equivalent to either halving the volume of traffic or doubling the distance of the observer from the road.

In the wet, the noise performance is further improved relative to that on impervious surfaces by lessening the incidence of 'splash' noise generated at the tyre–surface interface. For conventional non-porous surfaces, the tyre noise increases by between 5 and 10 dB(A) in wet weather, depending on the surface type and the speed of the vehicle.

7.3.4 Driver comfort and fatigue

This reduction in tyre noise is very noticeable inside as well as outside vehicles, making driving far more pleasant and less tiring. Although this improvement is quite dramatic, it seems to command little attention.

7.3.5 Spray

A substantial amount of spray can be generated by vehicles' tyres from even a fairly thin layer of water on a road surface. Transverse drainage from conventional impermeable road surfaces is too slow to prevent spray from being generated. The difference in the amount of spray generated on different surfacings is illustrated in Figures 7.8 and 7.9, which show the same vehicle travelling on rolled asphalt and then on porous asphalt.

It has been clearly demonstrated (Nicholls and Daines, 1992) that the amount of spray generated is directly related to vehicle speed. These measurements were supplemented by subjective and photographic assessments (Daines, 1992), and it was concluded that 'at 50 km/h (30 mph) spray is minimal, but at 110 km/h (70 mph) the vehicle ahead is virtually obliterated'. The general experience is that spray is only a serious problem at speeds above about 80 km/h (50 mph). Thus spray is a problem of high-speed roads rather than of city centres.

A new porous asphalt surface course, 50 mm thick, containing (say) up to 20% air voids, has the capacity to absorb up 11.5 mm of rain if it all fell instantly. In reality, that amount of rain would take considerable time to fall, even in very heavy rainfall. Also, drainage starts as soon as continuous films of water are established within the layer of porous asphalt. Consequently, a surface course of porous asphalt is able to prevent a

Fig. 7.8 Spray generated by a tanker travelling on rolled asphalt

substantial layer of surface water from forming for a considerable time, and greatly reduces spray during this period.

It would be reasonable to assume that clogging of porous asphalt with detritus would nullify this spray-reducing property. However, it has been observed (Please *et al.*, 1972; Daines, 1992) that even well-clogged porous asphalt, with low hydraulic conductivity, generates far less spray than a conventional rolled asphalt surface course. The reasons for this are far from obvious, but it is possible that the layer of porous asphalt remains permeable, even when the voids content is low, and that this allows a significant amount of water to flow through it, thereby reducing the amount of water lying on the surface and reducing spray.

It would seem to follow from this that hydraulic conductivity is a poor predictor of spray generation. Findings from TRL's trial on the A38 (Daines, 1992) support this in that:

Fig. 7.9 Spray generated by a tanker travelling on porous asphalt

- The porous asphalt (example) has a low hydraulic conductivity but is effective in reducing spray ... at approximately 90 km/h;
- This equation[2] implies that when the hydraulic conductivity has fallen to zero, and texture depth is still 2.0 mm (sand patch) then spray suppression is still 50% compared with hot rolled asphalt; and
- As hydraulic conductivity reduces with time there is no appreciable increase in the level of spray, until a relatively low hydraulic conductivity is reached.

7.3.6 Glare and illumination

The use of porous asphalt surfacing reduces the glare from headlights in wet conditions, as illustrated by Figures 7.10 and 7.11 (Daines, 1992).

Properties

Fig. 7.10 Glare on rolled asphalt

198   *Porous asphalt surface courses*

Fig. 7.11 Glare on porous asphalt

Although there are no statistics available to quantify this benefit, there can be no doubt that any such contribution to improved visibility is a contribution to safety. However, the low reflectance which reduces glare also reduces overall luminance, leading to an increased requirement for artificial illumination where such is provided.

7.4 APPLICATIONS

7.4.1 Reasons for using porous asphalt on highways

The main reasons for using porous asphalt are:

- to improve safety by reducing the risk of aquaplaning; and
- to reduce traffic noise.

The other benefits of using porous asphalt appear generally to have little influence on the decision-making of Highway Authorities. These other benefits are:

- improving visibility in wet weather by reducing spray;
- improving driver comfort;
- reducing glare in night driving in wet weather; and
- reducing fuel consumption and tyre wear.

However, reduced spray and improved driver comfort are conducive to its use on private-sector toll roads. They were major factors in the choice of porous asphalt to surface some privately operated toll roads in Italy because it was considered that these benefits would attract toll-paying drivers onto these roads.

The main disincentive for using porous asphalt is cost, although this does not appear to have inhibited its widespread use in several European countries, notably Austria, France, Netherlands, Italy, Spain, Sweden and Switzerland.

7.4.2 Hydraulic applications

Another, lesser reason for using porous asphalt on highways is to take advantage of its reservoir capability in order to either:
- compensate for slower than normal crossfall drainage; or
- reduce peak loadings on drains.

7.4.3 Suitability for sites

Porous asphalt may be used in almost any location. Its capability to reduce spray, aquaplaning and noise make its use particularly advantageous on bypasses, which frequently skirt well-populated areas, and on high-speed urban roads. However, its special properties confer little or no advantage where traffic speeds are low, as on most roads in town centres, because neither spray nor tyre noise is a serious problem at such speeds.

Porous asphalt may be used on either flexible or rigid substrates:

- in new construction; or
- in reconstruction; or
- as an overlay.

7.4.4 Use in the United Kingdom

7.4.4.1 Highways

The original purpose of porous asphalt was to prevent aquaplaning on airfield runways. Although spray is not an important consideration on runways, the porous character of the material also leads to a dramatic reduction in spray when it is used on highways, where many believe spray is important.

However, the main reason for its use in the United Kingdom is to reduce traffic noise. It is used for that purpose in badly affected areas when its use is calculated to be less costly than that of the alternatives of compensation to householders and the provision of double glazing and sound barriers. Hence, this has led to extremely limited use of porous asphalt in the United Kingdom. Table 7.1 shows that the usage of porous asphalt in the United Kingdom was very much lower than that of many continental European countries up to the end of 1995.

A porous asphalt surface course is sometimes the cheaper option for bypasses, which usually carry fast traffic around the outskirts of a town and, therefore, tend to subject large numbers of residents to high levels of traffic noise.

Although spray reduction is not normally the prime reason for using porous asphalt in the United Kingdom when it is used for noise-reduction, a minimum hydraulic conductivity (BSI, 1996a) is specified. It is presumed that this minimum is specified with the intention of ensuring a minimum degree of spray-reduction.

The use of porous asphalt in the United Kingdom is inhibited by its high cost compared with that of rolled asphalt and by concerns over its durability. The main factors contributing to the high cost of porous asphalt are:

- **Quality and cost of aggregate** All of the aggregate in a layer of porous asphalt must be of the same high quality. Generally, the polished stone value (PSV) (section 2.2.4.4) required for aggregate for porous asphalt is the same or similar to that for pre-coated chippings for rolled asphalt if laid on the same site. Aggregates of high PSV are relatively expensive because their sources are not widespread and tend to be in remote locations.

- **Size of aggregate** Currently, the UK specification allows only one grading with a 20 mm nominal aggregate size. Hence, porous asphalt is in competition with 20 mm chippings for rolled asphalt, which tends to increase its cost. Furthermore, the use of such a large aggregate size necessitates a minimum layer thickness of 50 mm, thereby preventing the use of a more economic thickness of 35–40 mm which would be possible if 10 mm, 14 mm or 16 mm aggregates were employed, as they commonly are on the European continent.
- **Mixing plant operations** When manufacturing porous asphalt, the mixing plant must be dedicated solely to that purpose. This means that:
  - the plant must be made ready, which may involve making a separate silo available for the hydrated lime filler and making a separate tank available if a modified binder is being employed; it will also necessitate emptying aggregate bins and obtaining and stockpiling the specific aggregate required;
  - a certain amount of production time will be lost during this preparation;
  - other asphalts cannot be manufactured in the normal manner during the period when porous asphalt is in production.

  All of these disruptions to normal operations create diseconomies. Much of such extra cost is incurred during preparation and start-up, and their effect is minimized by long production runs. Also, greater than normal technical control must be exercised to comply with the close tolerances for aggregate grading and binder content.
- **Laying and compaction** The laying contractor is normally required to lay the whole width of the carriageway in one operation so as to obviate the need for longitudinal joints. The reasoning is that longitudinal joints could impede the transverse drainage of water. However, this requirement normally necessitates the use of at least two pavers working in echelon and corresponding rollers. This increases direct laying costs because, say, any interruption of laying with one paver brings the other(s) to a halt. The requirement to avoid longitudinal joints means that the contractor may also be required to relay the whole width of the carriageway when material in only one lane is rejected. The contractor must allow for this risk when submitting his tender.
- **Compliance with specification** The specification is more difficult to comply with than for most other asphalts (including macadams). The main difficulties concern:
  - the analysis of samples taken from the laying site for binder content and aggregate grading, and
  - the values of hydraulic conductivity (BSI, 1996a) of the laid and compacted material.

The difficulties to comply with specification arise from the pronounced tendency of the larger aggregate particles to segregate from the fines during laying. Thus, the binder content and hydraulic conductivity vary across the laid mat, the latter tending to be lowest in the middle of the laid swathe. The properties of the 20 mm mixture greatly increase the risk of material being rejected as not complying with the specification. The contractor must allow for this risk, including the cost of following dispute procedures, when tendering. The *Specification for Highway Works* (MCHW 1) has been revised to recognize, and to some extent to allow for, this variation across the swathe. This should reduce the likelihood of material being rejected, but it may be some time before this reduced risk is reflected in tender prices.

The material supplier and laying contractor incur increased costs in attempting to cope with these difficulties, in negotiating with the client in cases of dispute, and in penalties when material is not accepted. Possible means of reducing costs include:

- **Reduce the specified PSV of the aggregate** There were some indications from the major UK trials on the A38 trunk road (Daines, 1992) that the skid resistance of porous asphalt is slightly higher than that of rolled asphalt at equal PSV of aggregate. Further (unpublished) work has shown that the aggregate PSV required of proprietary 'negatively textured' asphalt (Chapter 10) having surface characteristics very similar to porous asphalt is several points less than that required for chippings for rolled asphalt of equal skid resistance. This would suggest that the use of aggregates having a PSV of about 56 to 57 would be satisfactory. However, this offers only modest scope for reducing cost, because there are scant sources of such aggregates.
- **Allow the use of 10 mm or 14 mm aggregate** This could reduce the costs in three ways:
  - by reducing the cost of aggregate, because 14 mm aggregate is sometimes significantly cheaper than 20 mm material;
  - by allowing porous asphalt to be laid 35–40 mm thick instead of 50 mm; and
  - by facilitating compliance with specification because 10 mm and 14 mm materials segregate less than 20 mm material during laying.
- **Ease the specification for hydraulic conductivity** (BSI, 1996a) The specified values are high and are difficult to attain reliably at all test positions owing to the variability of the material due to segregation. The use of 10 mm or 14 mm material would ease this problem, but for 20 mm material a greater allowance is needed for variation from position to position. Further, it appears unnecessary to specify such high values of hydraulic conductivity because the reduction of spray is not the prime reason for using porous asphalt. The incongruity is exaggerated because there is evidence that hydraulic conductivity is not well correlated with spray reduction (Daines, 1992; Nicholls, 1997) and that significant spray reduction is obtained even when clogging with detritus has reduced the hydraulic conductivity to zero.

- **Allow single-lane laying and single-lane remedial work** The normal requirement to avoid longitudinal joints is based on a fallacy. In such an open material, a cut and 'painted' joint will not form a complete barrier to transverse flow. Even well-painted joints do not extend to the bottom of the layer where the water flows. Furthermore, a painted joint is unnecessary anyway because it has been found that a new swathe of this binder-rich material will join satisfactorily with the uncut edge of a previously laid swathe.

In summary, clients could bring about a significant reduction in cost by reducing the specification for hydraulic conductivity, by reducing the permitted aggregate sizes and layer thickness and generally reducing the anxiety of the contractor about the possibility of having material rejected.

Further reductions in cost will be possible if the current review of PSV requirements brings about more aggregate sources for the manufacture of porous asphalt.

7.4.4.2 Airfields

Porous asphalt has its origins in the development in the United Kingdom in the mid 1950s of a material which became known as 'airfield friction course macadam'. It was developed to prevent aircraft from aquaplaning and skidding on landing, and came into common use on both civil and military airfields in 1962. The maximum particle size of the aggregate was limited to 10 mm in order to minimize any damage from loose particles being ingested by jets (known as 'foreign object damage' or FOD). A layer thickness of 25 mm is normally employed.

It is interesting to note that, whereas noise reduction is now the main incentive for the use of porous asphalt on highways, it was of no consequence in its development and use on airfields. Further, spray reduction is of no importance on airfields because aircraft are well-spaced on landing.

7.4.5 Overseas

The use of porous asphalt in continental Europe varies considerably from country to country, but is generally at a much higher level than in the United Kingdom. Data for accumulated use are not available, but tonnages manufactured annually in European countries are collated by the European Asphalt Pavement Association (EAPA). The figures for 1995 are given in Table 7.1.

In **Austria**, much of the usage of porous asphalt is on trunk roads in valleys, where roads and residents are close together, and the disruption to life by traffic noise was greatly reduced by the use of porous asphalt. It has proved to very durable under very heavy traffic. Polymer-modified binders are employed almost exclusively. It is common practice to clean

Table 7.1 Estimated usage of porous asphalt in Europe to end of 1995

Country	Porous asphalt laid	Country	Porous asphalt laid
Austria	c. 8.0×10^6 m²	Italy	$>9.0 \times 10^6$ m²
Belgium	c. 1.0×10^6 m²	Netherlands	25.0×10^6 m²
Czech Republic	1.3×10^6 m²	Portugal	c. 0.4×10^6 m²
Denmark	0.2×10^6 m²	Spain	$>31.0 \times 10^6$ m²
France	25.0×10^6 m²	Sweden	0.1×10^6 m²
Germany	c. 0.8×10^6 m²	Switzerland	c. 3.9×10^6 m²
Hungary	0.1×10^6 m²	UK	$<1.0 \times 10^6$ m²

porous asphalt by pressure washing when the voids content has fallen by about 8%.

In **France,** porous asphalt has proved to very durable under very heavy traffic. Polymer-modified binders are employed almost exclusively.

In **Italy,** much of the porous asphalt in use is on toll motorways built by the private sector. This is a very significant demonstration of faith in the engineering and economic viability of porous asphalt. Again, modified binders are normally employed.

In the **Netherlands,** since 1988 all new or reconstructed trunk roads have been surfaced with porous asphalt. Porous asphalt remains in service after 10 years on some very heavily trafficked sections of road. Previously, porous asphalt had normally been manufactured with added fibres because of concern that modified bitumens may not be recyclable. However, these fears were resolved in 1996, and it now seems likely that modified binders will be commonly employed in the future. There are no upper limits to the traffic permitted on porous asphalt surfacings in Austria, France or the Netherlands.

Porous asphalt has also been used outside Europe, notably for airfields in the United States of America and on highways in Hong Kong. There is also considerable interest in the material in South East Asia and significant use in this region is anticipated in the near future.

7.5 COMPONENT MATERIALS

7.5.1 Aggregates and filler

7.5.1.1 General

The requirement for a high voids content means that the aggregate grading for porous asphalt must be relatively 'single sized'. Thus, traffic loadings are borne by the skeleton of the largest particles of aggregate, with very little support from the small amount of bitumen/fines/filler mortar. Therefore, the coarse aggregate must be resistant to both crushing and abrasion.

7.5.1.2 Coarse aggregate

For trunk roads and motorways in the United Kingdom, Clause 938 of the *Specification for Highway Works* (MCHW 1) allows the use of crushed rock or steel slag (but not gravel, whether crushed or not) possessing the following minimum properties, when tested according to the appropriate parts of BS 812:

10% Fines Value (section 2.2.3.2)	$\geq$ 180 kN	(dry);
Aggregate Abrasion Value (section 2.2.4.3)	$\geq$ 12; and	
Flakiness Index (section 2.2.1.3)	$\geq$ 25%	(mean of values for 20 and 14 mm sizes for 0 to 20 mm grading).

The polished stone value (PSV) (section 2.2.4.4) requirement for the coarse aggregate will depend on the nature of the particular site but, because porous asphalt is predominantly employed on busy high-speed roads, values less than 55 are rarely specified.

7.5.1.3 Fine aggregate

Clause 938 of the *Specification for Highway Works* (MCHW 1) for porous asphalt requires compliance with Clause 901, which permits the use of crushed rock or steel slag or natural sand or a blend thereof.

7.5.1.4 Filler

In an asphalt which is as permeable as porous asphalt, a strong resistance to stripping is clearly necessary. To ensure this, it is common practice to specify the inclusion of a minimum amount of an 'active' filler, for which hydrated lime is commonly employed. Clause 938 of the *Specification for Highway Works* (MCHW 1) requires a minimum of 2% of the total aggregate to be hydrated lime.

7.5.2 Binders

7.5.2.1 Unmodified binders

Bitumens are normally specified by penetration grade in the range 100 to 200.

With such an open asphalt, oxidation of the bitumen, leading eventually to embrittlement and failure, appears to be the main life-determining factor. It is considered that harder binders than 100 pen will reach the critical brittle state sooner than softer ones. Road trials in the United Kingdom (Nicholls, 1997) provide some evidence that this is the case. On the other hand, porous asphalt made with binders softer than 200 pen is likely to have insufficient cohesion and resistance to deformation.

In order to delay the onset of embrittlement, porous asphalt employs higher binder contents than would be the case for a 'normal' macadam having a similar grading. However, there is a natural maximum binder content that any particular aggregate blend will 'hold' without drainage. This drainage is due to the fact that unmodified bitumen has a zero yield stress, so that it will flow under its own weight and drain off the aggregate after coating, leaving only a thin film held on to the aggregate by physico-chemical forces. Consequently, it is common practice to take specific steps to increase the binder content above the natural maximum. These involve the use of either modified binders or fibres.

7.5.2.2 Modified binders

Although Clause 938 of the *Specification for Highway Works* (MCHW 1) refers to fibres as modifiers, they do not actually modify the properties of the bitumen (section 3.5.5). The draft CEN Terminology has been adopted in this publication, in which modified bitumen is defined as 'a bitumen whose rheological properties have been modified during manufacture by the use of a chemical agent'. An explanatory note specifically states that 'fibres and inorganic powders ("fillers") are not considered to be bitumen modifiers'.

Modifiers are employed to alter the rheological properties of bitumens in such a way that a thick film of binder will stay on the aggregate surface without draining under its own weight. Their primary role is to structure the bitumen so as to impart a yield stress, and thereby reduce the tendency for the binder to drain. The theory is that thick binder films will take longer to harden to the point of embrittlement than thin ones. There is now considerable evidence that this theory is sound, as porous asphalts having high binder contents have proved to be more durable than counterparts made with lower binder contents typical of normal macadams (Nicholls, 1997).

The modifiers employed all impart some degree of rubbery character to the bitumen. Natural (unvulcanized) rubber has been used in many of the trials conducted in the United Kingdom, and it has performed well under trial conditions. The rubber can be added either as a latex or as a powder (crumb rubber). However, it is little used commercially owing to its thermal instability: natural rubber has a tendency to break down at temperatures necessary for storage, coating, transport and laying of many asphalt mixtures. To overcome these problems, pre-blended binders containing thermally-stable polymers have been developed (section 3.5). Most of the polymers concerned could be described as synthetic rubbers. The main exception is the family of ethylene-vinyl acetate (EVA) co-polymers and others of related composition. These are better known as plastomers than rubbers. However, certain variants having appropriate vinyl acetate contents and molecular weights possess suitable properties and are among the modifiers in use.

At the present time, there are no UK or European (CEN) specifications for modified binders, although work is proceeding on the development of CEN specifications. Nevertheless, modified binders are almost exclusively used in France, Austria and some other European states in the manufacture of porous asphalt.

In the United Kingdom, certain modified binders are approved for use on trunk roads and motorways by the overseeing organization on a project by project basis, primarily on the basis of performance in trials. Paragraph 7 of Clause 938 of the *Specification for Highway Works* (MCHW 1) states that:

> Binder modifiers, including natural or man-made fibres, natural lake asphalt, natural rubber in latex or powder form, synthetic rubber, EVA or other similar materials, may be permitted, but only with agreement of the Engineer. Modifiers include any material added to or blended with the base bitumen.

A more general *Highways Authorities Products Approval Scheme* (HAPAS) (section 1.5.3) is in the process of being developed. A major component of this scheme is modified binders for porous asphalt (as well as for other applications).

7.5.2.3 Fibre additives

Certain organic and inorganic fibres are employed as alternatives to modified binders to increase the soluble binder content of porous asphalt. Both types have been tested in trials in the United Kingdom (Daines, 1992), and it was found that they were both effective in increasing the binder content of porous asphalt, although the durability of porous asphalt made with inorganic fibres was superior to that made with organic fibres.

In the case of the organic (cellulose) fibres, some of the binder will be absorbed into the fibres. Where the organic material is derived from reduced paper, a considerable amount of the additive is dust derived from non-fibre parts of the paper, which will act as additional filler and thereby impede binder drainage. It is questionable as to whether this form of binder 'consumption' increases the binder content as meaningfully as the inorganic fibres do. The limited evidence of the UK road trials (Nicholls, 1997) suggests that it does not.

Another consideration is the volume occupied by the fibre, because it will reduce the air voids content of the porous asphalt compared with the use of a modified binder. In the case of the paper-based additives, the use of 0.3% by mass of the total mixture will reduce the air voids content by approximately 1%, whereas the use of an inorganic fibre additive at 0.9% by mass of the binder will reduce the air voids by only 0.4%.

Fibre additives are commonly employed in the porous asphalt mixtures used in Holland. These are most frequently made with 0–14 mm aggregate, but smaller sizes, especially 0–10 mm, are also used. The 0–20 mm grading used in the United Kingdom is not used in Holland. Hitherto, the use of fibres rather than modified bitumens in Holland was based on concerns over the recyclability of the latter. These concerns have been resolved, and it now seems likely that Holland will follow the practice in most other countries of using modified bitumens.

7.6 MANUFACTURE, QUALITY CONTROL AND LAYING

Porous asphalt may be manufactured in either batch or continuous plants. Whichever is used, very close control of the aggregate grading and binder content is necessary in order to meet the 'tight' limits of the specification. Close control of mixing temperature is also required: it must be high enough to ensure that the aggregate is well-dried and well-coated but not so high that significant binder drainage occurs. The high-PSV aggregates required in the United Kingdom may fracture significantly in passing through the plant. Any consequential change of grading must be allowed for in order to ensure that the final product is within the accepted tolerances.

When porous asphalt is being manufactured, the mixing plant must be dedicated solely to that purpose (section 7.4.4.1) in order to ensure that a consistent material is produced which complies with the specification. In order to ensure that the narrow margins of the specification are met, it is also necessary for the manufacturer to employ high sampling and testing rates at the mixing plant and to make rapid adjustments when necessary. It is now common practice in the United Kingdom to take the main samples for analysis from lorries at the mixing plant because of the difficulty of obtaining representative samples of porous asphalt with 20 mm nominal aggregate at the laying site. Further samples are also taken at the laying sites, but less stringent tolerances are applied to them.

During laying, the client normally requires longitudinal joints to be avoided, to facilitate the passage of water through the finished layer. In order to achieve this when more than one lane is being surfaced, it is necessary to employ multiple pavers working in echelon with a maximum stagger of 20 m.

Care should be taken to avoid damaging the skeleton of coarse aggregate by over-rolling.

7.7 TACK COATS

A tack coat is generally required for porous asphalt both:

- with the intention of ensuring that there is an impermeable layer below the porous asphalt layer to stop water ingress into lower layers; and
- to improve the adhesion to the underlying layer, with which contact is not continuous because of the voids.

Varieties

The tack coat is usually a K1-70 emulsion in the United Kingdom, sprayed at a rate of at least 0.4 L/m². Currently, the tack coat is usually not modified.

7.8 VARIETIES

7.8.1 20 mm porous asphalt

While both 20 mm and 10 mm porous asphalts are specified in BS 4987 (BSI, 1993), at the time of writing the 20 mm grading is the only size of porous asphalt specified in the *Specification for Highway Works* (MCHW 1); this grading is not used in other European countries. The details of the mixture are given in Clause 938 of the *Specification for Highway Works* (MCHW 1); for full information, reference should also be made to the associated *Notes for Guidance* (MCHW 2) and advice in Chapter 5 of HD 27/94 (DMRB 7.2.4) in the *Design Manual for Roads and Bridges*. The salient features of the specification are as follows:

- it is essentially a recipe-type specification, with some prescriptive parts;
- the requirements for aggregates and binders are as discussed in sections 7.5.1 and 7.5.2, respectively;

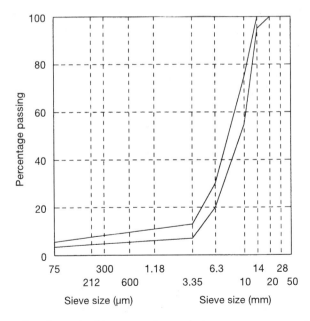

Fig. 7.12 Grading limits for 20 mm porous asphalt

- the aggregate grading (Figure 7.12) is a slightly modified form of the 20 mm open graded binder course macadam specified in BS 4987 (BSI, 1993);
- hydrated lime comprises a minimum of 2% of the total aggregate (section 2.5);
- the target binder content must be determined by the binder drainage test (BSI, 1996b) (a test in which the quantity of binder lost through drainage after three hours at an elevated test temperature is measured for mixtures with the same aggregate contents but with different binder contents; from the results, the maximum target binder content is determined as the value of mixed binder content 0.3% less than that at which 0.3% of the binder drains) with a minimum value of 4.5% mass being normally required;
- the maximum mixing temperatures are specified for all permitted binders in order to ensure that efficient coating takes place with a minimum of binder hardening or drainage;
- the minimum paver discharge temperature and the temperature(s) for substantial completion of compaction are also specified, for obvious reasons;
- the use of double-sheeted insulated vehicles is mandatory in order to guard against excessive cooling during transporting;
- a tack coat of K1-70 emulsion, sprayed at 0.4 L/m^2 minimum, is specified;
- the specified layer thickness is (50 ± 5) mm; and
- a minimum value of relative hydraulic conductivity (BSI, 1996a), measured in the near side wheel path of each lane before trafficking, is required.

7.8.2 10 mm, 14 mm and 16 mm porous asphalts

The use of 20 mm porous asphalt is limited to the United Kingdom; in continental Europe, a wide range of national specifications are in use. All of these employ smaller aggregate sizes than 20 mm, the commonest being 14 mm. Some variation, such as 16 mm, is due to the different sieve sizes currently in use in different countries. However, in France, one of the major users of porous asphalt, a 10 mm material is most widely used, apparently with satisfaction.

All of the mixtures based on these aggregate sizes are capable of having initial voids contents similar to those of the UK 20 mm mixtures at around 20%. However, the maximum pore sizes of the continental mixtures will be smaller than those of the UK 20 mm mixture. There is a view that such mixtures, as well as increasing the surface tension of the water passing through them, will become clogged more quickly than the UK formulation. However, there are other considerations, such as:

- clogging with silt may be related more to the finer voids than to the largest ones;
- larger pieces of detritus will be able to enter the larger surface voids of the 20 mm mixture; and
- the amount of fine and coarse detritus to which porous asphalt may be exposed is extremely variable.

There is limited evidence on which to base a general conclusion. In the United Kingdom, a trial section on the M1 motorway near Wakefield was replaced after two years (Nicholls, 1997) on the grounds of low hydraulic conductivity, that is it had become badly clogged. An adjacent section of 20 mm porous asphalt was still satisfactory. However, rapid clogging of 10 mm material is not normally experienced in France, and the reason for this disparity is not known. The failure has been ascribed to using a grading developed for 20 mm thick airfield pavements at double that thickness without checking for its suitability at that depth.

These continental mixtures are generally used in layers thinner than the 50 mm required in the United Kingdom, typically 35 mm to 40 mm for a 14 mm material. While this gives an immediate cost advantage, it provides a smaller water reservoir and thereby increase the risk that the layer will be flooded by prolonged heavy rain. However, even when flooded, porous asphalt which is still porous will reduce the risk of aquaplaning by dissipating pressure in the hydraulic wedge in front of the tyre. In any case, porous asphalt is no worse than a conventional dense surfacing with respect to surface water, and it still retains much of its noise-reducing property.

Porous asphalt with 10 mm nominal aggregate is widely used on airfields in 20 mm layers in the United Kingdom, where it is known as 'Friction Course', as well as other countries. The material has been used for this purpose successfully since the 1950s.

Generally, continental highway authorities and all airfield authorities have decided that the best balance of advantages and disadvantages lies with thinner layers than are used for UK highways, and this may be one of the reasons why porous asphalt is in greater use in many European countries than in the United Kingdom.

7.8.3 Very high void mixtures

A recent development in France and some other countries is to adopt aggregate gradings which result in materials having voids contents of about 30%, which is approaching the theoretical maximum for single sized tetrahedral particles. Such mixtures contain very small amounts of fine aggregate. Although conventional wisdom would suggest that such mixtures would have poor robustness and durability, their performance in trials in progress is very encouraging.

7.8.4 Two-layer systems

Trials have been carried out in Holland of a two-layer system of porous asphalt comprising:

- a 'reservoir' base layer, 55 mm thick, made with a 16 mm aggregate of undemanding quality; and
- a surface layer, 15 mm thick, made with an 8 mm high-quality aggregate.
  Both layers were manufactured using rubber-modified bitumen.
  The objectives of this system are:
- to provide a large reservoir for water in the base layer;
- to reduce the demand for high quality aggregate for the surface layer; and
- to provide a very quiet running surface.

Performance in these trial over five years (van Bochove, 1996) was excellent, and it was claimed that the system is ready for normal contract use.

The main characteristics of the system are:

- the fine (8 mm) surface is very quiet – quieter than conventional Dutch porous asphalt;
- the fine surface acts as a fine sieve, preventing ingress of coarse detritus so that the base layers remains quite clean;
- fine detritus is retained in the top layer, where it is accessible for complete removal using the 'Hydrovac' method; and
- the surface course can be removed and replaced when necessary without damaging the base layer.

7.9 PERFORMANCE IN SERVICE

Experience shows that porous asphalt is generally very resistant to rutting deformation.

The main difficulty is clogging with detritus, which reduces the spray-suppression. However, even when it is heavily clogged, considerable spray- and noise-suppression remains (Daines, 1992; Nicholls, 1997). The rate of clogging is very variable: a major factor affecting the rate of clogging is the amount of detritus prevalent, which is very variable – 10 mm and 14 mm varieties of porous asphalt in northern France and Holland have not become heavily clogged, which has been ascribed to the pumping action of heavy vehicles in wet wether. Some successful de-clogging of porous asphalt using a 'Hydrovac' machine has been carried out in Austria.

The durability of porous asphalt has been found to be considerably better than was expected. Long-term experience in the United Kingdom is very limited, but based on trials carried out in 1984, it was concluded (Daines, 1992) that:

Based on experience to date, 20 mm porous asphalt using 100 pen unmodified bitumen with 2% of hydrated lime as filler and a binder content of 3.7% will last in excess of 7 years under 4000 cvd with effective spray reduction. Porous asphalts with higher binder contents (which can be achieved with certain modifiers) will probably last significantly longer, but they will have reduced hydraulic conductivities.

In some continental countries, considerable amounts of porous asphalt have already been in service for more than ten years under unlimited heavy traffic. Thus, it is clear that the durability of porous asphalt is at least good enough to make it a viable surfacing material, with a service life comparable with that of dense asphalts.

7.10 RECYCLING

Owing to the relatively recent advent of porous asphalt, there is limited experience of recycling it. However, there have been verbal reports of successful plant recycling in Italy and the Netherlands. The recycling in Italy included material containing modified binder, which was recycled with no particular difficulty.

7.11 MAINTENANCE

7.11.1 Repairs

Pot holes and other localized damaged can be treated in one of two ways: by making an impervious plug or by replacing the damaged material by porous asphalt. In the former case, which is only appropriate for relatively small areas, the patch should be formed in a diamond pattern so that the rainwater can easily flow around it. Traditional rolled asphalt or dense bitumen macadam can then be used for the repair. If porous asphalt is being used as the replacement material, care must be taken to ensure that, in attempting to provide good adhesion between the old and new materials, the joint does not form a barrier to the passage of water. To this end, any tack coat on the vertical faces should be used sparingly.

Localized damage should be repaired as soon as practicable by routine maintenance to prevent such damage progressing. Routine measurements of texture depth are not generally necessary, but annual measurements in the spring on older porous asphalts using laser-sensor equipment may provide an early warning of failure by fretting. An increase in texture depth, particularly in the wheel-tracks during a cold winter period, may indicate that the binder condition is approaching the critical state. The porous asphalt will usually survive the following summer, but more-severe deterioration is likely during sub-zero temperatures the following winter.

Surface dressing and other veneer coats (Chapter 11) can only be used as maintenance treatments on porous asphalt if it is acceptable for the surfacing to become non-porous. If the noise- or spray-reducing properties need to be retained, then complete replacement will have to be considered.

7.11.2 Winter maintenance

7.11.2.1 Changes in procedures with porous asphalt

There are several differences in the characteristics of porous asphalt and conventional surfacing which affect 'winter maintenance', that is dealing with ice, snow, freezing rain, etc. The main differences in procedure are:

- **Freezing conditions after rain** In such conditions, there is less likely to be a serious problem with porous asphalt because there will be less water on the surface of the road.
- **Receptivity to salt** When salt is spread on porous asphalt for ice-prevention, a significant proportion will fall into the surface voids. This does not happen if salt is spread onto existing snow.
- **The use of grit** The use of grit must be avoided as it would fill voids in the porous asphalt.

7.11.2.2 Differences in thermal characteristics of porous asphalt and conventional dense surfacings

It has been found that the behaviour of porous asphalt differs from that of dense surface course materials in winter conditions. It is considered that there are four reasons for this (van der Zwan *et al.*, 1990):

- the roughness of the surface allows greater radiation to occur;
- the high voids content reduces the thermal conductivity of the material when dry;
- the interconnection between the voids allows the movement of fluids to take place; and
- the pores reduce the amount of de-icing material that is retained on the surface.

These points will be considered in turn.

The open nature of a porous asphalt surface increases the surface area exposed to the atmosphere, thereby allowing a greater exchange of heat with it, compared to a dense surfacing (Livet, 1989). Pronounced radiation of heat during the night leads to a greater and faster reduction in the surface temperature of the road in winter. (This increased surface area also leads to increased absorption of solar radiation in summer.)

The low thermal conductivity of porous asphalt slows the subsequent warming up in winter. Any resulting differences in temperature between porous asphalt and conventional surfacings will depend on particular circumstances; values of 0.5 °C (Koester, 1985) and 2 °C (Livet, 1989) have been reported.

The combination of increased radiation levels and reduced thermal conductivity implies that the critical temperatures for snow to settle or for ice to form will be reached sooner in the cooling phase and later in the warming phase on this material – that is, any hazardous conditions are likely to last longer. Measurements have shown that porous asphalt surfacings do stay below 0 °C for longer than conventional surface course surfacings, increasing problems with ice (van der Zwan et al., 1990). This difference will be reduced in wet conditions because the water in the voids will increase the thermal conductivity of the layer, bringing it closer to that of an impervious surfacing (Livet, 1989).

During cooling, in the absence of a freezing-point depressant such as salt, water will freeze at 0 °C. This releases latent heat, stabilizing the temperature at 0 °C until freezing is complete. During warming, the melting of ice will similarly absorb heat, stabilizing the temperature at 0 °C until melting is complete. Thus, the freezing and thawing of water will slow down the cooling and heating of the road surface and maintain it at 0 °C for a longer time than a conventional surfacing material (that is, the road surface will remain frozen for longer).

The above temperature-reducing effects are moderated by the heat-exchanging effects of the movement of air and water through the material (Livet, 1989). The effect is accentuated by the action of tyres successively pumping and sucking fluids into and out of the voids as they pass. The resulting constant circulation of air reduces the difference between the ambient air temperature and that of the pavement material (van der Zwan et al., 1990).

Because of the above difference in behaviour, and of the increased susceptibility of porous asphalt to damage, some modifications of normal procedures are required, as follows.

- Snow ploughs can be used successfully on porous asphalt, but a blade with a rubber front edge should be used to minimize damage to the material.
- The use of grit for winter maintenance is generally deprecated because it would reduce the porosity of the asphalt.
- De-icing salts when applied have to be used in greater quantities, although the extent of the increase varies. It has been suggested that, if a regime of regular applications of small doses is initiated, it should be effective with no overall increase in the quantity used. The reason for the additional salt is that, as well as the initial loss from salt falling into the pores, more is removed from the surface by the ice melting and the resulting solution draining away (van der Zwan et al., 1990).

There has been some use of brine, instead of salt, in the Netherlands. This proved quite successful, and applied less salt than when salt is used in the conventional manner. However, this practice has not become well-used because of logistical difficulties, such as the need to manufacture and store brine.

Critics of porous asphalt tend to regard its winter maintenance as presenting serious problems. However, these views are not shared in countries which employ considerable amounts of porous asphalt. Attitudes in such countries are indicated by comments such as 'If it is a problem, it is a small one which we can deal with' and 'We are not sure if it is a problem, but if it is, it is greatly outweighed by the advantages of porous asphalt.'

7.12 ECONOMICS

The cost of porous asphalt in the United Kingdom is currently higher than that for rolled asphalt or other surfacing material, both in terms of the material costs and the scheme costs. Some of the extra costs may be due to lack of familiarity with the material, but there are other contributory factors. The higher material cost is principally due to:

- the need for high PSV aggregate of 20 mm nominal size, which is a premium size because of its use for pre-coated chippings in rolled asphalt;
- the need for the premium aggregate through the 50 mm layer; and
- the tighter tolerances applied to the material in order to achieve the required porosity.

The reasons for the higher scheme costs include:

- the higher material cost;
- the need to provide positive drainage; and
- the need for extra total pavement thickness because of the relatively low structural strength.

Nevertheless, the environmental factors that are enhanced by the use of porous asphalt, not all of which can be calculated financially, means that the material has a wide group of supporters.

Notes

1. The author acknowledges and greatly appreciates the assistance of Cliff Nicholls in the preparation of this chapter, which has extended far beyond that required by his rôle of editor of the book.
2. $S = 270HC + 25SP$ where S = spray reduction relative to rolled asphalt; HC = relative hydraulic conductivity (s^{-1}); and SP = sand-patch texture depth (mm).

7.13 REFERENCES

Design Manual for Roads and Bridges, Her Majesty's Stationery Office, London.
  HD 27/94 – *Pavement Construction Methods* (DMRB 7.2.4)
Manual of Contract Documents for Highway Works, Her Majesty's Stationery Office, London.
  Volume 1: *Specification for Highway Works* (MCHW 1)
  Volume 2: *Notes for Guidance on the Specification for Highway Works* (MCHW 2)
British Standards Institution (1993) *Coated Macadam for Roads and Other Paved Areas; Part 1, Specification for Constituent Materials and for Mixtures; Part 2, Specification for Transport, Laying and Compaction.* BS 4987: Part 1: 1993, BS 4987: Part 2: 1993, British Standards Institution, London.
British Standards Institution (1996a) *Draft for Development: Method for Determination of the Relative Hydraulic Conductivity of Permeable Surfacings.* DD 229: 1996, British Standards Institution, London.
British Standards Institution (1996b) *Draft for Development: Method for Determination of the Maximum Binder Content of Bituminous Mixtures without Excessive Binder Drainage. DD 232: 1996.* British Standards Institution, London.
Colwill, D.M., G.J. Bowskill, J.C. Nicholls and M.E. Daines (1993) Porous asphalt trials in the UK, *Transportation Research Record 1427*, Transportation Research Board, National Research Council, Washington, DC.
Daines, M.E. (1986) *Pervious Macadam: Trials on Trunk Road A38 Burton Bypass, 1984*, Department of Transport TRRL Report RR 57, Transport and Road Research Laboratory, Crowthorne.
Daines, M.E. (1992) *Trials of Porous Asphalt and Rolled Asphalt on the A38 at Burton*, Department of Transport TRRL Report RR 323, Transport and Road Research Laboratory, Crowthorne.
Fabb, T.R.J. (1993) The case for the use of porous asphalt in the UK, *The Asphalt Year Book 1993*, Institute of Asphalt Technology, Staines, 46–58.
Jacobs, F.A. (1983) *M40 High Wycombe By-pass: Results of a Bituminous Surface-Texture Experiment*, Department of the Environment Department of Transport TRRL Report LR 1065, Transport and Road Research Laboratory, Crowthorne.
Koester, H. (1985) Porous wearing courses: observations of the behaviour under traffic, *3rd Eurobitume Symposium: Bitumen, Flexible and Durable, Volume 1*, The Hague, 512–17.
Livet, J (1989) *Les Enrobés Drainants et la Modification du Régime Thermo-Hydrique de la Surfaces des Chaussées* (Pervious Bitumen Coated Materials and Changes to the Hydro-thermal Behaviour of the Road Surfaces), Laboratoire Régional des Ponts et Chaussées de Nancy.
Nelson, P.M. and P.G. Abbott (1990) Acoustical performance of pervious macadam surfaces on high speed roads, *Transportation Research Record 1265*, Transportation Research Board, National Research Council, Washington, DC.
Nicholls, J.C. (1997) *Review of UK Porous Asphalt Trials*, TRL Report 264, Transport Research Laboratory, Crowthorne.
Nicholls, J.C. and M.E.Daines (1992) Spray suppression by porous asphalt, *The Second International Symposium on Road Surface Characteristics*, Berlin.
Permanent International Association of Road Congresses (1993) *Porous Asphalt.*
Please, A., B.J. O'Connell and B.F. Buglass (1972) *A Bituminous Surface-Texture Experiment, High Wycombe By-pass (M40)*, Ministry of Transport RRL Report LR 307, Road Research Laboratory, Crowthorne.

Roe, P.G., L.W. Tubey and G. West (1988) *Surface Texture Depth Measurements on Some British Roads*, Department of Transport TRRL Report RR 143, Transport and Road Research Laboratory, Crowthorne.

Roe, P.G., D.C. Webster and G. West (1991) *The Relation between the Surface Texture of Roads and Accidents*, Department of Transport TRRL Report RR 296, Transport and Road Research Laboratory, Crowthorne.

van Bochove, G. (1996) Twinlay, a new concept for porous asphalt, *Session 7, Innovations – Road Surfaces, Eurasphalt and Eurobitume Congress 1996*, Strasburg.

van der Zwan, J.T., T. Goeman, H.J.A.J. Gruis, J.H. Swart and R.H. Oldenburger (1990) *Porous Asphalt Wearing Courses in the Netherlands – A State of the Art Review*, Transportation Research Board 69th Annual Meeting, Paper No 890174, Washington DC.

CHAPTER 8

Mastic asphalt (and guss-asphalt) surface courses

I. J. Dussek, Wells (Trinidad Lake Asphalt) Ltd

8.1 CONCEPT AND HISTORY

The concept of mastis asphalt gussasphalt can be explained simply by the translation of the latter as 'poured asphalt'. It is thus, unlike most other surfacing asphalts, a material which is placed by means of its own mass, with a minimum need, theoretically, for compactive effort.

The basic components of the material are a binder mixed with an aggregate which is in powdered or finely crushed form. The two are normally heated and mixed together to form a pudding or porridge-like product, capable of being poured and trowelled or screeded out to level. Two principal product groups are:

- **mastic asphalt**, associated with British, French and Mediterranean practice, which tends to be voidless and, as described above, of the consistency of a pudding; and
- **gussasphalt**, associated with Germany, Northern Europe and Scandinavian countries, which relies on a graded aggregate structure but which, nonetheless, flows into place, albeit assisted by compaction.

Gussasphalt for road surfacing was developed principally by German engineers in the 1950s and was associated initially with the restoration of the Autobahn system constructed 20 years previously. However, it shares a common history with mastic asphalt, the oldest of all asphalt materials which dates back to biblical times and originated in the Middle East. Mastic asphalt was in use as early as the Fourth Millennium BC for a variety of purposes in connection with roads, building, waterproofing and sanitation (Forbes, 1936). The modern European mastic asphalt industry dates from the exploitation of the Neuchatel Rock Asphalt deposit in the Val de Travers, Switzerland (D'Eyrinis, 1721), although its use as a road surfacing material was developed nearly a century later through the chance observation that the lumps of bitumen-bearing rock, when crushed by wagon wheels, produced a surface course.

During the nineteenth century, the problems of horse-drawn traffic and horse effluent (section 1.1) resulted in the development of mastic asphalt suitable for city paving (Malo, 1898). The principal 'Natural Rock Asphalt' materials were supplied from European deposits, such as Neuchatel in Switzerland, St Jean in France, Limmer in Germany and Ragusa in Sicily; in 1859 these were supplemented by the introduction of Trinidad lake asphalt (section 3.3) and, subsequently, by the production of bitumen (section 3.4) as a by-product of the emerging petroleum industry. These latter two binders could be mixed with limestone and other powdered aggregates to produce a similar product to the naturally occurring material.

The use of mastic asphalt for very heavy duty pavements, notably in the United Kingdom in the cities of London and Liverpool, continued for a century but the slowness of output and labour intensive nature of its application, a craft skill based on apprenticeship, put limitations on its use. Rising labour cost and resistance to the introduction of mechanization did much to bring about its general demise in roadworks with the exception of such applications as guttering and footpaths. Where mechanization has been introduced (Figures 8.1 and 8.2), the cost of the specialist equipment has proved an obstacle and the availability of alternative heavy-duty asphalts and other materials has diminished its use throughout the construction industry.

Fig. 8.1 Machine laid mastic asphalt with a stirrer/cooker backing up to the paver, which is heavily shrouded to retain the temperature

Properties and applications 221

Fig. 8.2 Machine laid mastic asphalt in which the kneeling operative is sealing the longitudinal joint prior to application of the coated chippings

8.2 PROPERTIES AND APPLICATIONS

The design range of construction-related gussasphalts and mastic asphalts is very wide. These can be formulated to meet many requirements, based on the positive properties of this group of materials described in sections 8.2.1 to 8.2.2. However, there are several problems relating to the use of gussasphalt and mastic asphalt, as described in sections 8.2.3 to 8.2.5.

8.2.1 Waterproofing characteristics

The waterproofing characteristics were originally related to roofing, tankings and flooring applications, but they extended into paving work as water and corrosive chemical resistant protective layers on structures such as concrete bridge decks. For that situation, the waterproofing layer, which is typically 20 mm thick, must also be resistant to puncturing, particularly by traffic during construction of the structure, which will be the case with such bitumen-rich mixtures.

8.2.2 Durability

This group of asphalts have been shown to have exceptional life expectancy, particularly where they are applied in an enclosed situation such as a waterproof membrane between constructional or surfacing materials. In such situations, the hardening characteristics of the binder tend to be greatly inhibited; in such circumstances, the theoretical life of the asphalt is virtually endless. In exposed situations, such as on a roof, the voidless nature of a well-designed mastic asphalt should result in a life expectancy of up to a century, all other factors being equal.

Mechanical resistance also plays a major part in the assessment of durability. Clearly, a blend of binder and limestone powder would not be expected to provide mechanical strength, but the addition of hard road-stone, such as granite chippings, within the mixture or coated chippings on the surface will provide the properties required.

Steel bridge deck surfacings with a thickness of 38 mm have been shown to have a typical life expectancy of 20–30 years.

In the case of the gussasphalt produced to German designs, the mechanical strength of the mixture is achieved by means of the aggregate grading rather than the 'plum pudding' traditional mastic design. Gussasphalt surfaces have been shown to be effective in Northern Europe, especially where studded tyres have been permitted. In such instances, the wear is particularly intense and experience under heavy-duty conditions in Helsinki is that, whereas asphalt concrete has a life expectancy of three years, gussasphalt lasts for at least five years. It is estimated that some 45% of all existing autobahn surfaces in Germany are surfaced in gussasphalt and that many of them are 30 years old or more.

8.2.3 Cost

In many instances, the cost of material supplied and laid is comparatively high. Nevertheless, examinations of the whole-life costing frequently show compensation for the increase in initial cost, particularly where maintenance and re-surfacing present technical, political and environmental difficulties.

8.2.4 Manufacturing and application equipment and skills

The gussasphalt concept is different from that of other road surfacing materials. Mastic asphalt production tends to be low volume, from as little as 10 tonnes to 150 tonnes per day. At the lower end, this is sufficient for a gang laying by hand, but for machines capable of laying 100 tonnes or more per day, extensive modifications to standard road pavers are necessary.

For German gussasphalt, modified asphalt concrete plants are used, but the production rate is slower at approximately 1000 tonnes per day. Also, only a small proportion of German contractors have the necessary special paving trains capable of laying such large outputs.

Fig. 8.3 Gussasphalt texture (the coin diameter is 27 mm)

8.2.5 Surface texture

Because of the need to provide texture to what would otherwise be smooth and potentially slippery surfaces, gussasphalt is treated by means of small chippings (Splitt) rolled in using rollers. This gussasphalt texture (Figure 8.3) tends to be noisier than asphalt concrete, typically by 2 dB(A), but experimental work has shown that the use of 1 to 3 mm chippings provides a quieter ride.

8.3 COMPONENT MATERIALS

8.3.1 Aggregates and filler

8.3.1.1 Mastic asphalt

The principal component of mastic asphalt is limestone rock which is ground typically to the fine aggregate specification in Table 8.1.

The tests for the compliance are defined in BS 6925 (BSI, 1988b).

The coarse aggregate is defined as clean igneous, calcareous or silicious based aggregates passing a 14 mm sieve, retained on a 10 mm sieve and with an aggregate crushing value of not greater than 28.

The coated chippings for mastic asphalt are 20 mm or 14 mm nominal size with a flakiness index (section 2.2.1.3) not exceeding 25 and are typi-

Table 8.1 Fine aggregate specification for mastic asphalt

BS sieve	Proportion by mass of total aggregate retained on BS test sieve (%)	
	Minimum	Maximum
2.36 mm	0	3
600 μm	5	25
212 μm	10	30
75 μm	10	30
Passing 75 μm	45	55

cally granite or whinstone. Depending on the particular circumstances, a polished stone value (PSV) (section 2.2.4.4) may be specified. The chippings, for application purposes and especially to ensure they are firmly bonded in place, are pre-treated by coating them with 1.6 to 2.4% of 50 pen bitumen, applied in a suitable mixing plant at a maximum temperature of 185 °C.

8.3.1.2 Gussasphalt

The fine aggregate for gussasphalt is normally of limestone origin. The aggregate is principally 0 to 2 mm sand (replacing the larger end of the mastic fine grading). The sand particle shape is not as critical as for other asphalt mixtures. However, crushed sand is preferred, but both crushed and rounded materials are acceptable, usually blended in equal proportions.

The coarse aggregate is crushed basalt to the grading in Table 8.2.

The chippings for gussasphalt are typically 2 to 5 mm or 5 to 8 mm crushed basalt pre-coated with 0.8% binder spread at the rate of 15 to 18 kg/m².

8.3.2 Binders

8.3.2.1 Bitumen

The use of bitumen as the binder in both gussasphalt and mastic formulations is widespread, either on its own or in association with Trinidad lake

Table 8.2 Coarse aggregate specification for mastic asphalt

Sieve	Proportion by mass of total aggregate (%)
2–5 mm	22
5–8 mm	46
8–12 mm	32

asphalt. Choice of bitumen, particularly for mastic asphalt is comparatively more critical and some sources have been identified as being unsuitable for mastic production.

8.3.2.2 Trinidad lake asphalt

The rationale for the use of lake asphalt in mastic work is complex (Broome, undated). The consistency of the material, both chemically and physically, was advantageous in the manufacture of products made from natural rock asphalt, enabling natural variants to be adjusted. The composition of lake asphalt includes 36–37% porcellainite (silica) filler of which over 80% is below 60 μm particle size with 50% of this being below 10 μm. The filler has been blended in naturally over many thousands of years with the bitumen in the lake asphalt and is thus a 'premix' in the binder/aggregate blending process. Its primary use, however, in both gussasphalt and mastic surfacings is that it provides added mixture stability without the consequent disadvantages of low temperature cracking normally associated with hard grades of bitumen. Blends of bitumen and lake asphalt vary in proportion and, unlike some other binders, can be adjusted infinitely.

The binders in Table 8.3 are representative of those used.

8.3.3 Binder modifiers

8.3.3.1 Gilsonite

Gilsonite is a very hard, naturally occurring asphaltite mined in Utah and supplied in crushed form as an additive for increasing the hardness of the mixture. Gilsonite has a ring and ball softening point range of 150 °C to 200 °C and is principally used for increasing the hardness of flooring mixtures. Other similar materials include Manjak and Glance Pitch.

8.3.3.2 Polymers

The use of polymer-modified bitumen of many varieties, using both elastomers and thermoplastics, has been promoted in recent years and has

Table 8.3 Binders used in gussaphalt and mastic asphalt

Material	Bitumen content (%)	Lake asphalt content (%)	Penetration of blend @ 25 °C (dmm)
Mastic asphalt:			
Bridge deck	30–50	50–70	13–20
Paving	50	50	15–25
Gussasphalt	70–75	25–30	20–30

been shown to provide better low-temperature strain performance. However, the long-term performance of the polymer group of chemical is, as yet, not proven and, at 200 °C, both EPDM and SBS (section 3.5.2), two popular polymers, have been shown to reach their half lives within one minute (Kopsch, 1992). Oxidation inhibitors have been proposed. The use of polymer/lake asphalt binders for bridge surfacings has been successfully developed in Switzerland and Germany (Angst, 1992).

8.4 MASTIC ASPHALT

8.4.1 Manufacture

The traditional method of manufacture is by means of materials being fed in batches into a heated mixer with slowly rotating horizontal or vertical axis paddles. Such mixers may have a capacity of 1–8 tonnes and the mixing temperature is 200–230 °C. Once mixed thoroughly, after say five hours, the contents are allowed to pour out into moulds, producing 25 kg blocks. The blocks are tested and then taken to site for re-melting in mobile mixers. If necessary, grit is added at this stage.

More recently, the introduction of German plant has led to a more rapid process whereby the mastic asphalt is pre-mixed in a stirred tank (usually mounted at a height suitable for feeding trucks) or in a modified paving plant and is dropped into a mobile stirrer/cooker, together with grit as specified. Mixing continues en route to the site at which the asphalt is fed via mobile barrows or small insulated transporters onto the job; alternatively, the stirrer/cookers feed directly into the paving plant.

Where large quantities of gritted material based on blocks are required, such as on bridge decks, production can be accelerated by running the grit through an aggregate heater/drier.

Many sophisticated production methods have been produced over the years, but the traditional outlook and diminishing size of the industry has strictly limited development expenditure.

The principal test methods relating to mastic asphalt production include:

- binder penetration and softening point;
- aggregate grading and chemical type; and
- hardness number.

The hardness number is a measure of indentation by a given load at a given temperature for a given time through a flat-ended indentor pin of 6.35 mm diameter into a sample of mastic asphalt as made and the gritted materials as made. The test is defined in BS 5284 (BSI, 1993), in which a load of 311 N, giving a stress of 9.8 MN/m^2, is applied at 35 °C for 60 seconds.

8.4.2 Applications

The use of mastic asphalt as either a surfacing or waterproofing layer (or a combination of both) has a long history.

8.4.2.1 Waterproofing

The need to protect concrete structures, such as bridges, is important if damage, particularly to the reinforcement, by water or corrosive chemicals (including de-icing materials) is to be avoided. Mastic asphalt manufactured to BS 6925 (BSI, 1988b) is a well-proven means of protecting the deck. Once in place, it is overlaid with, say, a rolled asphalt surface course with or without a protective sand carpet layer.

The application procedure is straightforward. The surface of the concrete is swept clean and the waterproofing operation should be carried out under dry conditions. No preparation, such as scabbling or tack coating, is necessary for the concrete. The mastic asphalt is prepared in cookers (either by 'hot charge' or 'block') and laid in bays of approximately 3 m width. The material is laid in two courses of nominal 10 mm thickness, with the working bay joints overlapping. The application technique is governed by the laying of hot asphalt on cold, and normally damp, concrete. The result is the creation of bubbles of steam or 'blows' which permeate through the surface of the mastic asphalt. These have to be punctured, releasing the steam, by the mastic spreaders and then sealed up using trowelling techniques.

If possible, the mastic asphalt waterproofing layer should extend across the total width of the structure, as well as below kerbs and other details. Where this is not possible, care must be taken to ensure a complete seal is effected. In many cases, hot poured jointing compound or a bitumen based pre-formed strip system (such as Tokband) has been used.

The advantages claimed for the use of mastic asphalt waterproofing are:

- its thickness reduces problems arising from the puncturing of the layer by aggregate particles and from imperfections of the concrete surface; and
- once cooled, frequently within an hour or two, it can be trafficked by site vehicles prior to the application of the asphalt overlay – in many instances, the waterproofing is open to traffic for several weeks.

As mentioned above, care must be taken to seal 'blows' which, if not properly treated, can re-occur at a later date, particularly in hot weather. For this reason and due to the hand-laying process (Figure 8.4), mastic asphalt waterproofing is comparatively slow.

8.4.2.2 Tunnels and concrete bridge decks

Mastic asphalt provides a high-durability thin-section surfacing, which is of particular advantage in tunnels, where headroom restrictions are

Fig. 8.4 Hand-laid mastic asphalt on a bascule lifting bridge deck, hence the chevron ribs and studs to enhance bond

frequently encountered and where access maintenance activities are limited. The same conditions apply to mastic asphalt used on concrete bridge decks and similar structures.

Usually, mastic asphalt manufactured to BS 1447 (BSI, 1988a) is laid in a single course of 35–38 mm thickness directly on the concrete. Unlike the waterproofing grades, paving mastic asphalt is reinforced by the addition of suitable igneous or calcareous 10 mm grit, forming approximately 45% by mass on the final analysis. The resulting material has a high stability capable of resisting the deformation caused by the wheel-tracking of channelized vehicles in tunnel conditions which limit the weaving action obtained on normal road surfaces.

Typically, the mastic asphalt is laid in 3 m bays but, because it is a single course, particular attention must be paid to the working joints which, when properly carried out, are 'self-welding'. Pre-coated 20 mm chippings are rolled into the mastic asphalt at a rate of 11–12.5 kg/m^2.

8.4.2.3 Steel bridge decks

The use of mastic asphalt for the surfacing of steel (especially suspension) bridge decks came about in the United Kingdom following extensive research by the then Road Research Laboratory in the early 1950s. The need was to find a surfacing system which:

Mastic asphalt

- would bond to the steel deck and protect it;
- would move with the steel without problems of fatigue;
- would take heavy channelized traffic for many years; and
- was light.

Many materials were evaluated, but the design embodied into the Forth Road Bridge Construction in 1964 consisted of:

- a coat of solvent-based primer (Bostik 1255);
- a 3 mm layer of rubber-modified asphalt compound;
- a 35 mm single course of mastic asphalt paving, basically formulated to BS 1447; and
- pre-coated 12 mm chippings, applied at 7.7 kg/m^2.

The system, with hindsight of over 30 years' practical experience, proved very effective with working lives generally exceeding 20 years.

The history of these applications has been recorded (Dussek, 1992), but particular aspects of the various components and applications can be noted. The basic mastic asphalt formulation needs particular care in manufacture, the key consideration being to equate the penetration of the binder with the hardness number of the mastic as laid.

The two primer layers were modified in the late 1980s to allow the use of a methyl methacrylate system, Eliminator, to replace the asphalt based materials. Judging from the difficulty in removing the original system, it was most effective but the spray-applied build-up, including a calcined flint mechanical key to the mastic asphalt, presents a more technical approach to the problem.

Initially on such bridges as the Forth, Severn and Bosporus I, the work was carried out by hand but, for the Humber bridge (Robinson, 1982), a mechanical paver-operated laying technique was introduced. This technique worked well despite very taxing weather conditions but the output of up to 1300 m^2/day was limited by other operations being carried out on the bridge at the same time. In practice, mechanical laying tends to be of advantage on new decks while, on decks where an existing mastic deck has to be removed, the old hand-laying technique tends to be favoured by:

- the speed of removal of old material;
- the need to maintain the traffic flow; and
- the disruption to the weight distribution of the deck that the removal of the surfacing over a complete carriageway would cause.

During the removal operation, other repairs, such as welding or strengthening, may also have to be carried out, thus limiting progress.

The surface texture formed by 12 mm, and later 14 mm, pre-coated chippings has been employed generally, but there have been uses of resin-based high-friction surfacings (section 11.3). If such high-friction treatments are to work satisfactorily, the key factor is a complete bond between the mastic and the resin system. It has been suggested that laying a resin surface treatment over a binder-rich asphalt base can result in debonding, and thus great care is required to remove the surface layer by abrasion before the application of a resin-based system. One particularly successful application was made by laying the epoxy-resin/calcined bauxite over a coated chipping surface in which the chippings had embedded too deeply, being almost flush with the surface.

The natural cambers and cross-falls of bridge decks present serious application problems for mastic asphalt, particularly on exposed low temperature sites with high wind velocity. This can create variations of 20 °C across a 4 m screed bar in extreme cases and, therefore, temperature must be very carefully controlled to maintain a regular flow of material. This advice is also applicable to the feed mechanism in the paver itself, which is prone to damage if the mastic cools and blocks, or even shears, the feed screws.

The production of mastic asphalt for bridge deck work by mechanical means is plant intensive. On the Humber Bridge, eight 11 tonne stirred cooker tanks, of which five were mounted on truck chassis, were used for melting down blocks, mixing with grit and transportation to site.

8.4.2.4 Other paving

Mastic paving is a popular means of surfacing footpaths and similar surfaces, especially in urban areas. One major advantage is the ease of making, and subsequently repairing, utility excavations which can be profiled to precise level without subsequent need for compaction. This type of work is carried out directly onto the sub-base, typically lean-mix concrete, and is laid to kerb level with ramps to road level for pedestrian and mobility handicapped vehicles access. Texturing of the surface is achieved by use of a 25 kg crimper roller, run over the mastic surface while it is still warm. This results in a characteristic 'dimpled' surface.

Drainage systems for chemicals, car park decks and allied access ramps, loading bays and standing areas for heavy loads and vehicles are also often surfaced in mastic asphalt and, where necessary, heating cables can be embodied in the overall design (MACEF, 1991a).

8.4.2.5 Roofing mastic asphalt

Mastic asphalt is also a well established material for high durability surfacing of flat and similar roofs. It is seamless and can be worked in both horizontal and vertical planes as well as to complex profiles. It is also effective in sealing the joints to roof lights, vents and pipes for, for example, air-conditioning units and other roofing structures.

Roofing grade mastic asphalt, manufactured to BS 6925 (BSI, 1988b), can be laid onto concrete, timber (including wood wool slabs) and metal decking. Typically, it is applied in two coats to a total thickness of 20 mm. As the thickness of the mastic asphalt is not able to be adjusted to create falls, this function is carried out by means of a sand/cement screed, sometimes modified by the use of a foam-creating emulsion additive (such as 'Isocrete') which creates a lightweight material. Pre-cut boards are also used. Most roofing work is applied by hand in bays, using wooden battens which provide a guide for layer thickness. For roofing work, it is standard practice to lay the first course of the mastic asphalt on a separate layer of loose laid black sheathing felt (approximately 1.5 kg/m^2).

Roofing mastic asphalt is frequently laid in association with thermal insulation materials. The design of composite roof structures calls for great care because the rise in condensation from the structure of the building itself in use can result in a serious build-up of condensate. The end product of such a build-up is water and allied damage to the rooms immediately under the roof. The insulation can be laid beneath or above the mastic asphalt. Such insulating materials include polyisocyanate foams, cellular glass, expanded polystyrene, cork and fibreboard. Where the insulation is placed above the mastic asphalt (which is laid immediately above the roof concrete), the design is known as 'upside-down roofing'. The insulation is held in place and protected mechanically by means of a layer of, say, loose aggregate or paving slabs.

Detailing of edges, skirting, flashing, pipe-work and venting of the roofing structure is covered by standard drawings issued by the Mastic Asphalt Council and Employers Federation (MACEF, 1991b). Where vertical work is carried out above 300 mm from the horizontal, the mastic asphalt requires additional support such as provided by expanded metal tied to the base surfaces.

One problem relating to areas of the roof that are trafficked or subject to front loadings is the need to protect the surface against indentations from, say, stiletto heels or support legs to tanks. Mastic asphalt roofing can also be subject to ultra-violet attack and also to thermal shock resulting from rapid temperature changes. Depending on the use of the roof, tiles or slabs can protect the surface from mechanized attack while solar reflective coating treatments, white chippings or even aluminium paint helps prevent solar deterioration. The problem of thermally originated damage is principally resolved by the use of lake asphalt or polymer-based binders.

8.5 GUSSASPHALT

8.5.1 Manufacture

Gussasphalt production is carried out in modified road paving mixers. The binder may be pre-blended and mixed with the aggregate components for approximately two minutes at around 210–220 °C. Modifications to the plant are necessary to heat and dry the aggregate, as well as to heat the filler which, although lower in content than for mastic asphalt (20% as against around 50%), represents a substantial element in the mixture. At this stage, the gussasphalt batches are dropped into mobile 10–11 tonne stirrer/cookers ('Kochers'). In Germany, it is common practice on small-scale contracts to mix the bitumen and aggregate first and then to add lake asphalt to the cookers to blend for a least one hour before laying. However, by skilful scheduling and part loads, the turn-round time can be reduced to half that time. Daily production output of the order of 1000 tonnes per plant is obtainable.

Test methods for binders and aggregates are similar to those used for mastic asphalt but the DIN load pin indentation test, based on a load of 52.5 kg over an area of 5 cm^2 at 40 °C for 30 minutes, does not correspond to the BSI hardness number.

Gussasphalt, and mastic asphalt, can also be evaluated in terms of performance tests, notably in terms of wheel-tracking.

8.5.2 Application

Gussasphalt has been in use for some 40 years and is designed for laying on a wide range of highway and major road (autobahn/city ring roads) sites. This is in contrast to mastic asphalt, which is largely limited to applications on structures. Nevertheless, gussasphalt is also applied to steel and concrete bridge decks, tunnels and for car park surfacings.

In major roadworks, the gussasphalt is frequently laid on a binder course of comparatively high void material. This is important because the gussasphalt, being voidless, would otherwise entrap water vapour, causing 'blowing'.

The binder course 0/22 specification is based on 40/50 or 60/70 pen bitumen with a minimum 3.7% binder content. A typical aggregate formulation is given in Table 8.4. Forty-five per cent of the aggregate comprises crushed sand particles. The void content is approximately 6–7% and the layer thickness is a nominal 80 mm.

The gussasphalt 0/11 surface course is overlaid on the binder course. The surface course material is based on a binder of 40/50 pen bitumen and Trinidad lake asphalt in a 70/30 blend giving a softening point of 62 °C and a penetration of 26 dmm. The soluble content is 7.1% by weight, within a range of 6.5 to 8.0%. A typical aggregate formulation is given in Table 8.4. Sixty-six per cent of the aggregate comprises crushed sand particles. The pin indentation depth (section 8.5.1) is 1.0 to 3.5 mm which, for heavy duty pavements, may be reduced to 1.0–2.5 mm. The surface layer is laid to a nominal thickness of 35 mm.

As discussed in section 8.5.1, gussasphalt is delivered in stirrer/cookers to site, ready for feeding to the laying train (Figure 8.5). The laying equip-

Table 8.4 Typical aggregate formulations for gussasphalt binder and surface courses

Aggregate type	Size (mm)	Proportion by weight	
		Typical value (%)	Typical specification range (%)
Binder course			
Coarse aggregate	>2.0	68.1	65–80
Sand	0.09–2.0	27.8	17–32
Filler	<0.09	4.1	3–9
Surfacing course			
Coarse aggregate	>2.0	52.1	45–55
Sand	0.09–2.0	27.2	25–35
Filler	<0.09	20.7	20–30

ment is set up on steel rails or crawlers which are set to the desired height of the surface. The spread of the paver can be varied up to the total carriageway width, say 11.5 m, and is in the form of a train (Figure 8.6).

The stirrers/cookers feed into a heated non-vibrating screed bar, which spreads the gussasphalt to level, moving forward at a rate of 1 to 2 m per minute. This is followed by a distributor, spreading 2 to 5 mm coated chippings at a rate of 15 to 18 kg/m². The aim of these chippings is to provide a skid-resistant surface and to prevent a bitumen-rich skin

Fig. 8.5 Gussasphalt being fed by mobile stirrer/cookers in front of the screed bar with a feed paddle spreading the material evenly across the width

Fig. 8.6 11.5 m gussasphalt screed bar train

Fig. 8.7 Gussasphalt compaction with both rubber-tyred and steel-wheel rollers in action

Fig. 8.8 An early application of gussasphalt on the Avus motor racing circuit in 1955 showing hand finishing and the effect of crimping rollers

forming on the surface. The chippings are rolled into the gussasphalt by means of rubber-tyred rollers. As these tend to mark up the surface, the final compaction to eliminate such marks is completed by steel-wheel rollers (Figure 8.7).

For some time, it was common practice in Germany to texture the surface by means of crimping rollers (Figure 8.8). In such cases, chippings were forced into the surface by means of rollers which provided a textured surface of indentations, typically 20 mm wide and 5 mm deep. These were set in a diagonal pattern to minimize vibration effects on the traffic.

The use of gussasphalt has been widely adopted in Japan for the surfacing of major bridge structures, including the Honshu-Shikoku Link suspension bridge decks (Figure 8.9) and the surface of the Kansai Airport Approach Viaduct. The Japanese favour a combined surfacing design using the following procedure (Honshu-Shikoku Bridge Authority, 1983):

- preparation of the steel deck, removing rust by grit blasting;
- application of two coats of rubberized bitumen solution at 0.2 L/m² per coat;
- application of 40 mm thick gussasphalt binder course;
- application of bitumen emulsion tack coat at 0.3 L/m²; and
- application of 35 mm thick polymer-modified asphalt concrete surface course.

Fig. 8.9 Application of gussasphalt at Honshu-Shikoku, Japan

The DIN load pin test results (at 52.5 kg over 5 cm^2 at 40 °C for 30 minutes) fall within 1 to 4 mm.

The Japanese bridge deck gussasphalt contains a slightly higher filler content, at approximately 25%, and an increased binder content, to 8.1%, than its German counterpart. Manufacture is in a batch plant with an output of 200 tonnes/day suitable for bridge deck work. The gussasphalt is held in the cookers for at best one hour with the mixing temperature (220 °C) raised to 240 °C for application by means of a small non-vibrating screed working on a 3.5 m width.

The use of a 75 mm surfacing thickness represents a considerable increase in dead load compared to traditional steel-bridge surfacings but, among other advantages claimed, the structures and decks have to be resistant to earthquakes, high winds and major tidal movements. The design of the surface facilitates removal and replacement of the asphalt concrete without damage to the lower layer.

8.6 REFERENCES

Angst, Dr C. (1992) Gussasphalt im Brückenbau, *Proceedings of 10th International Natural Asphalt Congress*, Berne.

British Standards Institution (1988a) *Mastic Asphalt (Limestone Fine Aggregate) for Roads, Footways and Pavings in Buildings. BS 1447: 1988*, British Standards Institution, London.

British Standards Institution (1988b) *Mastic Asphalt for Building and Civil Engineering Work (Limestone Aggregate). BS 6925: 1988*, British Standards Institution, London.

British Standards Institution (1990) *Methods of Sampling and Testing Mastic Asphalt and Pitch Mastic Used in Building. BS 5284: 1993*, British Standards Institution, London.

Broome, D.C. (undated) *The Physical and Chemical Properties of Asphaltic Cements in Mastic Asphalt*, Baynard Press.

Dussek, I.J. (1992) Asphalt surfacing on steel bridge decks, *The Asphalt Yearbook 1992*, The Institute of Asphalt Technology, Staines, 44–7.

D'Eyrinis, E. (1721) *Dissertation sur l'Asphalte*, Paris.

Forbes, R.J. (1936) *Bitumen and Petroleum in Antiquity*, Brill.

Honshu Shikoku Bridge Authority (1983) *Deck Pavement Standards*, Honshu Shikoku Bridge Authority, Japan.

Kopsch, H. (1992) *Auswahlkriterien für Polymere zur Modifizierung von Bitumen*, Alterungs-verhalten Bitumen.

Malo, L. (1898) *L'Asphalte*, Paris.

Mastic Asphalt Council & Employers Federation (1991a) *Paving Handbook*, Mastic Asphalt Council & Employers Federation.

Mastic Asphalt Council & Employers Federation (1991b) *Roofing Handbook*, Mastic Asphalt Council & Employers Federation.

Robinson, P. (1982) The surface of the Humber Bridge with machine-laid mastic, *Proceedings of the 6th International Natural Asphalt Congress*. London.

CHAPTER 9

Stone mastic asphalt surface courses

C.A Loveday, Tarmac Quarry Products Ltd, and P. Bellin, State Highway Agency of Lower Saxony

9.1 CONCEPT AND HISTORY

9.2 General

Stone mastic asphalt (SMA) is a very high stone content, gap-graded mixture with a high binder content. A continuous coarse aggregate skeleton carries the traffic load giving very good deformation resistance. A bitumen-rich mastic mortar binds the aggregate skeleton together, resulting in both toughness and excellent durability. The projection of the aggregate skeleton presents a very uniform 'negatively textured' surface profile (section 10.4.8), combining low noise generation with good skid resistance.

9.1.2 History and development

Stone mastic asphalt was developed in the late 1960s in Germany. At that time, studded tyres were widely used in winter and the resulting wear was the critical factor determining the life of surface courses. The materials then available were asphalt concrete (Chapter 6) and gussasphalt (Chapter 8). Asphalt concrete was found to be prone to rapid wear. The binder content was relatively low and the binder tended to harden at the surface. The mortar then abraded away under the studs, soon resulting in the material unravelling. Attempts to overcome the problem by increasing binder contents resulted in instability and rutting by deformation as opposed to stud wear. Gussasphalt, on the other hand, performed very well. However, like UK mastic asphalt, it was expensive to produce and slow and labour intensive to lay, with the hand or mechanical addition of chippings or 'Splitt' to the poured gussasphalt on site. German asphalt companies responded to this situation by developing proprietary formulations of plant mixed stone filled mastic asphalt, for example 'Mastimac' developed by Deutag (Bellin, 1992).

Concept and history

Fig. 9.1 Stone mastic asphalt in Wilhemshaven, Lower Saxony, 24 years old

Fig. 9.2 Close-up of 24-year-old SMA showing excellent retained macro-texture

In 1975, the use of studded tyres was banned so that the original justification for the use of stone mastic asphalt disappeared. However, the stone mastic asphalt surfaces were performing much better than those with asphalt concrete. Asphalt concrete was prone to deformation with high binder contents while ageing and premature cracking developed with lower binder contents: in contrast, stone mastic asphalt seemed to have few vices and offered greatly improved durability (Figures 9.1 and 9.2).

Over the years the materials and techniques were progressively refined using feedback from successful (and unsuccessful) highway projects. The various Contractors' proprietary products were grouped together under the generic name 'Splittmastixasphalt (SMA)' and were standardized in 1984 into the German Federal Department of Transportation Supplemental Technical Specifications and Guidelines for the Construction of Asphalt Pavement *ZTV Asphalt – StB94* (Der Bundesminister für Verkehr, 1994). Stone mastic asphalt has steadily grown in popularity with over 100 million square metres laid and in several states is now the first choice surface course for autobahns and federal trunk roads.

Through the eighties, the use of stone mastic asphalt spread to many countries with variants adopted in Sweden, Denmark, Finland, Netherlands, Belgium, France, Switzerland and Japan. In the early 1990s, trials were conducted in the United States and use there has grown rapidly.

Interest in stone mastic asphalt in the United Kingdom was stimulated by exposure to the alternative technology through the involvement in the exercise to harmonize European Standards for asphalt mixtures. Laboratory studies and in-house full-scale trials were conducted by Tarmac during 1992 and 1993 resulting in their 'Tarmac Masterpave' variant of stone mastic asphalt. The first 'public' trial demonstration of stone mastic asphalt in the United Kingdom was, ironically, carried out by Tarmac on an airfield taxiway at RAF Lakenheath in July 1993 for the benefit of the United States Air Force as part of their research project to evaluate European pavement practices!

Evaluation of stone mastic asphalt was included in a TRL research project jointly sponsored by the Department of Transport, British Aggregate Construction Materials Industries and the Refined Bitumen Association. A study tour to Baden Wurtemburg in September 1993 was followed by a demonstration trial in October 1993 at Crowthorne (Figure 9.3). The trials proved very successful and are fully reported in TRL Project Report 65 (Nunn, 1994). Since then, the use of stone mastic asphalt, both as a standard generic product and as proprietary products, has grown rapidly with considerable success.

9.1.3 Concept of stone mastic asphalt

Stone mastic asphalt is based around a stone skeleton of inter-locking crushed rock coarse aggregate. This skeleton comprises largely single-

Fig. 9.3 SMA demonstration trial at TRL in 1993

sized stone of a size appropriate to the laying thickness and required surface texture. The single-sized nature of the aggregate skeleton leaves a relatively high void content between the aggregate particles. This void space is partly filled with a binder rich mastic mortar. The mortar comprises crushed rock fine aggregate, filler, bitumen or modified bitumen and a stabilizing additive. The mortar itself is voidless and has flow characteristics.

The mixture is designed so that, when fully compacted, the voids in the aggregate skeleton exceed the volume of the mastic by 3–5%. The structure of the material is then such that the traffic load is carried out almost exclusively through the interlocking aggregate skeleton. The mortar acts to bind the skeleton together and to make the layer impervious.

The composition of the mortar is very important in determining the performance of stone mastic asphalt. A very high binder content is essential to ensure durability and laying characteristics. Sufficiently high binder contents cannot be achieved using unmodified or unstabilized bitumens; drainage of bitumen or mortar would occur during transport and laying. Therefore, most stone mastic asphalt mixtures use a fibre stabilizer mixed with the binder in the mortar to keep it homogeneous. There are alternative means of stabilization using absorptive fillers and modified binders, but these are less widely used than fibre.

The nature of stone mastic asphalt is perhaps best understood by considering its volumetric composition. Figure 9.4 shows the breakdown by mass and by volume into its particular components. The coarse aggre-

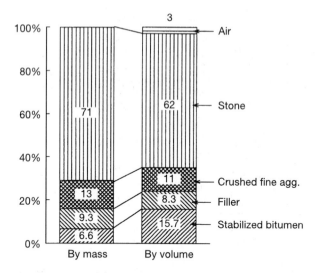

Fig. 9.4 14 mm stone mastic asphalt composition by mass and volume for total mixture

gate skeleton makes up over 70% of the total by mass, or just over 60% by volume. It needs to be sufficiently single sized in grading to leave about 40% voids, enough to accommodate 35% mastic mortar by volume, leaving sufficient air void space to ensure stability but not too much so that durability is compromised.

The process of designing a stone mastic asphalt mixture is very much one of adjusting the grading to accommodate the required binder and void content rather than the more familiar process of adjusting the binder content to suit the aggregate grading. Maintenance of the consistency of grading during manufacture is also of great importance as loss of volumetric balance can result in the mastic mortar 'fatting up' under compaction or traffic.

The richness of the mastic mortar can again best be understood from the standpoint of this volumetric composition (Figure 9.5). The approximate breakdown is 37% crushed fines, 26% filler and 37% bitumen. This compares with 22% bitumen by volume in a typical rolled asphalt (HRA) (Chapter 5) mortar (shown in Figure 9.5).

9.2 PROPERTIES

9.2.1 Deformation resistance

The high stone content and extreme gap grading ensure that loads are transmitted through the inter-locking aggregate skeleton rather than through the mortar. This imparts a very high degree of deformation

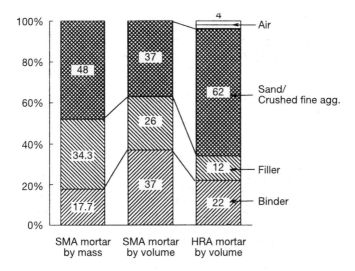

Fig. 9.5 Comparison by mass and volume of stone mastic mortar and comparison by volume of stone mastic and rolled asphalt mortar

resistance (TBV Consult, 1994), as is demonstrated by experience on the continent over many years.

Material from a number of trials and sites in the United Kingdom (Figure 9.6) has been subjected to the wheel tracking test, as described in BS 598: Part 110 (BSI, 1996a). The range of results has been generally between about 0.5 mm/h and 1.0 mm/h. These results represent excellent deformation resistance, significantly better than is achievable with even the best hot rolled asphalt.

9.2.2 Dynamic stiffness

Although not of prime significance, the structural contribution of surface courses has some relevance in pavement design. Testing has shown stone mastic asphalt to have generally very good dynamic stiffness. Specimens tested in the Nottingham Asphalt Tester in accordance with DD 213 (BSI, 1993) have given values of Indirect Tensile Stiffness Modulus up to 5 GPa.

9.2.3 Flexibility and fatigue life

The binder rich nature of stone mastic asphalt mastic mortar suggests that the material is likely to resist premature cracking and this is borne out by German practical experience over 25 years of use. At the time of writing, there is no definitive UK method for measuring fatigue or crack propagation properties but stone mastic asphalt specimens are being evaluated in those tests under development. There is every expectation that this will produce excellent results.

Fig. 9.6 Night laying of masterpave SMA at tool booths of Dartford river crossing to provide deformation-resistant surface

9.2.4 Durability

Experienced asphalt technologists, when they see stone mastic asphalt laid for the first time, tend to recognize very quickly that it will be very durable. It is 'richer' than any other asphalt material and the mastic mortar is effectively voidless. This means that 'ageing' through oxidation hardening of the surface binders will be very slow, which explains why surface cracking has been almost totally absent in Germany. It also means that the bonding of aggregate and binder is protected from damage by water ingress. Consequently, many examples of roads more than 15 years old without significant distress have been documented and typical lifespans in excess of 20 years can be anticipated.

9.2.5 Toughness

'Toughness' as a mechanical property is perhaps a little difficult to define. However, there are circumstances where surfacings are exposed to scrubbing, tearing and grinding much more severe than the normal rolling wear of traffic moving in straight lines. Given the origin of stone mastic asphalt as an antidote to the abrasion of studded tyres, it is hardly surprising to find that it performs particularly well under extremely tough conditions. The earliest Tarmac trials were laid on sites around quarry weighbridges, subject to constant turning traffic of fully loaded lorries with loose dirt on the surface. The material has been shown to succeed in such situations where all other solutions, including concrete, had failed. Stone mastic asphalt must, therefore, be considered a particularly tough surfacing.

9.2.6 Skid resistance

The skid resistance of a surfacing involves a range of factors: properties when first laid, macro-texture over time and micro-texture over time. When laid, stone mastic asphalt presents a very uniform negatively textured surface. The texture depth which can be achieved is very much dependent on the design of the mixture in terms of a sufficient gap grading and a suitable volume of mastic mortar. Sand patch texture depths in excess of 1.5 mm and up to 2 mm are possible with 14 mm nominal sized aggregate and 1.2 mm up to 1.5 mm with 10 mm. Thus, to be absolutely certain of meeting the 1.5 mm texture depth requirement usually imposed on high-speed roads in the United Kingdom, 14 mm aggregate is needed. This contrasts with German practice, where the largest aggregate size used is 11 mm and the trend is to use 8 mm or even smaller sizes, partly because it is considered that these provide better skid resistance (Figure 9.2).

Once the binder film has worn off, the surface presented to the tyre will be almost exclusively composed of the coarse aggregate. Given the anticipated lifespan of stone mastic asphalt in all other respects, the retained micro-texture under traffic could well become the critical factor in determining the life of a stone mastic asphalt surface. This has tended to be the case on the continent, where the practice has been to select very hard, wear resistant aggregates, usually igneous and invariably with a low polished stone value (PSV) (section 2.2.4.4). The UK practice of using generally much higher PSV aggregates with quality related to risk rating and anticipated traffic flow is better suited to ensuring a long life with good skid resistance. The surface presented to the tyre is very similar to that of porous asphalt (Chapter 7), so it is logical for PSV selection to follow the same rules.

Stone mastic asphalt surfaces in the United Kingdom have not generally been trafficked long enough for equilibrium values of mean summer SCRIM coefficient (MSSC) to be reached. Measurements have been made on new or fairly new surfaces from several sites. The demonstration trial at TRL generated SCRIM coefficients of 0.62 to 0.64, and these are fairly typical of values obtained on other sites.

However, there are some concerns over early life skidding resistance, as with porous asphalt. Potential problems are perceived of the performance under locked wheel braking conditions before the thick binder film has worn off, a process which is likely to take months rather than weeks or days. The phenomenon has been studied using the Braking Force Trailer both on the TRL trial sections and on various sites in Kent. The results of these investigations are somewhat inconclusive, largely because of the shortage of comparative data on other new surfaces and the absence of any established criteria for Braking Force Coefficient (BFC).

In Germany, a surface treatment is recommended for high speed roads in order to overcome any problem of early life skid resistance. Dust free, uncoated crushed grit of 1–5 mm size is spread at the light application rate of 0.6–0.9 kg/m^2 between the first and second rollers. The grit sticks to and penetrates the surface layer, enhancing initial skid resistance and speeding the erosion of the binder film.

9.2.7 Noise generation

Negatively textured surfaces are generally accepted to be less noisy under traffic than those with projecting chippings and this rule holds good for stone mastic asphalt. It is known to be a quiet running surface in all of the countries where it is widely used. Extensive studies of tyre/surface generated noise from a coasting vehicle were made on the demonstration trial sections at TRL. The results showed an overall reduction of noise levels of approximately 3 dB(A) when compared with conventional chipped rolled asphalt. This is a significant reduction, equivalent to a halving of traffic density or a doubling of distance from the road. Noise levels are even lower with the smaller aggregate sizes.

Fig. 9.7 Laying SMA in echelon on the A31 autobahn showing grit application from roller mounted spreader

Applications

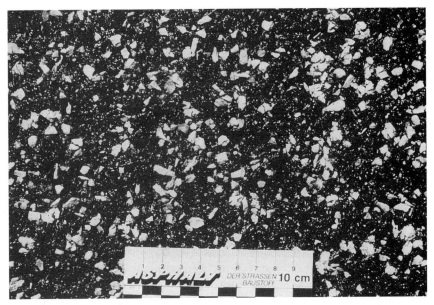

Fig. 9.8 SMA surface with application of 2–5 mm grit at 0.8 kg/m^2

9.3 APPLICATIONS

Stone mastic asphalt is sufficiently tough and durable for all types of highways and for a wide range of industrial paving applications:

- The stability and richness of the material in particular makes it suitable for laying at a wide range of thicknesses from thin surfacing overlays 20 mm or less to regulating surface courses 70 mm thick.
- Because of its stone skeleton structure, the majority of compaction is achieved under the paver screed and there is relatively little movement under subsequent rolling when compared with hot rolled asphalt or dense macadam. This means that it can be used as a regulating surface course to restore uneven profiles.
- The toughness of stone mastic asphalt makes it particularly suitable for stress sites, especially roundabouts.
- The mortar-rich nature of stone mastic asphalt makes it suitable for laying in fairly thin layers, although in this case, extra care needs to be taken over laying, compacting and tack coating.

Both 14 mm and 10 mm stone mastic asphalt have been laid in conventional 40 mm layers as a direct replacement for rolled asphalt surface course. However, the adoption of 40 mm or 45 mm as the layer thickness for rolled asphalt probably has more to do with the need for a sufficiently thick heat retentive layer to accept and retain pre-coated chippings than with an absolute need to lay surface courses at that thickness. Stone mastic asphalt, being rich and with a voidless mortar needing little compaction will clearly achieve durability in much thinner layers. This is important because the use of premium aggregate and stabilizing additives will tend to make the material fairly expensive.

A range of thickness has therefore been used in the United Kingdom from 35 mm down to 25 mm and 20 mm, approaching that of thin surfacings (Chapter 10). There is little to distinguish the surface appearance and apparent durability of the thinner layer stone mastic asphalts from the thicker alternatives but more care is required at the laying stage, both in terms of tack coating and in ensuring full compaction and inter-lock of the aggregate skeleton and some adjustment of composition may be required. Current experience suggests that 14 mm stone mastic asphalt can be laid down to 25 mm nominal layer thickness, 10 mm down to 20 mm and 6 mm down to 15 mm. Stone mastic asphalt has been used in the United Kingdom as a thin surfacing (Figure 9.9).

Moving in the opposite direction, the limitations on upper thickness are likely to be related to cost rather than engineering. The aggregate skeleton is extremely stable yet compacts well under the paver screed. Consequently, there is little movement of the mat under rolling so that an even riding surface can be achieved with thick layers and uneven substrates. Cost considerations apart, the material would also make an excellent binder course, given its deformation resistance, stiffness and impermeability.

Fig. 9.9 SMA as a thin surfacing – 25 mm thick masterpave on the A148 Holt Bypass, Norfolk

Stone mastic asphalt is very well suited for tough off-highway application. The combination of abrasion resistance, deformation resistance and general toughness means that it will perform where most other asphalt surfacing will be found wanting. When considering deformation resistance, however, serious consideration must be paid to the base on which it is laid. While the stone mastic asphalt will not deform in itself, because of the aggregate skeleton, transmitted loads may cause base deformation underneath. An example of industrial use is shown in Figure 9.10.

9.4 COMPONENT MATERIALS

9.4.1 Aggregates and filler

9.4.1.1 Coarse aggregate

There are two basic requirements for the intrinsic properties of coarse aggregate in stone mastic asphalt: strength and polishing resistance.

Polished stone value (PSV) (section 2.2.4.4) requirements will almost certainly dictate that stone mastic asphalt for highway use incorporates hardstone or steel slag coarse aggregate, although there is no reason why hard limestones should not be used in off-road paving. German specifications control crushing resistance using the Schlagversuch impact test (section 2.2.3.1). In the draft UK specification (Appendix A), this is trans-

Fig. 9.10 SMA as a tough industrial paving – masterpave on a brick stockyard in Leicestershire

lated into a requirement for 10% fines value (TFV) (section 2.2.3.2) greater than 180 kN. This strength requirement is necessary to ensure that the aggregate in the 'skeleton' does not break down at the points of contact.

The aggregate will also need to meet the standard requirements for polished stone value (PSV) (section 2.2.4.4) and aggregate abrasion value (AAV) (section 2.2.4.3) for the site. These values should be selected in the same way as for other asphalt surfacings. The aggregate should be totally crushed, to ensure optimum inter-lock, and will need to be single-sized and of good shape in order to ensure the correct volumetric proportions. The draft UK specification (Appendix A) calls for a maximum flakiness index of 30%.

9.4.1.2 Fine aggregate

The fine aggregate should be crushed rock or slag rather than natural sand. No other special requirements apply.

9.4.1.3 Filler

Added filler is normally limestone and the interim UK specification (Appendix A) incorporates requirements identical to those in BS 594: Part 1 (BSI, 1992).

9.4.2 Binders

German practice has been to use 80 pen binder for regular traffic and 65 pen for 'S' class stone mastic asphalt in heavily trafficked situations. UK bitumen grades do not correspond exactly with these and 100 pen and 50 pen grades have been used instead. Given the general growth in traffic densities the switch to the harder 50 pen grade is probably appropriate. The majority of UK stone mastic asphalt incorporates 50 pen binder and appears to be totally satisfactory.

9.4.3 Stabilizers, additives and binder modifiers

It is not possible to carry the high bitumen content of stone mastic asphalt in such a high stone content material without stabilizing the binder in some way. A range of materials has been used for this purpose.

9.4.3.1 Fibres

By far the most widely used form of mixture stabilization is fibre, in particular cellulose fibre. It seems to offer the best combination of economy and effectiveness. The breakdown of stabilization usage in Germany is roughly as follows:

Mixture composition

Cellulose fibre	90%
Mineral fibre	4%
Polymer	5%
Silicon	1%

Cellulose fibre mixes relatively easily with the binder and 'thickens' it almost totally overcoming any tendency to segregate or drain.

Cellulose fibre
This comprises finely milled cellulose fibre derived either directly from wood or more commonly by reprocessing newsprint. This is available loose in bulk, baled in pre-weighed melt down plastic covered bales or pelletized sometimes bound with a small proportion of oil or bitumen.

Mineral fibre
This comprises spun fibre produced by extruding molten mineral through fine spinnerets. The fibres are broken down into short lengths and packaged in a similar way to cellulose.

9.4.3.2 Polymers

A wide range of polymers has been used in stone mastic asphalt formulations including styrene-butadiene-styrene (SBS), styrene-butadiene rubber (SBR), ethylene vinyl acetate (EVA) and natural rubber (section 3.5). Polymers are much less effective at preventing segregation than fibres and are more likely to be used to enhance the material in other ways. Because of this, it is likely that polymer-modified binders would be used in conjunction with fibre. Given the dependence of stone mastic asphalt performance on the aggregate skeleton rather than the mortar, the improvement resulting from polymer use is likely to be rather less than in rolled asphalt. However, there are circumstances where extra flexibility is required and elastomeric binders will be of benefit. Polymers may also assist in resisting the tendency to 'close up' under very heavy traffic, an increasing concern as traffic density continues to rise.

9.4.3.3 Special fillers

Special fillers include silicic acid and diatomaceous earth.

9.5 MIXTURE COMPOSITION

The most important aspect of stone mastic asphalt is that it is designed to meet the volumetric criteria described elsewhere in Section 9.1.3. This means a gap grading with a coarse aggregate sufficiently single sized to accommodate the required volume of mortar.

Specifications are expressed as a grading 'envelope' within which the target grading must lie, together with requirements for binder and stabilizer content. Examples of the German specifications (Der Bundesminister für Verkehr, 1994) are shown in Table 9.1. It must be stressed that these are not recipe composition specifications in the conventional sense. The producer must use his skill to design a mixture within these constraints which will perform satisfactorily. In the case of aggregates of unusual density, such as steel slag, this will even require departure from the basic specification requirements to adjust to a suitable volume composition.

There is no specific laid down mixture design method to parallel the BS 598: Part 107 (BSI, 1990) procedure for rolled asphalt. However, the following steps are usually followed to arrive at a Job Mixture Formula:

Step 1
Having chosen aggregates, binder, filler and additive and based on feedback from former Job Mixture Formulae, a trial mixture grading is chosen complying with the specified target envelope.

Step 2
Trial mixtures are made in the laboratory at the target grading and with the required minimum binder content and two or three additional higher binder contents.

Step 3
An assessment is made of the tendency for the mixture to drain. This tests the effectiveness of the stabilizing additive. Requirements for such

Table 9.1 German specification for stone mastic asphalt

	Mixture types 'S' signifies heavy traffic			
	0–11 S	0–8 S	0–8	0–5
Sieve size:	Proportion passing (%)			
16 mm	100			
11 mm	90–100	100	100	
8 mm	max 60	90–100	90–100	100
5 mm	30–40	30–45	30–55	90–100
2 mm	20–25	20–25	20–30	30–40
Filler (passing 90 micron)	9–13	10–13	8–13	8–13
Stabilizing additives	0.3–1.15% organic or mineral fibre, silica or polymer			
Binder grade (pen)	65, PMB 45	65, PMB 45	80	80
Binder content (%)	6.5–7.5	7.0–7.5	7.0–7.5	7.2–8.0
Thickness (mm)	25–50	20–40	20–40	15–30

drainage tests are not generally formalized. In Germany a simple procedure, known as the Schellenberg/von der Weppen test (Schellenberg *et al.*, 1986) is used. This involves storing 1 kg loose samples of stone mastic asphalt in 800 mL glass beakers for 60 minutes at 170 °C. At the end of this period, the mixture is tipped out by rapidly upending the beaker. The loss in mass resulting from material adhering to the beaker is determined. Any value greater than 0.3% of the original is considered unacceptable. The UK interim specification suggests the use of the Binder Drainage Test described in Clause 939 of the *Specification for Highway Works* (MCHW 1). In practice, it is found that an addition of 0.3% cellulose fibre is so effective at binder stabilization that binder drainage and mixture segregation is not a problem.

Step 4

Marshall specimens are prepared in accordance with the appropriate procedure and at the target composition. The void content is determined from the bulk density of the Marshall specimens and the maximum mixture density. For maximum mixture density, the 'Rice' method to DD 228 (BSI, 1996b) is currently recommended but it must be appreciated that breaking the mixture into sufficiently small pieces is a laborious process for such a rich and sticky material. The void content of the Marshall specimens needs to be in the range of 2–4%. If the required air void content is not achieved then adjustments should be made to the trial mixture compositions against the following priority list:

Priority 1   Change the coarse aggregate grading or coarse aggregate content;
Priority 2   Change the content and/or the type of filler;
Priority 3   Change the content and/or the additive; and
Priority 4   Change the binder content (the last resort!).

Step 5

The proven target composition is then adopted as the Job Mixture Formula and tolerances are applied about it for compliance assessment.

9.6 PRACTICAL ASPECTS

9.6.1 Mixing

The mixing of stone mastic asphalt is conventional in all respects other than the incorporation of fibre. High mixing temperatures are needed to ensure an appropriate mastic viscosity at the laying stage. For 50 pen binders, this means a temperature in the truck of 175–180 °C and, hence, a maximum mixing temperature of 190 °C. While care must be taken to avoid excessively high temperatures, it must be appreciated that the high film thickness and the voidless nature of the mastic mean that binder hardening will be much less pronounced than in dense macadams or rolled asphalt at the same temperature. The low fine aggregate content of stone mastic asphalt will also result in a low overall moisture content in the cold feed and asphalt plant driers will have no problems achieving the required temperature.

The difficulties in production are those of mechanical handling and blending in of the fibre additives. Cellulose fibre is a low density, bulky material which is added to the mixture at very low proportions. Mechanical handling of the loose fibre is possible but this requires blowing, and subsequent proportioning into the mixture is a problem because, in this divided state, it is tricky to weigh and will not flow easily under its own weight.

Therefore, most fibre is added to the mixer in pre-weighed compressed packages wrapped in thin, low melt plastic film. These can be added manually through suitable openings in the mixer box. It is also possible to use semi-automatic conveying systems linked to the plant control system and which count in the requisite number of bags per batch (Figure 9.11).

To simplify mechanical handling, a number of fibre manufacturers now produce pelletized fibre. This is made by mixing the fibre with a small amount of hard bitumen or oil and then compressing it and extruding it. The resulting pellets are sufficiently dense to flow under their own weight and can be conveyed more easily. The fibre content does, however, seem to be slightly less effective at stabilizing the binder than is loose fibre.

Mixing stone mastic asphalt involves blending as little as 0.3% of fibre into the mixture. It is important that this small amount of fibre is evenly

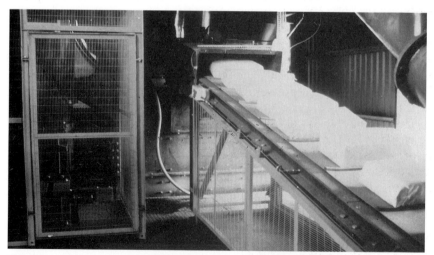

Fig. 9.11 Automatic system for delivering fibre pillows to the mixer at Tarmac's Ettingshall plant

distributed in the aggregate before the binder is added or an inhomogeneous product with 'balling' will result. Therefore, it is necessary in pug-mill mixers to introduce a dry mixing period into the mixing cycle. This must be long enough to distribute the fibre but not so long that the individual fibres are broken down by attrition. It is also necessary to extend the wet mixing cycle slightly to ensure thorough blending. The combined effect of this is roughly a 50% increase in the mix cycle time and a consequent reduction in the rate of output from the plant.

Most stone mastic asphalt has been produced in conventional batch type asphalt plants with pug-mill mixers and there is little information on the use of drum mix or continuous mix plants. It will almost certainly be essential to use pelletized fibre and considerable care will be needed to ensure thorough and even distribution if using such plants.

Quality control of production needs to be to a very high standard. Close control of the grading of feed aggregates and of all proportioning is essential in ensuring that the important volumetric relationships are maintained so that the stone mastic asphalt performs to its best.

9.6.2 Laying

Laying of stone mastic asphalt requires no special plant but there are differences in handling and some special skills need to be acquired. A tack coat should be used under stone mastic asphalt. For conventional layer thicknesses, say in excess of 30 mm, a normal tack coat, such as that specified in BS 594: Part 2 (BSI, 1992), is adequate. For much thinner layers, modified tack coats incorporating polymers, such as the Mastertack product developed by Tarmac for use with their Masterpave system, have been found to be more effective.

Stone mastic asphalt must be considered as a material for machine laying. Hand laying is possible and, in small areas where this is unavoidable, it can be done very effectively. However, it is undeniably very hard work as stone mastic asphalt is very stiff to rake at anything but the very highest laying temperature and the proportion of hand work should be kept to a minimum.

There are no special requirements for pavers, but the best results are achieved if the maximum possible compaction is reached under the paver screed. Therefore, the temperature should be kept up, preferably above 150 °C in the paver hopper, and the tampers and screed set to optimize the compaction of the mat before rolling. If this is done correctly, there is very little subsequent movement of the mat during rolling so very little surcharge is needed. There is also much less ability to roll out level errors, such as poor joint matching, so that great care is required in getting everything right first time.

The best roller combination is at least two steel wheeled rollers of 8–12 tonnes deadweight, one of which should be a tandem. The first

Fig. 9.12 Versatility of SMA – laying operations on the 1 : 4 gradient Porlock Hill, Somerset

roller should be kept as close to the paver as possible. The use of vibration should be kept to a minimum, probably only on the first pass, if at all. Vibration is not advisable on thin layers of 20 mm or less or at layer temperatures below 100 °C. Pneumatic tyred rollers are not generally recommended because of the risk of mortar flushing to the surface. However, there are instances where special techniques with pneumatic tyred rollers have been used to enhance surface texture and assist with gritting.

A reasonable joint can be made in stone mastic asphalt without cutting, especially if the second mat is laid before the first is fully cold. For the very best results, however, the joint should be cut and the vertical surface painted with bitumen. It is a matter of skill judging the best time to cut the joint, a slight delay will leave a very tough job.

When required, the early life skid-resistance can be improved by a light application of grit. Various grit sizes can be used but they should be clean, that is substantially retained on a 1 mm sieve and with a top size between 3 and 5 mm. The grit is spread between the first and second rollers and then lightly rolled in. Application rates are variously quoted as between 0.5 kg/m^2 and 2 kg/m^2. However, there is no standardized procedure for this gritting and clean sand and sealing grit have also been used.

One important advantage of stone mastic asphalt is that it can take traffic very soon after laying. Once rolled, with the skeleton inter-locked, it appears to withstand traffic without deformation, tearing or marking almost immediately and free standing joints are little damaged by traffic overrun.

9.7 COST AND MAINTENANCE CONSIDERATIONS

The client must consider the overall economics of a surfacing material in terms of some whole life costing assessment. Tonne for tonne, stone mastic asphalt will be more expensive than most other asphalt surfacings, including rolled asphalt, for the following reasons:

- high percentage of premium high PSV aggregate;
- high binder content;
- use of stabilizing additives;
- high temperatures;
- lower production rate; and
- extra attention to quality control.

However, the material does offer a considerable range of advantages including:

- high resistance to permanent deformation;
- high wear resistance;
- low noise generation;
- good structural contribution;
- excellent resistance to ageing and cracking;
- good low temperature performance;
- toughness and impact resistance;
- ability to be laid thinly;
- reduced traffic disruption during laying; and
- long service life, possibly in excess of 20 years.

Stone mastic asphalt thus offers a more expensive but longer lasting surface at the same thickness as rolled asphalt (Chapter 5). It is, however, possible to reduce the initial cost at lower thicknesses without compromising the life expectancy.

9.8 REFERENCES

Manual of Contract Documents for Highway Works, Her Majesty's Stationery Office, London.
  Volume 1: Specification for Highway Works (MCHW 1), Clause 939 Binder Drainage Test.
Bellin, P.A.F. (1992) *Use of Stone Mastic Asphalt in Germany: State of the Art,* 71st Transportation Research Board Annual Meeting, June 1991, Washington DC.
British Standards Institution (1990) *Method of Test for the Determination of the Composition of Design Wearing Course Asphalt. BS 598: Part 107: 1990,* British Standards Institution, London.

British Standards Institution (1992) *Hot Rolled Asphalts for Roads and Other Paved Areas; Part 1, Specification for Constituent Materials and Asphalt Mixtures; Part 2, Specification for Transport, Laying and Compaction of Rolled Asphalt. BS 594: Part 1: 1992, BS 594: Part 2: 1992,* British Standards Institution, London.

British Standards Institution (1993) *Method for the Determination of the Indirect Tensile Stiffness Modulus of Bituminous Mixtures. DD 213: 1993,* British Standards Institution, London.

British Standards Institution (1996a) *Sampling and Examination of Bituminous Mixtures for Roads and Other Paved Areas; Part 110, Method for Determination of Wheel Tracking Rate of Cores of Bituminous Wearing Course. BS 598: Part 110: 1996,* British Standards Institution, London.

British Standards Institution (1996b) *Methods for Determination of Maximum Density of Bituminous Mixtures. DD 228: 1996,* British Standards Institution, London.

Der Bundesminister für Verkehr, Bonn (1994) *Zusätzliche Technische Vertragsbedingungen und Richtlinien für den Bau von Fahrbahndecken aus Asphalt. ZTV Asphalt – StB 94, Ausgabe 1994,* Bonn, Germany (The German Federal Department of Transportation, *Supplemental Technical Specifications and Guidelines for the Construction of Asphalt Pavements, ZTV Asphalt – StB 94).*

Deutscher Asphalt Verband (1991) *Leitfaden Splittmastixasphalt,* Offenbach, Germany (German Asphalt Pavement Association, *Guidelines on Stone Mastic Asphalt).*

Nunn, M.E. (1994) *Evaluation of Stone Mastic Asphalt (SMA): A high stability wearing course material,* Project Report 65, Transport Research Laboratory, Crowthorne.

Schellenberg, von der Weppen, Rotweil (1986) Verfahren zur Bestimmung der Homogenitäts – Stabilität von Splittmastixasphalt, *Bitumen* 1/1986 Seiten 13 & 14, Baden-Württemberg. (Bitumen drainage test for the evaluation of the stability of stone mastic asphalt homogeneity, *Bitumen* 1/1986.)

TBV Consult (1994) Rutting resistant asphalt mix trial – stone mastic asphalt, *USAFE R & D Project European Pavement Practice 1994.*

CHAPTER 10

Thin surface course materials

D.F. Laws, W.S. Atkins, East Anglia

10.1 CONCEPT AND HISTORY

10.1.1 Definition

Thin surface course materials are defined as being of intermediate thickness between veneer coats (Chapter 11), which are very thin surfacing systems such as surface dressings and slurries, and materials laid as surface courses at the more traditional thicknesses of 40 mm to 50 mm, such as rolled asphalt (Chapter 6) and coated macadam (Chapter 5). Therefore, the materials are those laid 15 mm thick and upwards to a maximum of 35 mm, but more generally 30 mm. In practice, such materials will generally be paver-laid when installed in the mat, although similar materials can be laid with surface dressing equipment (section 11.1.2.2).

10.1.2 History

Until at least 1991, the thin surface course materials laid in the United Kingdom, as defined above, were restricted to either:

- coated macadams of generally 10 mm nominal size and below, including fine cold asphalt (which is neither laid cold nor is an asphalt in terms of the traditional, more restricted, meaning of a gap-graded mixture with a binder/filler/fine aggregate mortar) to BS 4987; or
- sand-carpet type rolled asphalts to BS 594;
  or, less usually, to:
- dense tar surfacing; or
- mastic asphalt.

Figure 10.1 illustrates some aspects of this past usage.

Increasing traffic levels, the speed of private vehicles and the increasing power and weight and changing configuration of commercial vehicles have progressively overtaken the physical capacity of many of the above materials.

In recent years in the United Kingdom, apart from in the Housing Estate Road context, coated macadams have tended to be used as a relatively low-cost expedient for restoring surface and riding qualities of deteriorating minor roads. In this context, due to a combination of vary-

The types and scope of coated macadam	Model specification for roads and footways on Housing estates
An ACMA Advisory Service publication Third, revised edition, 1975	An ACMA Advisory Service publication Second, revised edition, 1975
Function of the constituents	Construction of wearing course
Each of the constituents exerts an effect on the coated macadam which is a function of the quantities present. In general these functions are:	The wearing course shall consist of:
	* Rolled asphalt to BS 594: 1973, Tables 4, 5 or 6; Coarse aggregate content: 30 per cent; Thickness: 35 mm; 14 mm coated chippings at 8.0–9.3 kg/m^2.
Coarse aggregate Provides the main structure of the material and forms a layer of interlocked material which distributes the loads imposed by traffic.	* Dense coated macadam of 10 mm nominal size to BS 4987: 1973, Tables 45–48; Thickness: 25 mm.
Fine aggregate According to the quantity present, fills or partially fills the voids which there would otherwise be in coarse aggregate. Also controls the finished surface texture.	* Fine cold asphalt to BS 4987: 1973, Tables 53–56; Thickness: 20 mm 14 mm coated chippings at 8.0 kg/m^2.
Filler In coated macadams the filler combines with the binder to increase the thickness of coating film on the aggregates, thereby assisting to fill small voids. The addition of filler also has the effect of increasing the viscosity of the binder and so reduces any run-off of binder from the aggregate. In dense tar surfacing, filler is used in greater quantities when it combines with the binder to form a strong mortar substantially filling all the voids.	* Dense tar surfacing to BS 5273: 1975; 10 mm nominal size material; 35 per cent stone content; Thickness: 30 mm
	* Open-textured coated macadam of 10 mm nominal size to BS 4987: 1973, Tables 33–36; Thickness: 20 mm
Binder The binder acts as a lubricant when hot, thereby assisting compaction. When cold it acts as a bond, giving stength with some flexibility to the material.	

Fig. 10.1 Extracts from 1975 ACMA publications

ing thickness and variable and often poor compaction, the materials have too often suffered from poor performance with fretting and pot-holing. To achieve better workability, softer binders were often used with resultant softening and rutting in hot weather. Despite improvements in later editions of BS 4987 where thicker layers in relation to aggregate size are now specified, and the use in some enlightened counties of voids specifications to achieve good and consistent compaction, the cost of coated macadams is significantly greater than surface dressing and the service life is generally perceived as only moderate. In the footway and for other off-road purposes, 6 mm dense and 3 mm close graded surface courses are still in general use, as also are fine-graded rolled asphalts.

Surface texture and consistency of finish could be variable with the above mixtures and sometimes 'fine cold asphalt' and sand-carpet rolled asphalt would be finished with rolled-in coated chippings, but this added to the cost, and such relatively fine graded materials could suffer from chipping submergence and rutting.

Tar, for dense tar surfacing, is no longer widely available (section 3.2) and is now recognized as a health hazard, while mastic asphalt is labour intensive and expensive (section 8.2.3). Some cold screeded systems such as Ralumac, Reditex and Nimpactocote have proved successful on some sites, but fall at the lower thickness limit of this category and are covered in Chapters 11 and 12.

In Europe, the development of thin surface course materials has occurred in several countries (AASHTO, 1991; Litzka et al., 1994). Rut-resistant long-life stone mastic asphalt (Chapter 9) has been used in some form in both Germany and Switzerland (Partl et al., 1995) for over 20 years, but the lack of a skidding resistance or texture requirement obscured its effectiveness as far as the United Kingdom was concerned until recently. Thin surfacing techniques have been under development in France for some 10 years, with the Laboratoire Central des Ponts et Chausées (LCPC) collaborating with major national manufacturers and contractors to evolve cost-effective rapid laying systems with low-noise characteristics and reasonable life expectancy.

In 1991, Associated Asphalt invited a mixed group of engineers, including Department of Transport, Transport Research Laboratory and County Council representatives, to meet their counterparts on French soil and witness the Euroduit system, developed by LCPC and SGREG Ouest, in operation. As a result of this expedition, trials of the system were arranged in the United Kingdom under the revised tradename 'Safepave' on sites ranging from the A1 in Cambridgeshire laid directly on PQ concrete to rural roads in Cumbria, Lancashire and Wiltshire (Nicholls et al., 1995). Interest in, and trialling of, the new generation of thin surfacing systems for use on UK roads has expanded from that beginning.

10.1.3 Categories

Thin surface course materials, or thin surfacings, can be categorized by several means, but the simplest approach is to consider the material type from which they were developed. In this way, the categories currently available (Nicholls, 1995) are:

- thick slurry surfacing;
- multiple surface dressing;
- paver-laid surface dressing;
- thin polymer-modified asphalt concrete; and
- thin stone mastic asphalt.

Thick slurry surfaces (also known as micro-surfacing or micro-asphalt) and multiple surface dressing are covered in sections 11.2 and 11.1 respectively and are not generally regarded as 'thin surface course materials' as described in this chapter.

Paver-laid surface dressings have a single size coarse aggregate in a mortar and are laid on to a hot-applied polymer modified spray-coat. This type, known as ultra thin hot mixture asphalt layer (or UTHMAL) using the France classification system, will not be included in the European Standard for thin surfaces because, at present, there are only two such products, and they will be dealt with under the European Technical Approvals (ETA) procedure.

Thin polymer-modified asphalt concretes are mixtures developed from asphalt concrete and macadams which have been designed not to require much post-paver compaction, a necessary attribute for thin layers when there is not much time for compaction (Nicholls and Daines, 1993). They tend to be gap graded and therefore are not true asphalt concretes. The materials of this type tend to be proprietary, but this may change if the proposal to prepare a European Standard specification for this category of material is pursued, probably under the title, derived from the French classification, of very thin surface layers (or VTSL).

Thin stone mastic asphalt (thin SMA) is a variant of SMA, which is covered in Chapter 9, but has many similarities with thin polymer-modified asphalt concretes. It is uncertain whether thin SMAs will be classified as stone mastic asphalts or very thin surface layers (or both) under the European specifications.

Some indication of the interest both of the manufacturers and of their potential clients in the categories of modern thin surfacing materials can be gleaned from the rapid proliferation of proprietary materials, some examples of the trade names offered are given below:

- Paver-laid surface dressing – ultra thin hot mixture asphalt layer
  - Safepave (called Euroduit in France and Novachip elsewhere) (Associated Asphalt)
  - Combifalt (Colas Denmark)
- Thin polymer-modified asphalt concrete – very thin surfacing layer
  - UL–M ('ultra thin' from the French 'ultra-mince') (Tilcon, White Mountain and others)
  - Masterflex (previously called Wimphalt) (Wimpey Minerals, now Tarmac)
  - Axoflex (Redland Aggregates)
  - Hitex (Bardon Roadstone)
  - Brettpave (Brett Asphalt)
  - Viatex (RMC)
- Hybrids of the two above
  - Tuffgrip (ARC)
  - Colrug (Colas Ltd)
  - Thinpave (Bardon Roadstone)
  - Euro-Mac (McSweeney/Murray Brothers)
- Thin stone mastic asphalt
  - Masterpave (Tarmac Quarry Products)
  - SMAtex (Bardon Roadstone)
  - Brettmastic (Brett Asphalt)
  - Megapave (Mid-Essex Gravel)

And there are others.

The Highways Agency have now issued a draft working specification (Appendix B) which, it is hoped, will enable manufacturers to tender against client specified generic requirements subject only to additional site specific requirements. Details of these follow in the text.

10.2 PROPERTIES AND APPLICATIONS

10.2.1 Properties

Modern materials in this category are expected to perform to a high standard. Current expectations of properties for thin surface course layers for use in the United Kingdom cover the following:

- acceptable texture and skid resistance;
- good rutting resistance and load-spreading ability;
- good ride;
- consistent appearance;
- long service life with resistance against crack growth, stripping and delamination;
- reasonable cost;
- relative ease and speed of installation.

Different mixtures exhibit varying degrees of noise- and spray-reduction by comparison with chipped rolled asphalt.

Without exception, the currently evolving range of mixtures and systems embracing most of the above goals are:

- hot-laid;
- paver-laid; and
- indented- (or negative) textured surface (section 10.4.8).

All rely on inter-lock of particles and stone-to-stone contact for in-service stability and resistance to rutting.

10.2.2 Applications

Currently, thin surface course materials are mainly used in the maintenance context where the speed and simplicity of operations can match that of surface-dressing:

- without the need for after-care;
- without the likelihood of loose, flying chippings; and
- with a lessening of the uncertainty factor of varying weather conditions affecting the integrity of the end product.

Further, it is becoming apparent that the life expectancy of many of these materials is likely to match that of chipped rolled asphalts, with some developments exceeding it, and it is now found that intermediate thickness surfacings have become part of the armoury for new-build designs including the Design Build Finance and Operate (DBFO) scene.

The performance of conventional surfacings with rolled asphalt design mixtures 30/14 and 35/14 is now suspect under heavy traffic conditions and Specification Trials have been arranged for the Highway Agency to assess a performance-related approach based on wheel-tracking rate and void content (Figure 10.2). If adopted, this type of specification clause will result in greater use of modified rolled asphalt mixtures in order to meet the wheel-tracking rate requirements with resulting cost implications.

By comparison, thin surfacing materials and stone mastic asphalts already exhibit good resistance to rutting and their relative cost-effectiveness will make them more attractive, subject to satisfactory texture, skid resistance and layer adhesion, while the limiting wheel-tracking requirements in the draft rolled asphalt table (Table X in Figure 10.2) could equally be applied in specifications for thin surface course materials as appropriate.

Given adequate applications of compatible tack-coat, thin surface course materials appear to be equally effective over pavement quality (PQ) concrete (Figure 10.3), continuously-reinforced concrete pavement (CRCP), and a variety of flexible constructions. Nevertheless, thin surface course materials cannot be expected to remedy basic structural deficiencies and adequate preliminary repair procedures must be carried out prior to laying.

10.3 COMPONENT MATERIALS

10.3.1 Aggregates and filler

Because thin surface course materials are all self-texturing, the coarse aggregate must be of a character to provide the long-term resistance to wear and polishing appropriate for the traffic it has to carry. Size (section 2.2.1.1), shape (section 2.2.1.3), polished stone value (PSV) (section 2.2.4.4) and aggregate abrasion value (AAV) (section 2.2.4.3) have to be specified to provide the required characteristics, and these aspects are covered in the descriptions and specifications which follow, both in the text and in the Appendices. For long-term durability, the crushing strength or impact resistance of the aggregate may also need to be considered, particularly where heavy traffic and high PSV aggregates coincide.

The fine aggregate has to be covered by the specification as appropriate to the material type. Filler of crushed limestone is generally required.

10.3.2 Binders

Penetration grade binders are necessary with grades as specified or as formulated by the producer to give the performance characteristics required.

Combinations of Factors which Promote Wheel-Track Rutting

23. The majority of permanent deformation occurs during the summer months when the hot rolled apshalt is at higher temperatures, particularly when there are slow heavy-goods vehicle movements, such as climbing lanes. This is especially the case when newly laid material is not yet fully stable and it is less resistant to deformation under wheel loads. This can occur on road construction sites during summer months when partially completed pavements are re-opened to highway traffic under contraflow arrangements. The added factor of concentration of traffic can provide the worst combination of factors to cause permanent deformation. For this reason, Table Y* has special categories I, II, III and IV, to cater for schemes where such conditions can be anticipated in the early life of the wearing course to be laid. Special categories I, II, III and IV are also applicable to locations which can be regarded as "sun traps", in particular south facing cuttings where vehicles are travelling uphill.

24. In assessing the appropriate category, other local factors may also influence the choice, including areas which have previously demonstrated high surface temperatures and the use of aggregates with particular characteristics such as dark or light colouring. The problems of high surface temperatures can also be exacerbated on elevated structures which have less thermal capacity than where there is ground support of the pavement, and consequential higher temperatures.

TABLE X: Limiting Wheel-Tracking Requirements for Site Classifications

	Classification		Test Temperature (°C)	Maximum Wheel-Tracking	
No.	Description			Rate (mm/h)	Rut Depth (mm)
0	Lightly stressed sites not requiring specific design for deformation resistance		Not required (Shall comply with the requirements of BS 594: Part 1)		
1	Moderately to heavily stressed sites requiring high rut resistance		45	2.0	4.0
2	Very heavily stressed sites needing a premium modified rolled asphalt		60	5.0	7.0

Trafficking Newly Laid Hot Rolled Asphalt Surfacing

25. In addition to the requirements of BS 594 that "newly laid sections of asphalt shall not be opened to traffic until all pavement layers have cooled to ambient temperature", in hot weather, sites classified 1 in Table X should not be trafficked until at least 24 ahours after paving and sites classified as 2 should not be trafficked until at least 48 hours after paving, especially after contraflow working.

* not reproduced here

Fig. 10.2 Extract from the Highways Agency draft performance-related specification clause for rolled asphalt

10.3.3 Binder modifiers

Binders for these thin surface course materials can utilize a variety of modifiers ranging from polymers to stabilizing additives, such as cellulose or mineral fibre, and these are covered in the specifications which follow and in the Appendices.

Fig. 10.3 Laying 14 mm SMA over pavement quality concrete

10.4 VARIETIES

10.4.1 Ultra thin hot mixture asphalt layer – paver-laid surface dressing

This type of system essentially comprises a hot paver-laid mixture of high quality stone of either 10 mm or 14 mm nominal size carrying a sand/filler mixture in penetration grade binder. This matrix is laid directly onto a hot-sprayed polymer-modified binder film and forms a closely packed mosaic of stone with a relatively high level of surface voids, securely held in place. The original variety is known in France (where it was developed as the only product in the French classification of ultra thin hot mixture asphalt layer (UTHMAL)) as Euroduit, in the United Kingdom as Safepave (Heather, 1992) and elsewhere as Novachip.

This system produces a relatively open-textured surface (Figures 10.4, 10.5 and 10.8) and originally required a specialist paver incorporating the pre-spraying mechanism, although now conventional pavers can be modified for use.

10.4.2 Very thin surfacing layer – thin polymer-modified asphalt concrete

There are several proprietary very thin surfacing layer (VTSL) materials that have been developed in France and, more recently, elsewhere. The first one to be introduced into the United Kingdom was UL–M (Parkinson

Varieties

Fig. 10.4 10 mm Safepave after four years' trafficking (background) and 10 mm untrafficked UL–M (foreground)

and Lycett, 1993) (Figures 10.4, 10.5 and 10.9) although now Masterflex, Axoflex, Hitex and others are also available. These proprietary materials generally comprise a 10 mm nominal size through graded mixture with high-quality stone coated in a polymer modified penetration grade binder, paver-laid onto a tack-coat, which may be polymer modified.

10.4.3 UK specification for ultra thin and very thin surface layers

With the introduction of thin surface course materials into the United Kingdom, there became a need to specify what is required of them. The British Board of Agrément, on behalf of the Highways Agency, other overseeing organizations and the County Surveyors' Society, are setting up a certification scheme for thin surfacings under their *Highways Authorities Products Approval Scheme* (HAPAS) series (section 1.5.3). For use until the scheme is operative, a draft specification has been prepared by the Highway Agency which covers both paver-laid surface dressing (or UTHMAL) and thin polymer-modified concrete (or VTSL) materials.

The UK draft specification for thin surfacings was developed around Safepave and UL–M, the first material of each type to be introduced into the United Kingdom. An extended extract from the specification is given in Appendix B, and the draft will evolve as more experience is gained. A check should be made with the Highways Agency regarding the current version prior to use.

Fig. 10.5 Untrafficked 10 mm UL-M (left) and 14 mm Safepave (right)

This draft specification has enabled a range of materials to be produced, many of which are hybrids between the original Safepave and UL–M products, and will be formally included in the 1997 revision of the *Specification for Highway Works* (MCHW 1).

Varieties

10.4.4 Thin stone mastic asphalt

Stone mastic asphalt is covered in Chapter 9. It is a gap-graded mixture of high-quality coarse aggregate filled and cemented by a rich sand/filler mastic with penetration grade binder and stabilizing additives, which may consist of fibres or polymer modifiers or combinations of both, to produce a comparatively thick binder film. The material is paver-laid onto a tack coat which may be polymer modified, and can be laid as a thin surfacing course (Figures 10.3, 10.6 and 10.7).

10.4.5 UK specification for stone mastic asphalt

Stone mastic asphalt surfacing course is covered as a generic material (Nunn, 1994) by the Highway Agency's Draft Specification (Appendix A), which gives a dense lower portion with the top surface exhibiting moderate to relatively high surface voids depending on the gradation selected. There are national specifications for stone mastic asphalt elsewhere in Europe (SSO, 1994). In the United Kingdom, several manufacturers are producing materials complying with the basic requirement of the draft specification for stone mastic asphalt and some proprietary thin surfacing course materials, such as Masterpave and Smatex, have been developed based on the principles of stone mastic asphalt.

The draft specification will evolve as more experience is gained and a check should be made with the Highways Agency regarding the current version prior to use.

Fig. 10.6 14 mm stone mastic asphalt over concrete with a saw-cut joint

Fig. 10.7 14 mm stone mastic asphalt on A1, Cambridgeshire

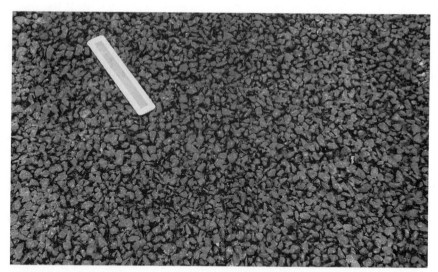

Fig. 10.8 14 mm Safepave on A1, Cambridgeshire

10.4.6 Eire provisional specification for thin bituminous surfacings

In the Republic of Ireland, the National Roads Authority has issued a provisional specification for trials of these materials (O'Driscoll, 1995). This specification, which is reproduced in the appendix, adopts a performance approach and would allow a range of thin surface course materials which encompass any of those previously mentioned. This type of specification may become more widely used in future. Its requirements

Varieties 271

Fig. 10.9 10 mm UL-M on A1, Cambridgeshire

cover skid resistance, texture depth after three years and resistance to disintegration after five years. A material named Euro-Mac has been observed being laid successfully in County Cork and others are understood to be on trial elsewhere.

10.4.7 Thoughts for specifiers

Layer thickness
Beware the temptation to specify too thin a layer. The specifier should be aware of the contract tolerances normally applied during the laying process and specify either with a minimum thickness requirement or allowing sufficient leeway on the nominal thickness to accommodate the thinnest variant taking account of the permitted tolerance.

Texture depth
Some manufacturers of thin surfacings may tend to formulate mixtures with a coarse texture in order to attain a specified minimum texture depth (say 1.5 mm for major roads). Excessively coarse textures are likely to be detrimental to overall service life and it is suggested that adoption of a specified texture range (say 1.5 mm to 2.0 mm or 1.2 mm to 1.8 mm to suit the circumstances) might control such extremes.

Production tolerances
Many reputable materials suppliers currently aim to control binder variations to within ± 0.3%. With modern plant, this is a practical measure

which increases consistency and hence enhances confidence in the performance, workability and layability of a given material. It is suggested that working to a binder tolerance of ± 0.3% is now a practical and desirable proposition which could be applied with benefit to these mixtures.

10.4.8 Surface finish and texture

These paver-laid mixtures are usually capable of providing a degree of regulating ability in a single pass and are generally perceived as producing a high standard of longitudinal profile and 'ride quality'. With all of the system types described above, the relatively thick binder films in the mixtures, achieved in their different ways, promise high levels of cohesion and good durability, but the running surfaces prior to trafficking also appear to be rich in binder (Figures 10.7, 10.8 and 10.9). This binder film wears away with trafficking, but concern has been expressed over the possibility of short-term shortfalls in skidding resistance and, in some cases, a light spread of 3 mm grit has been applied prior to final rolling to enhance the initial performance. Research is continuing to establish whether this is necessary, either generally or only in high-stress areas, and comparisons with chipped rolled asphalt will be an important factor in this matter.

All thin surface course systems exhibit a texture radically different from the chipped finish produced by rolled asphalt or surface dressing. With thin surface course materials, the finish is an integral part of the formed surface and is referred to as an indented, or negative, texture (Figure 10.10). Sand patch values exceeding 1.5 mm can be achieved. Production controls on these materials tend to be relatively tight and this leads to generally consistent surface finishes. Sensor-measured texture values on indented-textured materials tend to be lower than those recorded on positive-textured surfaces despite similar sand-patch values and work is in hand to establish credible correlations for such in-service measurement techniques. Such correlations may vary with material type, binder source and aggregate source.

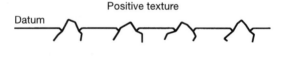

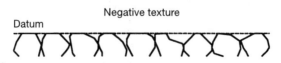

Fig. 10.10 Schematic representations of different types of texture

Varieties

Indented textured materials, as covered in this chapter, give a noticeably quieter ride inside a vehicle and external measurements are indicating that the noise generated by passing vehicles is measurably less than with chipped rolled asphalt, surface dressing or textured concrete but not as low as can be achieved by porous asphalt. Reductions of the order of 2 dB(A) and sometimes more have been claimed and research work is in hand which may enable this benefit to be more confidently quantified.

A degree of spray suppression has sometimes been claimed, but this will vary from material to material and will not approach the level of benefit in this respect which can be derived from use of porous asphalt (albeit at a far higher cost).

10.4.9 Skidding resistance

Comparative SCRIM investigations have been undertaken on the thin surfacing and stone mastic asphalt trial sites in Cambridgeshire (Laws, 1996) on the A1 at Eaton Socon and on the A10 at Littleport utilizing materials complying with the Highway Agency draft specifications (Appendices). These trials incorporate records of SCRIM coefficient values obtained from July 1995 within three weeks after opening to traffic (hereafter referred to as initial values), another run in September and then through to May 1996 (A1 only), which include 'control' values on lengths of existing rolled asphalt (together with a length of 10 mm Safepave laid on A1 in 1991) which have enabled comparisons of relative skidding resistance to be made with some confidence. In addition, a length of new rolled asphalt with coated chippings was incorporated in the A10 trial site in order to compare the development of skidding resistance of conventional HRA with that of thin surfacing materials.

The emerging conclusions, as of summer 1996, are broadly as follows:

- within 15 weeks of opening to traffic, all systems trialled exceeded the investigatory levels in Table 3.1 of HD 28/94 (DMRB 7.3.1);
- after initial trafficking all systems trialled, including new rolled asphalt with chippings, gave initial values close to the investigatory levels;
- stone mastic asphalt initial values were only marginally lower than those on the new rolled asphalt with chippings;
- the initial values for gritted stone mastic asphalt were at the same level as those on the new rolled asphalt with chippings;
- the thin surfacings showed skid resistance increasing at a faster rate than that for new rolled asphalt with chippings; and
- the stone mastic asphalt showed skid resistance increasing at a rate not less than that for new rolled asphalt with chippings.

In Autumn 1996, the Highways Agency's new high-speed pavement friction tester, operated by TRL, was run at 130 kph (80 mph) over the north bound lanes of the A1 trial site and the results on all these thin surfacing systems, including stone mastic asphalt, were not dissimilar to those obtained on rolled asphalt with chippings, with only minor variations between the three different systems.

These materials are expected to be incorporated into a revision of *Specification for Highway Works* (MCHW 1).

10.5 TACK COATS

Thin surface course materials depend critically on good bonding with the underlying support layers in order to achieve their full life potential. Therefore, a tack coat is always required, but the type and spread rates used in the United Kingdom vary depending on the system in use. It should be noted that all French contractors known to lay thin surfacings now use polymer-modified tack coat as standard and this is likely to be a sound investment for long-term durability given that adhesion to the support layer is likely to be a limiting factor.

10.6 MAINTENANCE

Rutting and fall-off of skid resistance are unlikely to be major problems with these systems provided that they are competently formulated and that coarse aggregate with PSV adequately matched to the traffic has been incorporated, although the AAV and other durability-related aspects may need to be carefully considered to control overall wear and attrition. Some closing-up of the surface may occur in the wheel-tracks in the early life of the materials but again, if competently formulated, this should not be of a scale to seriously affect the skidding resistance. However, should the texture deteriorate, these materials would lend themselves to re-texturing techniques such as high pressure water jetting because they are composed of skid-resistant aggregate throughout their depth.

Generally, these systems appear to promise long relatively trouble-free service, but at some stage in their lives, hopefully far removed, these materials may suffer local deterioration and material suppliers need to give thought to the evolution of suitable repair techniques for future use.

Some repair materials, such as Safepatch, are becoming available in bagged or tubbed form but have yet to be proved in long-term use.

10.7 ECONOMICS

The speed and simplicity of installation of the current range of thin surface course materials, coupled with the greater surface coverage per tonne compared with conventional rolled asphalts, combine to reduce the cost per square metre and hence appear to more than offset the extra costs of high-quality aggregates, the tight control of production and the incorporation of various binder modifiers. Properly designed and installed, these materials also promise good-life expectancy and hence will be attractive on a whole-life cost basis.

However, there is some contention over whether or not a thin surface layer contributes to the road structure. In a new-build situation, with the surfacing an integral part of the design, it is the overall thickness of bound materials which is counted and, in that context, the surfacing adds to the structure. Provided that a thin surfacing is formulated so as to spread the load onto its support layer, and is bonded to it, it would seem logical to assume that some structural benefit is gained by adding a layer of this range of thickness in a maintenance context – the thicker the layer, the greater the gain.

Experience in Europe with thin surface course materials appears to strongly support their economic viability and the materials covered in this chapter are in widespread and successful use in various countries, in some cases as part of a national specification.

10.8 REFERENCES

Manual of Contract Documents for Highway Works, Her Majesty's Stationery Office, London.
  Volume 1: Specification for Highway Works (MCHW 1).
Design Manual for Roads and Bridges, Her Majesty's Stationery Office, London.
  HD 28/94 Skidding Resistance (DMRB 7.3.1).
American Association of State Highway and Transportation Officials (1991) *Report on the 1990 European Asphalt Study Tour*, American Association of State Highway and Transportation Officials, Washington DC.
British Standards Institution (1992) *Hot Rolled Asphalts for Roads and Other Paved Areas; Part 1, Specification for Constituent Materials and Asphalt Mixtures; Part 2, Specification for Transport, Laying and Compaction of Rolled Asphalt. BS 594: Part 1: 1992, BS 594: Part 2: 1992*, British Standards Institution, London.
British Standards Institution (1993) *Coated Macadam for Roads and Other Paved Areas; Part 1, Specification for Constituent Materials and for Mixtures; Part 2, Specification for Transport, Laying and Compaction. BS 4987: Part 1: 1993, BS 4987: Part 2: 1993*, British Standards Institution, London.
Heather, W.P.F. (1992). Safepave – a process for the nineties, *Highways & Transportation*, **39**(1), Jan., Institution of Highways and Transportation, London.
Laws, D.F. (1996) 10 years of highways trials in Cambridgeshire, *Highways & Transportation September 1996, East Midlands Special Issue*, Institution of Highways & Transportation, London.
Litzka, J.H., F. Pass and E. Zirkler (1994) Experiences with thin bituminous layers in Austria. *Transportation Research Record 1454*, National Research Council, Washington DC.
Nicholls, J.C. and M.E. Daines (1993) *Acceptable Weather Conditions for Laying Bituminous Materials*, Department of Transport TRL Project Report 13, Transport Research Laboratory, Crowthorne.
Nicholls, J.C. (1995) Thin surfacings, *Surface Treatments* Seminar, Dec., Transport Research Laboratory, Crowthorne.

Nicholls, J.C., J.F. Potter, J. Carswell and P. Langdale (1995) *Road Trials of Thin Wearing Course Materials*, Department of Transport TRL Project Report 79, Transport Research Laboratory, Crowthorne.

Nunn, M.E. (1994) *Evaluation of Stone Mastic Asphalt (SMA): A High Stability Wearing Course Material*, Department of Transport TRL Project Report 65, Transport Research Laboratory, Crowthorne.

O'Driscoll, D. (1995) *Provisional Specification for Hot Laid Bituminous Thin Surfacings*, National Roads Authority, Republic of Ireland.

Parkinson, D. and P.J. Lycett (1993) A new wearing course from France, *Highways & Transportation*, **40**(6), June, Institution of Highways and Transportation, London, 5–10.

Partl, M.N., T.S. Vinson, R.G. Hicks and K.D. Younger (1995) *Experimental Study on Mechanical Properties of Stone Mastic Asphalt*, Report No. 113/8 (EMPA No. FE 155'176), Transportation Research Institute, Oregon State University (Swiss Federal Laboratories for Materials Testing and Research).

Swiss Standards Organisation (1994) *Splittmastixasphalt Deckschichten – Konzeption, Anforderungerungen, Ausfuehrung, Swiss Standard SN 640432*, Swiss Standards Organisation.

Van der Heide, J.P.J. (1992) Materials and mix design practices in Europe, *Association of Asphalt Paving Technology Journal*, **6**, Minnesota.

CHAPTER 11

Veneer coats

S. St John, Colas Ltd

11.1 SURFACE DRESSING

11.1.1 Concept and history

11.1.1.1 History

There is no doubt that the advantages, principally of comparative dustlessness, and the immediate relief which the tarring of an existing macadam surface will give, have caused this method of road treatment to be regarded in many quarters, particularly by automobilists, with considerable favour. There is no doubt also that it tends to economy and maintenance by prolonging the life of a macadam by acting as a waterproof seal to prevent access of water to the core of the road, and by preventing surface abrasion of the metal. (Walker-Smith, 1909)

While today's 'automobilists' may not regard surface dressing with considerable favour, highway engineers recognize the need for this treatment on many types of road. Public ignorance of the significant cost benefits of using surface dressing may be part of the cause of public disfavour, but there is no doubt that poor quality work has led to understandable criticism. This is unfortunate because surface dressing, properly designed and carefully executed, has benefits which no other treatment can offer.

Although elimination of dust was the original reason for developing surface dressing using tar as an adhesive with stone chippings as a running surface, the waterproofing benefits were quickly recognized; surface dressing preserved the structural integrity of the road. Initially, the binder was applied by can and then tar-brushed, with chippings applied by hand, but soon mechanical methods of application (Figure 11.1) were developed using horse power or steam. While much has changed in the types of binders and equipment used, including their sophistication, the concept remains basically the same; a bituminous binder is applied to a suitably prepared road surface, followed by chippings which are rolled to press them into the binder film. Subsequent traffic forms the dressing into a mosaic.

Fig. 11.1 Patent automatic tar-spraying machine outside Buckingham Palace in 1927

There are five reasons for using surface dressing:

- to provide a non-skid surface;
- to seal the road surface against ingress of water;
- to arrest disintegration;
- to provide a coloured surface; and
- to provide a uniform appearance to a patched road.

The binder used for many years was tar (section 3.2), produced as a by-product at the many municipal gas works across the country, and refined and blended into a binder that was heated so that it could be sprayed. The change to natural gas in the seventies led to bitumen replacing tar so that now it is virtually the only binder used. Some tar is still produced during smokeless fuel production, but is almost invariably combined with bitumen in a blend for surface dressing use. This represents a very small proportion of the market.

11.1.1.2 Basic principles

There are two forces at work on a surface dressing, as shown in Figure 11.2. If these forces are balanced, the dressing stays in place giving many years of life, assuming a reasonably even film of binder under the chippings.

Surface dressing 279

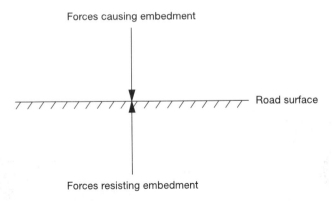

Fig. 11.2 Basic principles of surface dressing

The forces causing embedment are the wheels of traffic, especially commercial vehicles. These forces can be increased by:

- slow moving traffic;
- traffic running in channels; and
- traffic going uphill.

The forces will be decreased by the opposite actions:

- fast-moving traffic;
- wide roads or traffic lanes; and
- traffic running downhill.

The forces resisting embedment can be divided into two:

- the substrate of the road; and
- the size of chippings used in the dressing.

The substrate hardness can be classified as being from 'Very hard' to 'Very soft' (determined using the relevant graph as reproduced in Figure 11.3 for one UK latitude and one altitude category). Concrete (unless it is defective) is always 'Very hard', and will completely resist the forces causing embedment. Asphalt materials, including old surface dressings, can range from 'Very hard' to 'Very soft'. The softer the substrate is, the easier it is for chippings to be pushed into the surface (embedment). However, asphalt materials are thermoplastic; that is, they get softer as they become warmer and harder when cooled. Therefore, it is essential to consider the temperature when measuring hardness. Meanwhile, chippings resist embedment, with larger chippings being more resistant to being pushed into the substrate.

Once a dressing has been laid, it needs to stabilize with initial embedment taking place typically over a few weeks. For this to happen, the road needs to be trafficked, and it needs to be warm. If it is not warm, chippings remain like peas on a drum, held only by the binder film. When the temperatures drop as winter approaches, the binder becomes

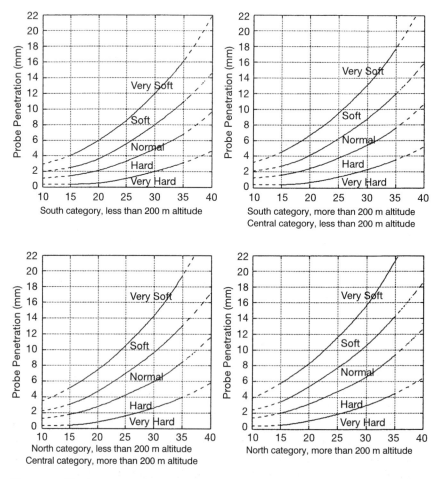

Fig. 11.3 Hardness categories from depth of penetration and road surface temperature for one latitude and one altitude category

more brittle (because it is thermoplastic) and the chipping can break away. If roads do have to be dressed later in the year, the risk of insufficient embedment can be reduced by going for a smaller chipping size and/or more binder. Nevertheless, it must be stressed that surface dressing after the end of August becomes increasingly risky. The circumstances of each site need to be analysed in the light of general principles and experience (surface dressing has been successfully laid in December and January on appropriate sites exercising sensible care and controls).

Areas of the road which are shaded will be measured as having similar hardness to areas in full sun but the shade will effectively increase the forces resisting embedment because it remains cold, and therefore more binder must be applied, as required in Road Note 39 (Nicholls, 1996) and

Table 11.1 Local factors influencing the rate of spread of binder

Influence	Property	Effect	Comments
Surface condition *	Very binder rich	0	The road condition will affect how much the binder is required to provide similar conditions at the interface
	Binder rich	1	
	Normal	1	
	Binder lean	2	
	Very binder lean	3	
Gradient	10% and over uphill	0	The gradient affects the stresses applied to the surfacing
	Less than 10%	1	
	10% and over downhill	2	
Shade	Unshaded	0	Shaded areas are cooler and therefore harder to average
	Partially shaded	1	
	Fully shaded	2	
Local traffic	Design range	0	Sizeable areas with hatched lines are effectively untrafficked
	Effectively untrafficked	2	

* Local small-area variation in the general condition, not the general surface condition used to select the type of surface dressing.
Source: From Road Note 39, TRL.

reproduced as Table 11.1 in which a unit of 'effect' can increase the binder content by up to 0.1 L/m^2.

The thickness of binder required will be related to the forces causing and resisting embedment, the size and type of chippings, and the time of year. Road Note 39 (Nicholls, 1996) gives useful guidance, and if followed would eliminate many mistakes. For multiple designs, the design steps described in Road Note 39 can be carried out using a computer program developed by TRL. The designs derived from Road Note 39 are not automatically the only designs that will work in that situation, so there is some room for error, but when two or three factors are wrong there are likely to be problems. If a dressing is going to fail, it will fail where it is weakest, such as areas light on binder. The root cause, however, may be wrong specification on chipping size.

The binder is more critical when the surface is very hard, and where no embedment takes place. Polymer addition can be advantageous in these circumstances in that it can:

- help the binder to hold on to the chippings;
- prevent the binder from becoming very brittle or very soft; as well as
- enable the binder to absorb some traffic stress.

It has often been said that surface dressing is both art and science, perhaps because it has frequently seemed unpredictable, with apparently random failures. In order to overcome this, several attempts have been made to develop a rational design approach. Some work carried out using this approach has produced spectacular failures, all from the application of too much binder. It is interesting that much surface dressing theoretical research dates from more than 25 years ago. One paper on rational design (Hanson, 1934) was written by a New Zealand highway engineer, who claimed success. In the United Kingdom currently, however, no consistently successful rational design approach is available.

Surface dressing failures are not random, and detailed analysis of the root causes of many surface dressing problems has revealed a lack of sufficient knowledge on the part of first and second line supervisors; that is, those who normally draw up specifications and execute the work. The Road Surface Dressing Association has produced a number of publications (RSDA, 1995a, 1995b) to help improve the quality of work, and is heavily involved in training local authority personnel as well as its members' employees. Successful surface dressing requires a wide range of knowledge and experience, and the development of National Vocational Qualifications (NVQs) may help to recognize this.

11.1.1.3 Types of specification

Full members of the Road Surface Dressing Association need to have Third Party Quality Assurance to BS EN ISO 9000 (BSI, 1994). In the past in the United Kingdom, it has been common for surface dressing tenders to be let on a 'Recipe Specification', with the client telling the contractor what is to be done in some detail. Arguments can and do arise between client and contractor if something goes wrong as to whether the root cause is specification or workmanship. It can be argued that the contractor should not have done work which he believed to be flawed, but commercial judgements are made, and no-one is satisfied, least of all the public, and the image of surface dressing is damaged. With the increasing use of quality assurance, and with moves to harmonization within the European Union, it is becoming more common for contracts to be let on an 'End-Product Performance Specification'. For this approach to work, it is essential for both parties to cooperate, so that all necessary information to bid is available to the contractor, and the client has assurance that he will get value for money.

In this type of contract, the choice of binder, chipping source, chipping size and all other matters are the responsibility of the contractor. In making his choice of materials and rate of spread, he will take into account what the client requires concerning skid resistance and texture depth. It is essential to specify how and when these are to be measured. When letting such a contract, it is extremely helpful for the client to supply the following basic information to those contractors selected to bid:

- details of the type of existing surface;
- hardness measurements;
- detailed traffic data; and

- any other relevant data such as altitude, gradient and road geometry.

Supplying this information avoids the waste of several contractors either making the necessary measurements, or pursuing the client's organization for data. When such a contract is let to a quality-assured contractor, the client should also save on the costs of supervision and testing which are often felt necessary under recipe type specifications.

11.1.2 Properties and applications

Surface dressing can provide a waterproof, non-skid surfacing. It can arrest disintegration, and can also be used for aesthetic reasons, either to provide a uniform appearance on a patched surface, or, when used with suitable coloured aggregates, as a more pleasing alternative to dull, asphalt black, or harsh, concrete white.

11.1.2.1 Preparation

Roads need to be properly prepared prior to surface dressing, and this includes siding out where necessary to remove encroaching vegetation, pre-patching, and pre-sweeping. The latter needs to be done immediately before the dressing is laid, but if there is excessive detritus, this does not normally form part of the surface dressing contractor's responsibilities. The responsibility for removing any excess should be defined in the contract.

Ideally, pre-patching should be done months before the dressing, to avoid the problems of binder being soaked up in the patch. It is essential that dense material with a low binder content is used for patching. If the material is fluxed in any way, sufficient time must elapse before the dressing for the flux to evaporate. Sound dressings can be spoiled by chippings disappearing into soft patches, or coming off where the porosity has resulted in insufficient binder remaining to hold the chippings. If porous material, such as dense bitumen macadam, is newly laid, it should be sealed with slurry seal or cement slurry a few days before the dressing takes place.

Surface dressing has no shaping or profiling ability, merely following the contours of the existing surface. When surface dressing is used as a final running surface in new construction, the surface which will take the dressing should be laid to final levels because the thickness of the surface dressing will be negligible. To illustrate this, it needs to be understood that the chippings settle themselves so that their least dimension (LD) is vertical. The typical least dimension of a 14 mm chipping is between 8 and 10 mm, for a 10 mm chipping the least dimension is 5–6 mm, and for a 6 mm chipping it is 2–4 mm. Because the chippings will embed into most surfaces, the actual thickness of the dressing above the surface on which it is laid is only a few millimetres (Figure 11.4).

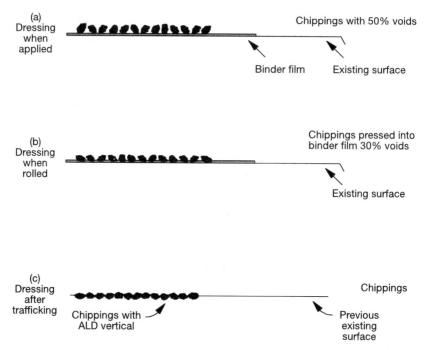

Fig. 11.4 Orientation and embedment of chippings

The rate of progress from (b) to (c) in Figure 11.4 will be dependent upon traffic volumes (especially commercial vehicles) and temperature.

11.1.2.2 Types of dressing

The past 15 years have seen an increase in the types of surface dressings that are available. Some companies have developed their own variations and techniques to provide specific solutions to certain problems. For example, the 'Hot Chip' process was for many years the only way to get good textured dressings on hard, high textured surfaces, and 'Fibredec', a surface dressing incorporating chopped glass fibres, is used on cracked and crazed surfaces, or as a Stress Absorbing Membrane Interlayer.

Single surface dressing
The single surface dressing is still the most widely used and cheapest surface dressing because it has the least number of operations. However, even when modified binders are used, there is a limit to the amount of traffic stress it can withstand.

Inverted double dressing
An inverted double dressing (also known as a pad coat followed by a single dressing) is not used now as frequently as it once was. It is a dressing with a small chipping which, after trafficking, is overdressed with a larger chipping. It was used where the surface was too hard or rugous to take a large chipping immediately, which was required to provide adequate texture and wear, but racking-in is now more common in such instances.

Racked-in surface dressing
The 1980s saw the introduction from France of racked-in surface dressing, where two sizes of chipping are used (typically 14 mm and 6 mm or 10 mm and 3 mm) with one application of binder. After applying the binder, the larger chippings are laid at about 90% of the coverage used for single dressings (the precise spread rate depending upon the shape of the chippings), followed by the smaller chippings. The stable dressing produced leads initially to a reduction of loose large chippings, and to texture depths of the final mosaic higher than can be achieved with single dressings. It is usually laid using polymer modified binders on fast, heavily trafficked roads.

Double surface dressing
The double surface dressing uses two layers of binder and chippings laid in one pass by two sets of equipment. As with racked-in surface dressings, two sizes of chippings are used. It is suitable for binder lean surfaces, and produces few loose chippings.

Sandwich surface dressing
In sandwich surface dressing, two sizes of chippings are used with one layer of binder. The larger chippings are laid before the binder is applied, and the smaller chippings are then applied over the binder. The larger chippings provide a raft, or capping layer, and resist embedment. This type of dressing, more commonly known by the trade name Mosaic, is particularly suited to roads of variable hardness, such as those heavily patched, or where fatting up has occurred in the wheel tracks.

Multiple-dressings
A proprietary system, Surphalt, was introduced from Scandinavia with trial sections in 1995 with multiple dressings that become of sufficient thickness to be almost a thin surface course material (Chapter 10). Surphalt consists of a 14 mm/10 mm 'racked-in dressing' followed by a 6 mm 'single dressing' using emulsion binders, with the 10 mm being pre-coated and the rate of spread of the 'single' binder layer being selected to fill the voids in the 'racked-in' dressing. The resultant system not only has the skid-resistant properties of a conventional surface dress-

Table 11.2 Surface dressing operations

Number	Operation	Description
1	Identify	Identify roads to be dressed. Advise statutory undertakers that any works they might foresee should be completed prior to the dressing being laid.
2	Repair	Repair potholes and damaged areas.
3	Assess	Measure the hardness of the road surface and assess the number of commercial vehicles. Note areas of high stress – roundabouts, sharp bends, traffic lights, steep hills, altitude, etc.
4	Design	Using above data, select type of dressing, binder and chipping type and size.
5	Materials	Decide who should be responsible for ordering chippings* and binder with the appropriate properties.
6	Contract administration	Contracts should be let well in advance of the work to secure skilled contractors and the best equipment. Select tenderers. Evaluate tenders, including discussions on binder types, aggregate sources, methods, programmes of work and resources to be used.
7	Method of working	Agree method of working and traffic control between client and contractor. Where necessary, give advance warning to those likely to be inconvenienced by the work.
8	Site preparation	Remove vegetation from the road edge, sweep, mask reflective studs and ironwork, note and record road markings.
9	Traffic control and execution	Implement warning signs and traffic control, sweep, apply the binder and chippings and roll
10	Aftercare	Remove surplus chippings, control the speed and path of traffic until the dressing has stabilized. Replace stop, give way and other road markings. Dust the dressing during periods of hot weather.
11	Record	Keep a record of the materials used, including their application rates, and the weather conditions, including air and ground temperatures together with relative humidity at the time of dressing if an emulsion binder is being used. Note any unusual occurrences during the work.
12	Inspect	Inspect the work regularly during the early life and record any deficiencies. Progressively extend the period between inspections, but inspect after the first frost after periods of sub-zero temperatures.
13	Investigate	Where defects have occurred, assess the reason and, in extreme cases, consider the need for remedial works.

* chippings need to be ordered well in advance.
Source: From Road Note 39, TRL.

not only has the skid-resistant properties of a conventional surface dressing but also the noise-reducing properties of thin surface course material.

11.1.2.3 Surface dressing operations

Table 1 of Road Note 39 (Nicholls, 1996) gives a helpful overview of the various activities surrounding surface dressing, and is reproduced as Table 11.2. The responsibilities for the various activities will vary from client to client, and whether a recipe or end-performance approach is followed.

11.1.3 Component materials

The materials used in surface dressing are binder and chippings. The chippings provide the running surface for traffic, and the binder is the adhesive to hold the chippings in place.

11.1.3.1 Chippings

There are a number of requirements for chippings which need to be considered and defined to ensure their suitability because not all aggregates make suitable surface dressing chippings. These requirements are size and grading (section 2.2.1.1), flakiness (section 2.2.1.3), binder affinity (section 2.2.5), polished stone value (PSV) (section 2.2.4.4), colour, coating, and aggregate abrasion value.

Selecting the chipping size is covered in Road Note 39 (Nicholls, 1996), where road hardness, traffic volumes (particularly of vehicles over 1500 kg unladen weight) and the type of dressing being laid are considered. In selecting the chipping size and dressing type, it should be borne in mind that the texture depth has a significant bearing on both skidding and non-skidding accidents, whether the road is wet or dry (Roe *et al.*, 1991). Accidents of all types are rare where the macro-texture is high and more common where macro-texture is low.

BS 63: Part 2 (BSI, 1987) sets out the requirements for grading, flakiness, and resistance to crushing of the chippings, including maximum dust content. Dust can be a source of adhesion problems, although opinions vary on this. Some aggregates are known to generate further dust in transit and use, and may therefore be requested to be 'double- washed' in order to ensure that they are still fit for purpose when used. If chippings are stockpiled for a long period and subject to rain, any dust in the stockpile will wash down. This can lead to concentrations of dust that can cause problems.

There are a number of possible tests for binder affinity, such as the Immersion Tray Test and the Total Water Immersion Test (or TWIT) developed by Shell. Other, more informal tests can be used to ensure the proposed binder and aggregate will work together, and binder suppliers can advise which aggregates are suitable. The Vialit adhesion test is also a test for binder/aggregate adhesion and compatibility; the test is being considered for inclusion in the *Specification for Highway Works* (MCHW 1) in the surface dressing clauses.

The polished stone value (PSV) of an aggregate defines its resistance to polishing and is found in BS 812: Part 114 (BSI, 1989). In selecting the required PSV, account needs to be taken of HD 28/94 (DMRB 7.3.1) from the *Design Manual for Roads and Bridges*. These take account of the relationship between skid resistance, traffic, and the required PSV. Chippings with a high PSV are a finite resource, and care needs to be taken in avoiding over-specifying the requirements.

The colour of the aggregate can sometimes be of particular significance. While bright colours, apart from shades of grey, are not available from natural sources, it is possible to obtain a variety of red and green shades, as well as several suitable types of gravel (section 12.8). Because the colours are natural, natural variations occur, and misunderstandings can happen if this is not borne in mind.

In order to promote more rapid adhesion between the aggregate and cutback bitumen binders, lightly coated chippings can be used (although they should not be used with emulsions because they actually inhibit the break); the binder film also eliminates surface dust on the aggregate. The binder used for this is generally 50 pen bitumen, with the film being as thin as possible; indeed, it does not need to be continuous. When the coating is too thick, the chippings can stick together during laying and cause considerable problems. The amount to be used is a matter of experience, and tends to vary between aggregate types; Table 7 of Road Note 39 (Nicholls, 1996) gives a guide.

Other types of coated chippings that are available are those chemically coated to remove dust and promote adhesion. This type of coating is degraded when exposed to the elements and, because they are transparent, it is impossible to see whether the coating is still in place.

In the 1950s and 1960s, various trials were conducted where doping agents were sprayed on to cutback binders before the chippings were applied in order to help promote adhesion. This approach has not been pursued in the United Kingdom, but is used in other countries, with the dope being applied by the chipping spreader immediately in front of the chipping bar. Ironically, perhaps, some of the equipment to do this is manufactured in the United Kingdom.

11.1.3.2 Binders

The purpose of binders used in surface dressing is to seal the existing road surface, and act as a bond between the surface and the chippings. As the dressing stabilizes and the chippings embed into the road surface, its role depends upon the type of surface and its traffic levels, and the types of traffic stress. The changing requirements from initial laying through to years of life present a challenge, because the best formulation to give years of durable life may be impossible to lay!

Resin thermosetting binders will be dealt with separately in section 11.3. For the moment we will consider two main types of binder, cutback bitumen and bitumen emulsion. Cutback bitumens (section 3.4.3) are often termed 'hot', and are usually sprayed at temperatures in excess of 150 °C. Bitumen emulsions (section 3.4.2), which contain water, are termed 'warm', and are usually sprayed at 85 °C. Both these types of binders can be further sub-divided into modified and unmodified. The approximate relative market share (1993) of each binder type is:

- Conventional bitumen emulsion   85%;
- Conventional 'hot' cutback bitumen   3%;
- Modified bitumen emulsion   9%; and
- Modified 'hot' cutback bitumen   3%.

Cutback bitumen (unmodified)
This is usually a 100 pen or 200 pen bitumen to which is added a diluent such as kerosene to reduce its viscosity. This enables the binder to be sprayed and also helps wetting of the chippings to take place. Cutback (or fluxed) bitumen is defined by the Standard Tar Viscometer viscosity specification in seconds. Typically, early season work in March, April and early May will be carried out using 100 seconds material. As ambient temperatures rise, a higher viscosity can be used, such as 200 seconds, and the freshly laid dressing in warmer conditions is more secure with a stiffer binder. Measuring these properties in use is difficult, and there is nothing as good as observation and experience. As road temperatures rise on a newly laid surface, it is possible to see the chippings begin to turn, presenting sticky binder to vehicle tyres. In these situations, it is essential immediately to dust the dressing or close the road, or failure will occur as the chippings are plucked out.

Many cutback bitumens are supplied with a specially formulated passive adhesion agent (section 3.5.4). These agents not only assist wetting of the chippings but also help to prevent stripping of the binder from the chippings in the presence of water. If the weather is uncertain, the use of active wetting agents can assist. Unlike the passive agents, these tend not to be heat stable and degrade within a few hours of being in binder held at spraying temperature. This means a treated tank of binder, which is not used on the day, must be treated again the following day if marginal weather conditions persist.

Another way of promoting immediate adhesion between the binder and the chippings is to use lightly coated chippings (section 11.1.3.1). The temperature at the interface between binder and chippings necessary for adhesion to take place is considerably lower when using lightly coated chippings (Gomersal, 1966). The wetting of chippings with a cutback becomes more difficult the higher the binder's viscosity, hence the fact that modified cutbacks are invariably used with lightly coated chippings.

Modified cutback bitumen

All the modified binders available today are proprietary materials. They have been developed for use where the site to be dressed poses particular difficulties, such as high stress on bends, very hard surfaces (like concrete), fast roads where early life strength and resistance to traffic stress is essential, and where wide varieties of temperature are experienced. They can provide assurance against failure at a reasonable premium on cost.

Some examples of use may help. The first dressing of a UK motorway at night was achieved by Colas Roads in 1990 using Surmac (Figure 11.5), a polymer-modified cutback. At the same time, a trial undertaken using emulsion was unsatisfactory due to high humidity. Lightly coated chippings with a high-viscosity polymer-modified binder on a concrete surface gave rapid stability and cohesive strength allowing sweeping to take place before the motorway was opened to traffic. When traffic was allowed on at normal speeds, chipping loss was virtually zero. Similar results have been achieved using other modified cutbacks, such as Shellphalt SX.

Surface dressing in urban areas, with heavily trafficked inter-urban roads frequently having footpaths where loose chippings can be dangerous as well as unpleasant, has led some engineers to specify polymer-

Fig. 11.5 Surface dressing of the M25 in 1991

modified cutbacks, using the early 'grab' and rapid stability as a means of shortening the time between laying the dressing and removing the last loose chipping from the footway.

Although modified and unmodified cutback bitumens have a number of significant advantages, it must be stressed that they are all sprayed at temperatures well above their flash point. While fires involving the use of cutbacks are rare, the author has direct knowledge of five incidents on site during a 20-year period, three of which led to the total destruction of a spray tanker, one on a public road. The irony is that, in experiments when attempts have been made to ignite spraying cutback, it proved virtually impossible, yet a tanker spraying near a roofer's asphalt boiler has ignited, and in the incident on a public road referred to earlier, no source of ignition was ever established.

In addition to the risk of fire, research has established the presence of significant amounts of airborne hydrocarbons in the immediate vicinity of the spray bar. In the past, operators have been observed wearing sandals, shorts and nothing else – apart from a cigarette. With present knowledge, a spray-bar operator spraying cutback bitumen should wear a heat resistant boiler suit, safety boots, protective hood, gloves, suitable face mask and a hard hat. Additional protection is needed when transferring cutback from one tanker to another.

Bitumen emulsions

Bitumen and water are immiscible; that is, they do not mix. Emulsification technology enables globules of bitumen to be dispersed throughout the continuous water phase. The globules of bitumen in emulsions vary from 0.1 microns up to 5 microns in diameter, and an emulsifier holds the bitumen in suspension through stabilized electrostatic charges. All binders used in surface dressing are cationic; that is, the bitumen particles are positively charged. Because mineral aggregates are naturally negatively charged, and unlike poles attract, the positive ions surrounding the globule of bitumen are drawn to the aggregate when bitumen emulsion and aggregate come into physical contact. The attraction starts the emulsion breaking process, and the bitumen particles then begin to coalesce. This coalescing can be encouraged by rolling the dressing or by the use of breaking agents.

The initial condition of a surface dressing laid with an emulsion can present a dilemma; not until the binder is fully broken is there much cohesive strength in the binder, so traffic stress can easily disturb the dressing. On the other hand, rolling provided by moving traffic can rapidly accelerate the break, and very rapid stabilization and mosaic formation can be achieved. Control of traffic immediately after laying is essential with all binders, but perhaps most of all with unmodified bitumen emulsion.

Unlike cutback bitumen, the viscosity of bitumen emulsions does not need to be modified according to the time of year. The viscosity of the emulsion is, in fact, unrelated to the viscosity of the bitumen used, which is typically 200 pen, and usually produced from selected sources of crude oil, because some make better emulsions than others. Unmodified emulsions for surface dressing are type K1-70, as specified by BS 434: Part 1 (BSI, 1984a). The designation 'K' shows that it is cationic, '1' that it is fast breaking and '70' that it has approximately a 70% bitumen content.

The standard allows for a viscosity range of 25–35 seconds. If it is higher than 35 seconds, there may be problems getting good transverse distribution. If it is lower than 25 seconds, there may be problems with run off of the binder on steep slopes. There may be problems in such instances, but viscosity has no direct bearing on the adequacy or otherwise of the emulsion. Storage of bitumen emulsion is even more critical than for cutbacks, and must be properly controlled. Most spray tankers and mobile storage tanks use oil-fired burners to heat the binder. Bitumen emulsion that is physically against the burner tubes is therefore subjected to considerable heat which will break down the binder, so circulation during heating is necessary to avoid this problem. Also, the closer the temperature comes to 100 °C, the more rapid the evaporation of the water phase.

BS 434: Part 1 (BSI, 1984a) allows for some diluent to be added to the bitumen in an emulsion, and manufacturers will normally produce different grades during the year. This needs to be borne in mind when assessing the likely success and performance of a dressing when done at the extremities of the season. For example, a dressing may be carried out in March, and the rate of spread of binder increased to ensure the dressing is held when there is little warmth in the road to enable embedment to take place. In July, the road is subjected to road temperatures of over 50 °C, and the winter grade binder is very soft due to the cut in the bitumen. A traffic diversion leads to a sudden flow of heavy traffic, and the lively binder is picked up on vehicle tyres, with disaster being the result. Conversely, a dressing done in a September heat-wave uses an emulsion with no cut at all and the subsequent road temperatures will not allow the dressing to stabilize. Sharp frosts in late November make the binder extremely brittle, and the action of traffic causes the chippings to break away, with loose chippings suddenly appearing.

Modified bitumen emulsion

As with cutbacks, emulsions can be modified with a variety of polymers (section 3.5). These can either be added to the base bitumen before emulsification, or blended in the water phase and added during the milling process. Adding polymers to an emulsion alters its characteristics in several ways, including its viscosity and break performance. Care is needed by the contractor to ensure the material is stored and applied correctly so that a successful dressing is achieved.

The performance of modified emulsion binders is similar to that of modified cutbacks: greater cohesive strength, improved temperature susceptibility of the binder, making it less brittle at low temperatures, and more viscous at higher temperatures. Unlike a polymer modified cutback, emulsions cannot be claimed to give a good initial 'grab' of the chippings because the benefits of the polymer cannot be experienced until the emulsion is broken. However, the early life vulnerability of bitumen emulsion surface dressings can now be overcome by adding a breaking agent during the spraying of the emulsion, which will make the emulsion break immediately. 'Emulcolas' from Colas Ltd and 'Nybreak' from Nynas UK AB are two systems in the market in the United Kingdom which are claimed to give the benefits of a cutback with early life stability without the health and safety drawbacks. Both require modifications to the tanker spray bar, and the fitting of ancillary equipment.

11.1.4 Equipment

However good the material that is being used, little will be achieved unless it is correctly laid. Equipment plays a vital part in this, but only if it is maintained and used correctly. Tankers and chipping spreaders must be properly calibrated, and appropriate checks and tests need to be carried out to ensure the calibrations are still sound. Manufacturers and contractors have sought to improve equipment, but some aspiring technologies have been abandoned. One of the major problems faced by contractors is the cost of new plant when set against the limited season for surface dressing. Because the latest equipment is intended for premium work, and the season for premium work is between late May and August, funding such purchases in this country, where the prices for surface dressing are considerably lower than continental Europe, is sometimes difficult to justify on purely commercial grounds.

Spray tankers
There are a variety of spray tankers currently used in the United Kingdom. During the last ten years, a number of tankers have been brought in from France, Germany and Denmark. They vary in the type of spray bar fitted, although all of the foreign made tankers have slot jets fitted, whereas UK machines usually come with swirl jets. Some developments in spray tank technology have solved some problems, but generated others. Contractors have learned to be cautious over some of the claims made by manufacturers, and a pragmatic approach is advised. There is little point having sophisticated machinery unable to spray an even mat consistently, even though it may provide impressive data print-outs.

Chipping spreaders
Chipping spreaders have tended to be home produced. Tailboard gritting boxes fitted to four- or six-wheeled tippers are still more than adequate for many types of work, although the basic design has remained unchanged

for many years. Learning from some continental imports, some UK machines have been modified so that the roller which meters the chippings is driven hydraulically, providing a more even and consistent spread rate. Care needs to be exercised when working on steep hills, with careful planning to avoid the tipper becoming unstable during gritting. Most work is now done using self-propelled chipping spreaders. These have variable heads to match the variable spray bars on tankers, and some can chip across 4 m in one pass, coinciding with traffic lanes, and avoiding joints in the dressing being made in wheel tracks. These spreaders have evolved, with incremental development, and the latest machines are now hydrostatically driven, with four-wheel drive enabling them to pull six-wheeled tippers with ease and safety, even on steep inclines. Recently German machines have entered the market, with the ability to deliver even rates of spread to a precision not before attainable.

Rollers
Three types of rollers are used in surface dressing. Most widely used are rubber-coated steel-drum vibrating rollers which, by the vibration, not only press the chippings into the binder film but also begin the orientation of the chippings. Because the drum is rigid, there can be areas untouched by the roller if the surface is barrelled or deformed. This is not the case with pneumatic-tyred rollers (PTRs) which have individual wheels which can roll every chipping within the width of the roller in one pass. Steel rollers, either small vibrating or 6 tonne deadweight, are still used but have much slower speeds than the other types. This can be a significant factor, not so much in rolling the work, but in the time taken to track between sites on a widespread programme. Each type of roller has its detractors and exponents. The mechanical effect of rolling helps speed the break of emulsions. In terms of obtaining a good mosaic, subsequent traffic is of far greater significance. Whichever type of roller is used, they need to be properly maintained and driven by experienced operators so that loose chippings are not generated by excessive speed and so that, in hot weather particularly, the wheels do not begin to pick up chippings.

11.1.5 Traffic control and aftercare

Apart from the obvious need to ensure the safety of the crew and members of the public, it is essential to consider the quality of the dressing when deciding on the appropriate traffic management both during the work and subsequently until the road can be fully open to unrestricted traffic. The County Surveyors' Society and the Road Surface Dressing Association have produced a code of practice for traffic safety and control of surface dressing operations (CSS, 1993). Each surface dressing site is different, and thorough training of those responsible for signing is essential.

Following completion of the work, the dressing needs to stabilize. On low speed, lightly trafficked roads little effort may be required. However, as traffic volumes and/or speeds increase, greater planning and care is required. Convoy systems, the use of cones and mandatory speed limits can each be used when suitable.

Once stability has been achieved it is essential to remove loose chippings. Initially, vacuum only may be used, but as the strength of the dressing increases, a full width sweeper with suction is very effective. Some loose chippings are inevitable, but laying the correct amount in the first place, and planning and executing an adequate sweeping regime, considerably reduces the problem. It should be noted that high temperatures when laying a dressing can require higher than normal application of chippings in order to avoid the chippings picking up on the machines doing the work. Some binders are more susceptible to this than others, and the significance will be related to the type of dressing being applied and ambient weather conditions.

11.1.6 Maintenance

Provided a sound approach has been followed when determining the specification and the work has been carried out with care, a surface dressing will give many years of life. Dressings have lasted in excess of 15 years and look set to last many more. It is usually possible to redress a road if the dressing is becoming unsatisfactory through aggregate polishing or loss of texture or because the road has been dug up and patched. Many satisfactory dressings have been removed as a result of reconstruction necessary because the underlying structure of the road has failed.

When End Product Performance Specifications are used, the measurement of the dressing's condition must be carried out. It would be helpful if recording and preserving data had a more significant and permanent place in surface dressing practice, so that learning from success and failure was an easier exercise.

If a dressing does fail, the remedial work needs to be approached with care because some remedial works can look worse than the failure. Unless immediate remedial action is absolutely necessary, remedial work should be carried out within the normal seasonal constraints for surface dressing, which usually means the following summer. This does not apply if the remedial action is something other than a further surface dressing.

11.1.7 Economics

Surface dressing can be used as part of new construction. At one level, this can take the form of constructing lightly trafficked areas with Type 1 material, to provide strength, and double dressing with a glass fibre reinforced surface dressing, to provide waterproofing and a running surface;

296 *Veneer coats*

if the Type 1 is laid with a paver, then the ride quality can be very good. On heavily trafficked roads, construction can be undertaken with normal black material. Rather than provide a surface course of rolled asphalt with pre-coated chippings rolled in (Chapter 5), it is feasible to surface dress a surface course of high stone content asphalt to provide the required texture for the running surface without having to roll in pre-coated chippings.

When surface dressing is used as preventive maintenance, it should be applied to the road before it has begun to deteriorate, thereby saving on patching costs, and prolonging the life of the road. It is often possible to surface dress more than once to prolong its life even further. The cost of initial construction, or reconstruction, is high, particularly when the delays to traffic are included. When a sound road is constructed, it represents a valuable asset which should be preserved.

When a heavily trafficked road is constructed, it will start to deteriorate before this is obvious, and requires an experienced eye to ensure that it is surface dressed before hairline cracks start to let in water, or before the skid resistance or lack of texture becomes a problem (Figure 11.6). If this is done, the costs of patching are eliminated. If only patching is done, problems can multiply until reconstruction is inevitable.

Surface dressing is so fast that it is not unknown for the pre-patching of a road to take two weeks, and the dressing over the completed patching to take half a day. As well as the cost per square metre of the patching, the cost of preparing documents and letting the patching work, as well as supervising the work, need to be considered. Many highway authorities have followed a policy of requiring roads which have been built by developers to be surface dressed before the authority adopts them, so that they are acquiring a preserved asset. Similar thinking needs to be applied to other new roads.

Some roads consist of little more than surface dressing on surface dressing, and they provide excellent service in many parts of the road

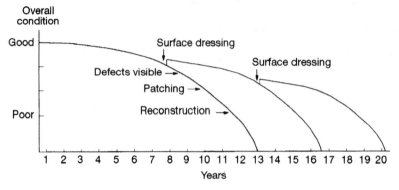

Fig. 11.6 Surface dressing to preserve the heavily trafficked highway asset

Slurry surfacing

network (Figure 11.7). Here the gain from each dressing is very marked. It may be that a time comes when another surface dressing is not possible, and another surface treatment, such as micro-asphalt, is required. The main point to bear in mind is the need to preserve rather than reconstruct. This approach of preservation is not a new idea. Few homeowners allow their window frames to rot and then buy new ones; a regular coat of paint every few years will preserve them for decades.

11.2 SLURRY SURFACING

11.2.1 Concept and history

'Schlämme' in German means slime or mud. This is the name given in the early 1930s in Germany to a mixture of sand, filler, clay, water and straight-run bitumen which was laid on a road surface. The particles of water and bitumen were separate from each other, and the aggregate particles were uncoated until the material was laid; evaporation of the water and coating of the aggregate by the bitumen resulted in a 'slurry seal'. The result of applying a thin slurry seal to an old, weathered asphalt surfacing, is to replace the binder and fines lost over the years.

The process was introduced in the United Kingdom in the late 1950s. In the late 1960s and early 1970s, further development work in Germany produced a material which could be laid in thicker applications, and this material is known as micro-surfacing, or micro-asphalt. Possible applications now range from ready mixed tubs of fine material suitable for a garden path, to micro-asphalt materials laid through specially designed mixer/applicators, able to re-profile a heavily trafficked carriageway.

11.2.2 Properties and applications

The following are the four main areas where slurry surfacing can be used, enabling the process to meet a number of needs:

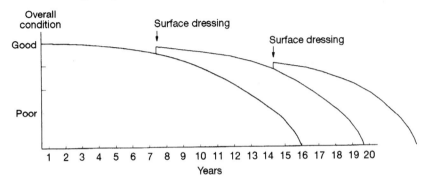

Fig. 11.7 Surface dressing to preserve the lightly trafficked road

- footway slurry: thick or thin application
- carriageways: thin slurry surfacing
- carriageways: thick slurry or micro-asphalt
- airfield surface treatments

Footway slurry
Usually this process is mechanically mixed and hand applied, with brush finishing to provide surface texture. Many footways have been neglected for years and thorough preparation is vital to ensure the slurry is effective. For example, power washing ensures good adhesion. By the addition of differing aggregate sizes, and binder modifiers (including polymers and fibres of various types), it is possible to regulate up to 50 mm in depth. It is essential for both client and contractor to be clear which of the following is required:

- a thin seal to a flat surface where no preparation is required; or
- the reshaping and patching of a deformed footway which is breaking up and full of weeds.

Both thick and thin slurry surfacings can be done, with an aesthetically pleasing dense material as the result. It may also be necessary to raise ironwork, depending on the thickness laid. As well as modifiers to give the slurry greater strength and cohesiveness, pigmented binders can produce coloured slurries which can be used to denote cycle tracks and the like.

Thin carriageway slurries
This is usually a 3 mm application laid by a continuous-flow machine. The speed of laying is such that traffic disruption can be minimized. The most widely used material used to be an anionic system where the cure time was extremely rapid; with the cationic systems now used, the set is slightly slower but hand work in restricted areas is easier. Adjustment of ironwork is not normally required with thin slurries, but preparation work such as patching and weed killing needs to be considered.

Thick slurries
Thick slurries are usually one-coat treatments which use fast-breaking, polymer-modified emulsions with 6 mm graded aggregate and which give a more durable finish than thin slurries. These are more appropriate on more heavily trafficked sites which require little re-profiling. Weed killing and patching may be required prior to the application of the slurry.

Micro-asphalts
Micro-asphalts (Figure 11.8) typically use aggregates up to 10 mm and are usually laid in two layers; the first layer regulating and re-profiling while the second layer provides a dense, textured riding surface. Many imperfections can be corrected during the application, filling ruts and potholes.

Any ironwork will almost certainly need adjustment, and the areas around the raised ironwork need to be applied carefully using the same material as the surfacing. Unfortunately, in the past, some excellent work has been spoiled by poor ironwork adjustment.

Airfield surface treatments

Slurry sealing of runways and taxiways has been widely used for many years. Slurry sealing seals the surface extremely effectively providing good skid resistance. The aggregates used are fine enough to prevent loose chippings being a hazard to jet engines and, most significantly in such a facility, the process is very fast. Unlike the other slurry surfacings, this application is usually rolled with a pneumatic-tyred roller. Aprons, taxiways and maintenance areas of an airfield can be subjected to fuel spillages. In such areas a fuel resistant slurry can be applied.

All slurry surfacings and micro-asphalts are cold processes. It is unnecessary to use large amounts of scarce hydrocarbons heating up a material which then has to cool before it can be trafficked. As awareness grows of the significant environmental footprint resulting from the hot mix processes, slurry surfacing and micro-asphalts are likely to be a preferred option for many clients.

11.2.3 Component materials

Slurries and micro-asphalts are a blend of coarse and fine aggregates, mineral filler (usually cement), and bitumen emulsion. BS 434: Part 2

Fig. 11.8 Micro-asphalt applicator with articulated laying box

(BSI, 1984b) and the *Specification for Highway Works* (MCHW 1) both include two grades of slurry seal, the thickest being laid 3 mm thick. K3-60 emulsion is the binder most frequently used in standard footway and carriageway slurries. The K3-60 designation means it is a stable emulsion, with 60% binder content.

The earliest slurry seal emulsions were stable, slow setting anionic systems. They relied upon evaporation of the aqueous phase to initiate the break and could only be applied in a thin layer. For some time, the carriageway market was dominated by an anionic system, which set extremely quickly. This material required considerable knowledge and experience to lay satisfactorily. The speed of set was fast, varying between 30 and 60 seconds under normal conditions. This made it extremely useful on airfields, where stories of a slurry machine pulling off a runway as a plane landed have passed into contractor folklore. On the other hand, this speed of set made handwork impossible, and hand applied cationic material had to be applied in areas where the machine could not lay.

With K3 emulsions, the setting times are about 10 minutes and rely on the electrostatic attraction of the positively charged emulsified bitumen particles to the negatively charged aggregate particles. The speed of break can be controlled as the process is applied by the use of doping agents, but care needs to be taken in case there are factors affecting the break which the doping agent masks. In this respect, it is vital to select the right aggregates in terms of reactivity with the binder. If the match between binder and aggregate is poor, there will be problems when laying with material either going off too fast or too slowly, or never achieving acceptable levels of strength and stability.

The difficulties of finding suitable aggregates is demonstrated by Ralumac, which was the first micro-asphalt introduced into this country. The fine aggregate used originally in Germany was a 0–2 mm graded crushed glacial moräne sand, and this has been used extensively in this country because of its excellent characteristics. No directly equivalent material is available from UK sources.

The polished stone value (PSV) (section 2.2.4.4) of the coarse aggregate fraction is important because this affects the final skid resistance of the finished surface. It is the proprietary systems which are able to use the coarser aggregates, thereby providing better texture and skid resistance, but lasting texture depths in excess of 1.5 mm are difficult to achieve with attempts to use larger aggregate in order to achieve such textures leading to unacceptable chipping loss. However, the critical point where texture becomes a problem, in terms of its relationship to all types of accidents in wet and dry conditions, is about 0.7 mm SMTD (sensor-measured texture depth) (Roe *et al.*, 1991), and this can be achieved and maintained by micro-asphalt systems.

The market has many proprietary processes. As well as using various polymers to enhance the performance of a binder to coat the aggregate

Slurry surfacing

and to give greater cohesive strength in the finished surface, the material characteristics are changed through different sources of aggregates, the addition of various doping agents, and the use of fibres.

Coloured slurries can be produced in one of two ways (section 12.8). In the first method, pigment is added to the mixture in sufficient quantities to achieve the desired colour. This can be satisfactory for either untrafficked or very lightly trafficked areas but, because the coloration relies on the pigment alone, the colour is lost when the pigment wears off the exposed surface of the slurry. The choice of colours is also very limited. The effect of the pigment on the make up of the mixture needs to be taken into account because, being made up of a very fine powder, it can adversely affect the grading of the fine fraction of aggregate, and possibly weaken the slurry. The other way is to use a colourless or albino binder. Such binders are readily pigmentable and, when used with aggregate which is sympathetic to the desired shade, a truly coloured slurry can be achieved. These materials are considerably more expensive than conventional slurries.

11.2.4 Contract types

Using BS 434: Part 2 (BSI, 1984b) or the *Specification for Highway Works* (MCHW 1) enables a client to let a contract on a recipe specification basis, and this still does happen on some contracts. However, this can prevent the use of materials offering better value for money. Alternatively, the *Highway Authority Standard Tender Specification for Slurry Surfacing* (CSS, 1995) follows the European Union approach of end product performance specification, and can be applied to both footways and carriageways, covering both thin and thick systems. The document emphasizes the need for the client to measure the end performance, which is defined in terms of:

- trafficking time;
- transverse surface regularity;
- deformation;
- macro-texture (on carriageways);
- abrasion loss; and
- weed control.

11.2.5 Maintenance

One of the benefits of using slurries and micro-asphalts is their pleasant appearance. This is not immediately apparent on carriageway work because the surfacing often relies on the action of traffic to enable it to settle down. Persistent uneven appearance is usually due to inconsistent mixing, and is difficult to rectify other than by going over the entire surface again. The use of an end-product performance specification requires monitoring by the client, and the use of a suitably quality assured contractor. This should lead to a job being right first time, and consequently meeting expected performance requirements.

11.2.6 Economics

The variety of systems available means that there is a wide variation in prices. The client needs to be clear just what the surfacing is required to provide and then select a material accordingly, remembering that value for money is not just a matter of price. Apart from clarifying what is included, consideration also needs to be given to how much time and effort the client may need to invest if the contractor is poorly organized with inadequate resources. Poor quality does not come cheap.

11.3 HIGH-FRICTION SURFACINGS

11.3.1 Concept and history

Immediately after the creation of the Greater London Council (GLC) in 1965, much thought was given to the methods by which road accidents could be reduced. It was realized that about three-quarters of all the road accidents occurred at or within 15 m of road junctions, pedestrian crossings or other similar conflict locations. Les Hatherly, then Chief Engineer of the GLC, approached Shell Chemicals to enquire whether they had any system suitable for use on the approaches to junctions. Shell's response was that they had a bitumen-extended epoxy-resin system which, in association with calcined bauxite aggregate, might be a possible solution. Experiments were carried out in 1966 on a very heavily trafficked section of road in London, and it was apparent that the system was viable, provided equipment could be designed and built to apply it.

The first six junctions were treated in 1967, and within a few months it became clear that a very large reduction in accidents had been effected. Until this time, it had been assumed that skidding did not play a large part in the creation of accidents in a major city, and the reported level of skidding accidents had always been remarkably low. It is believed that this was because few people, including the driver, were aware they were in a skidding condition during an accident. The term skidding is frequently taken to apply to a vehicle wildly out of control, but this rarely happens in the relatively low-speed traffic patterns of a major city. What is needed in this situation is to ensure the shortest possible stopping distance by increasing the skid resistance.

From this collaborative approach between local authority and industry, 'Shellgrip' (and the technically equivalent 'Spraygrip') was born (Figure 11.9). To date, this surface treatment alone has saved staggering amounts of money. For example, in London alone £3 million had been

High-friction surfacings

Fig. 11.9 Spraygrip application at Blackwall Tunnel

spent by 1981, and £24 million saved through accident prevention. More recently, alternative systems have been developed. These utilize alternative binder types. The key examples to date are:

- straight epoxy resin;
- polyurethane resin;
- acrylic resin; and
- thermoplastic rosin ester materials.

The original systems are applied by machine but the more recent systems are typically hand-applied.

11.3.2 Properties and applications

High-friction surfacings have played a well-documented role in accident reduction when used at critical sites where sudden braking or turning is likely. Approaches to roundabouts, traffic signals, pedestrian crossings and railway level crossings are designated categories J and K in Table 2 of HD 28/94 (DMRB 7.3.1). These are the sites where research has shown that 80% of London's accidents occur. Because the government has set a national target for the reduction of road casualties, the need for such systems remains.

The performance of the original systems, using bitumen-extended epoxy-resins and calcined bauxite aggregate, has been shown to be sustained over long periods of time. This is despite not meeting the 1.5 mm texture depth requirement for high-speed roads in the *Specification for Highway Works* (MCHW 1). Westminster Bridge had a Spraygrip surface for 19 years, still performing satisfactorily when other works led to its removal. Calcined bauxite is very resistant to polishing, and thermosetting binders are unaffected by changes in temperature, particularly when compared with thermoplastic or bitumen-based systems. Embedment of the chippings does not take place. The binders used are fuel-resistant as well as extremely strong, which has made them ideal for use with red aggregate to designate bus lanes, including bus stops, in London. Although high skid-resistance is not necessary in these situations, this application of part of the high-friction course technology is worth bearing in mind.

The bitumen-extended epoxy-resin systems can be specified using clause 924 of the *Specification for Highway Works* (MCHW 1), but this clause is not appropriate for the other binder types. The British Board of Agrément, on behalf of the Highway Agency, other overseeing organizations and the County Surveyors' Society, are setting up a certification scheme for thin surfacings under their *Highways Authorities Products Approval Scheme* (HAPAS) series (section 1.5.3) as an alternative to preparing a series of clauses for each binder type that are technically equivalent. Under HAPAS, both the product and the applicator will need to be certified, and any specification can then require that the system is certified to the relevant level and applied by a certified applicator. The high-friction systems will be tested using laboratory tests (Nicholls, 1997), the primary ones having been calibrated against road experience (Nicholls, to be published).

The overall rankings for the various resins trialled by TRL with calcined bauxite as the aggregate were found to be as follows:

High-friction surfacings 305

	SCRIM Coefficient	Texture depth	Visual assessment	Overall
• Epoxy resin	1 =	1	1	1
• Polyurethane resin	1 =	2	2	2
• Acrylic resin	1 =	3 =	4	3
• Rosin ester	4	3 =	3	4

Combining the rankings for the individual properties gives the overall ranking, with the systems that were tested in descending order of performance being epoxy resin, polyurethane resin, acrylic resin and rosin ester. This overall ranking might be affected by the addition of other factors, such as damp susceptibility. In particular, a defect observed at the polyurethane resin sites soon after laying was that some areas had a yellow appearance in which the aggregate was not retained properly as a result of a reaction between the curing agent and moisture, causing gassing and a breakdown of the system's adhesive and cohesive strengths. The most likely source of the moisture was from the filler component, which it is assumed had become contaminated. This problem demonstrates the particular care that needs to be taken with the component materials of these systems prior to mixing.

11.3.3 Component materials

11.3.3.1 Binders

Bitumen-extended epoxy resin
Thermosetting polymers are produced by blending two liquid components, one containing a resin and the other a hardener, which react chemically to form a strong three-dimensional structure. When two component epoxy resins are blended with bitumen, they display the properties of modified thermosetting resins rather than those of bitumen.

A specially designed machine heats, meters, mixes and sprays the material on to the road surface; the calcined bauxite is then applied to excess. The curing time on the road is dependent on ambient temperature, typically being between 2.5 and 4 hours, after which the excess bauxite is swept off and the road opened to traffic. Even though the binder has then cured sufficiently to hold the aggregate, further curing continues. Indeed, the testing required in Clause 924 of the *Specification for Highway Works* (MCHW 1), which is described in BS 2782: Part 3: Method 320A (BSI, 1976), requires seven days to elapse between taking samples of the mixed material and testing them.

Adhesion is excellent (other than to cement concrete, which is poor and the use of standard systems is not recommended) provided the road surface is not contaminated, although more care needs to be exercised when applied to freshly laid asphalt and dense bitumen macadam. These types of systems are normally applied by machine as required in Clause 924 (MCHW 1). Once applied, the surface needs to be protected from traffic throughout the curing process.

Epoxy resin
Systems using this binder, which gives faster setting times than the bitumen-extended resin, are comparatively recent. Machine application is more difficult, but the material can also be applied by hand. The material offers good adhesion to most surfaces, including cement concrete. Health and safety considerations are similar to bitumen-extended processes. The system does not comply with Clause 924 of the *Specification for Highway Works* (MCHW 1).

Polyurethane resin
The polyurethane resins used are typically two- or three-part binder systems with good adhesion to all surfaces, although pre-treating with a primer may be necessary with some. The systems are hand applied with brushes, rollers or spray. Polyurethane contains isocyanates, and has significant health and safety risks to those using it, as well as those nearby. It, too, does not comply with Clause 924 of the *Specification for Highway Works* (MCHW 1).

Acrylic resin
Acrylic resin is a very fast setting two-part material which will adhere to all surfaces without pre-treatment. Normally, machine-sprayed application is used with high productivity possible. The clear binder makes pigmentation possible. However, care is needed when applying due to the toxic nature of materials. Again, this system does not comply with Clause 924 of the *Specification for Highway Works* (MCHW 1).

Thermoplastics
The binder is generally rosin ester, but these materials are generally supplied pre-mixed with calcined bauxite. The thermoplastic mixture is heated and then screeded on to the road by hand, giving a characteristic patchwork appearance. Setting times are fast, and the operation requires little skilled knowledge or investment in equipment to apply. The thermoplastic systems will also fail to meet the specification laid out in Clause 924 of the *Specification for Highway Works* (MCHW 1). The material can be easily pigmented, although the colour is not durable and tends to discolour under traffic. Although many sites have demonstrated very poor durability, those which have been satisfactory have shown good skid resistance.

Surface dressing binders
Successful work with calcined bauxite as the aggregate and polymer-modified bitumen emulsions and cutbacks has been undertaken.

Provided stress levels are not too high, and, more importantly, the aggregate does not embed into the underlying surface, these may prove of benefit in certain limited areas. However, they are not suitable for use in very high stress areas and will not comply with Clause 924 of the *Specification for Highway Works* (MCHW 1).

11.3.3.2 Calcined bauxite

All the above are adhesives for the application of high polished stone value (PSV) aggregate, generally calcined bauxite although there are other artificial aggregates such as 'Dynagrip'. Calcined bauxite is a by-product of aluminium production, and the major supplier is Guyana. Supplies are also available from Brazil and China, but colour and other characteristics are more variable. Supplies can be unpredictable with consequent effect on prices. Typical figures for Guyanese bauxite are a PSV of 75 and an aggregate abrasion value (AAV) of 4. Both these figures reflect the superb performance of the aggregate. It has a high specific gravity in comparison with natural aggregates. The characteristics of calcined bauxite cause high wear rates on the machines that handle it, especially suction sweepers which remove the surplus aggregate once the binder has cured.

11.3.4 Maintenance

The various binder and calcined bauxite systems can have local adhesion problems causing patches to plate off. In such circumstances, it is comparatively simple to repair the surface. With thermoplastic systems, reinstatement is also straightforward, and traffic management is likely to be the main difficulty.

11.3.5 Economics

While initial costs per square metre of road treated are high with high-friction surfacings, it is essential to assess them in terms of whole life cost and to include the savings made through accident reduction. Fatal accident costs are put at £750 000 per accident, and the cost of injuries and damage to vehicles and property produces a huge, although comparatively hidden, cost to the economy each year.

Some of the systems now available are cheaper to apply, and time will confirm whether or not they give good value.

11.4 REFERENCES

Manual of Contract Documents for Highway Works, Her Majesty's Stationery Office, London.
　Volume 1: Specification for Highway Works (MCHW 1).

Design Manual for Roads and Bridges, Her Majesty's Stationery Office, London. *HD 28/94 Skidding Resistance* (DMRB 7.3.1).

British Standards Institution (1976) *Methods of Testing Plastics; Part 3, Mechanical Properties; Method 320A, Tensile Strength, Elongation and Elastic Modulus. BS 2782: Part 3: Method 320A: 1976,* British Standards Institution, London.

British Standards Institution (1984a) *Bitumen Road Emulsions (Anionic and Cationic); Part 1, Specification for Bitumen Road Emulsions. BS 434: Part 1: 1984,* British Standards Institution, London.

British Standards Institution (1984b) *Bitumen Road Emulsions (Anionic and Cationic); Part 2, Code of Practice for Use of Bitumen Road Emulsions. BS 434: Part 2: 1984,* British Standards Institution, London.

British Standards Institution (1987) *Road Aggregates: Specification for Single-Sized Aggregate for Surface Dressing. British Standard 63: Part 2: 1987,* British Standards Institution, London.

British Standards Institution (1989) *Testing Aggregates; Part 114, Method for the Determination of the Polished-Stone Value. BS 812: Part 114: 1989,* British Standards Institution, London.

British Standards Institution (1994) *Quality Management and Quality Assurance Standards. BS EN ISO 9000: 1994,* British Standards Institution, London.

County Surveyors' Society (1993) *Code of Practice for Traffic Safety and Control of Surface Dressing Operations,* Wiltshire County Council.

County Surveyor's Society (1995) *Slurry Seal Surfacing (Works). Highway Authorities Standard Document* and *Notes for Guidance,* Cheshire County Council.

Gomersal (1966) *Report 0361,* Coal Tar Research Association.

Hanson, F.M. (1934) Bituminous surface treatments of rural highways, *Proceedings of New Zealand Society of Civil Engineers,* Vol 21.

Nicholls, J.C. (1996). *Design Guide for Road Surface Dressing, Department of Transport TRL Road Note 39* (4th edn), Transport Research Laboratory, Crowthorne.

Nicholls, J.C. (1997) *Laboratory Tests on High-Friction Systems for Highways,* TRL Report 176, Transport Research Laboratory, Crowthorne.

Nicholls, J.C. (to be published) *Trials of High-Friction Systems for Highways,* TRL Report 125, Transport Research Laboratory, Crowthorne.

Road Surface Dressing Association (1995a) *Code of Practice for Surface Dressing,* Road Surface Dressing Association, Matlock (01629 732259).

Road Surface Dressing Association (1995b) *Guidance Note on Surface Dressing Aggregates,* Road Surface Dressing Association, Matlock (01629 732259).

Roe, P.G., D.C. Webster and G. West (1991) *The Relation Between the Surface Texture of Roads and Accidents,* Department of Transport TRL Research Report 296, Transport Research Laboratory, Crowthorne.

Walker-Smith, J. (1909) *Dustless Roads Tarmacadam.*

CHAPTER 12

Specialized materials

C.A. Catt, Consultant

12.1 INTRODUCTION

This chapter covers a variety of materials and processes which do not fit into the categories of the preceding chapters. With the exception of cold-laid surface courses (which in one or other of its forms is a material used every day by many) and re-texturing (which uses no new material) the materials described are low tonnage, niche products for particular purposes. Most of these specialist materials are produced by a limited number of suppliers and are often only available by prior arrangement. This is particularly so for those which use special binders like the coloured macadams and dense tar surfacing or special aggregates like Delugrip RSM and, again, the coloured materials. Some of the materials are proprietary, like Delugrip RSM and some grouted materials, and are licensed to particular manufacturers or paving contractors.

12.2 COLD-LAID MATERIALS

12.2.1 Fluxed bitumen macadams

There are three groups of cold-laid macadam surface courses:

- those made with fluxed bitumen;
- those made with deferred set binders; and
- those, mainly still being developed at the time of writing, made with flux free bitumen emulsion.

These latter meet defined engineering properties with respect to elastic stiffness and deformation resistance. They are also mixed cold. A summary of the desirable properties for cold-laid materials and how they are met by these three categories is given in Table 12.1 (after Smith, 1995).

The traditional cold-laid materials are based on fluxed bitumen which may be delivered to the manufacturer already cutback (section 3.4.3) or based on a penetration grade bitumen, usually 300 or 450 pen (section 3.4.1). Traditionally, the flux used has been creosote but now the safer kerosene is used. The flux is added at the macadam mixing plant to increase workability and storage time. The gradings used are normally the open textured ones from BS 4987: Part 1 (BSI, 1993a); usually 6 mm medium graded or 10 mm or 14 mm open-graded materials.

These fluxed materials are normally ordered with a required storage life such as 3, 7, 14 or 28 days. Producing these materials is a very inexact science which depends on many factors, not least of which are the weather during storage and the quality of their storage, neither of which is within the control of the producer. There is, of course, no particular cut-off between usable and unusable and this is often a source of much discussion between supplier and customer. The traditional practice of adding more flux in winter must be resisted as over-fluxed material may appear to take traffic satisfactorily while the weather remains cold but will deform rapidly (and possibly dangerously) as soon as the temperature rises. The materials with the longer storage times are often unusable for the first third of their nominal storage life as they have insufficient strength to carry traffic and can deform dangerously even under quite light traffic. The more heavily fluxed materials are also subject to binder drainage.

Taking all the above into consideration, it can be seen that fluxed bitumen macadams have little or nothing to recommend them. Better materials are now available at little or no extra cost, and usually at less cost when all factors including less wastage, better performance and longer life are taken into consideration.

Although they are members of this group of fluxed materials, macadams made with 200 or 100 sec cutback bitumen behave somewhat

Table 12.1 Desirable properties of cold-laid materials

Property	Fluxed	Deferred set	PCSM cold mixture
Minimum stiffness	no	rarely	yes
Good rutting resistance	no	rarely	yes
Long storage life	yes	yes	yes
Low wastage	no	yes	yes
Safe handling	probably	probably	yes
Safe laying	probably	probably	yes
Ability to stock on site	yes*	yes	yes
Use un-insulated vehicles	yes	yes	yes
Full load delivery	usually	yes	yes
Simple storage requirement	yes	yes	yes
Good workability	yes	yes	yes
Early stability after compaction (30–60 min)	no	doubtful	usually
Durable	doubtful	doubtful	not known†
Skid resistant	yes‡	yes‡	yes‡
Conforms to Montreal Protocol (no solvents)	no	no	probably

* '7-day cutback' or shorter period materials only.
† The materials are too new to be certain but, because of the design and compaction requirement, they probably will be when properly installed.
‡ Provided the correct aggregate is used.

better, particularly on footways. They should be delivered directly to site and laid warm on the day of delivery. For more information on the manufacture and use of this grade of material, see BS 4987: Part 1 (BSI, 1993a), BS 4987: Part 2 (BSI, 1993b) and BS 3690: Part 1 (BSI, 1989).

12.2.2 Deferred set macadams

Deferred set macadams are designed to have a useful storage life with adequate initial strength to carry traffic. This is achieved by using either a cutback bitumen with a less volatile and higher viscosity flux than normal kerosene or a bitumen emulsion containing a flux as well as water. Macadams made with either of these binders are mixed hot and the bitumen emulsion breaks during or shortly after mixing and the flux is necessary for workability during laying. No flux should be added at the time of mixing and, therefore, the materials produced have consistent workability and storage properties although, as with all bitumen based materials, these properties vary with temperature.

Deferred set macadams are available in both open and close textured gradings from BS 4987: Part 1 (BSI, 1993a) and also some proprietary gradings intermediate between the two. Not all suppliers supply all gradings as some combinations of binder and aggregate perform less well with some gradings than others. Deferred set materials are also available in either buckets or bags which extend the storage life of the materials to six months, or even longer if both are used. This makes them very suitable for rapid reaction emergency patching as a few buckets or bags can be carried by, for example, maintenance staff or safety audit teams.

All the cold-laid materials dealt with so far in this chapter have been mixed hot in order to dry the aggregate and make the binder sufficiently fluid in order to coat the aggregate, thus necessitating the inclusion of some sort of flux to give sufficient workability at the time of laying. The flux is very slow to evaporate from the laid material particularly if there is a recently laid cold-laid binder course under the surface course. Of course, this also means that stiffness and resistance to deformation are very slow to develop and will be very unlikely to reach the levels of even a 200 pen bitumen for several years. One source of 6 mm surface course in a bucket has come close to achieving the Highway Authority and Undertakers Committee (HAUC) stiffness requirement for 200 pen material within two years of laying, but this is unusual for a deferred set material; it was also laid on a binder course that was laid hot several years previously so there was no flux migrating from below. The fact the flux evaporates means that these materials contravene the *Montreal Protocol* which calls for the removal of organic solvents from the atmosphere and all fluxes used in asphalt materials are of course organic solvents.

A trial of deferred set material carried out in the mid 1980s showed that even the surface hardness (using a surface dressing hardness probe) took about two years to develop sufficiently for surface dressing to take place with a reasonable degree of confidence that the dressing would behave similarly on the existing road and the trial patch. This led to one particular authority changing to hot materials (45/10F rolled asphalt with 100 or 200 pen bitumen) for all permanent patching and reinstatements (Catt, 1994).

12.2.3 Cold mixed materials

12.2.3.1 General

For many years, it has been appreciated that the production of cold-laid material with the structural properties of a hot-laid material would be very useful. Various manufacturers have carried out small-scale trials at various times but there was little incentive to put a lot of effort into them as there were no defined engineering parameters to meet and there was no evidence that customers would be willing to pay a higher price than for deferred set materials which, while acknowledged not to be ideal, rarely failed catastrophically.

The catalyst for change was the publication of the *Specification for the Reinstatement of Openings in Highways* (HAUC, 1992), produced as a result of the *New Roads and Streetworks Act* (House of Commons, 1991). The Act and the Specification between them contained two fundamental changes from what went before in the context of surface courses. The Act required the Utilities to be responsible for two years (or in some cases three years) for the quality of their reinstatement and the Specification laid down the engineering properties required of cold-laid materials if they were to be acceptable in lieu of hot materials. These engineering properties are set out for asphalt materials in Table A10.1 of the HAUC Specification. Different levels of stiffness are given for equivalence to 50, 100 and 200 pen bitumen macadams. These cold-laid materials have given rise to the acronym 'PCSM' (permanent cold-laid surfacing materials) and it is these materials that this section deals with.

It rapidly became clear that, with the possible exception of bucket stored 6 mm deferred set macadam surface course which might possibly reach 200 pen equivalence, no macadam containing flux would meet any of the criteria. It also became clear that, because the equivalencies are based on dense hot materials, it is necessary for the PCSMs to be laid as dense as possible. This is also desirable on the grounds of durability as prevention of binder oxidation is one of the most important factors in prolonging the life of asphalt materials.

In order for a PCSM to be laid and compacted cold, it must contain a sufficient quantity of a fluid with a low enough viscosity to give the mixture adequate workability, and this quantity is likely to be close to the optimum quantity for maximum density. This may not be the same quantity as, but is analogous to, optimum moisture content for maximum dry density in BS 1377: Part 4 (BSI, 1990). The simplest way of making bitumen fluid at ambient temperatures for a long period is to emulsify it with water (section 3.4.2 and Whiteoak, 1990).

The bitumen emulsion in the PCSM must not break until completion of rolling, when the break should take place rapidly at least sufficiently completely for the material to have reasonable resistance to traffic induced stresses. Hot-laid materials usually need about half an hour cooling before they can safely be opened to traffic and it would seem reasonable to allow cold materials the same time period to gain strength. Certainly, anything longer than about an hour (except in unusual circumstances or weather conditions) would not be acceptable to a laying contractor.

In the longer term, it may be possible to devise a roller with a normal vibration mode to compact the material and a different one, possibly very high frequency, to break the emulsion. It is known from surface dressing that additional passes of vibrating roller can assist the speed of breaking emulsion binders, although there is no known formal testing of this experience.

Up to mid-1995, no PCSM had gained formal approval from HAUC at any equivalence level. Several companies have formal trials under way and at least some were giving early indications that they would meet the 200 pen equivalence. It is acknowledged by a number of suppliers that it will be necessary for materials to reach 100 pen equivalence before there will be wide acceptance of them; there will be reluctance for utilities to place the additional thickness needed for 200 pen equivalence material and it is easier for a 100 pen material to meet the deformation resistance requirement than a 200 pen material. It is, perhaps, surprising that the deformation resistance required is not different for different traffic levels as they are for wheel-tracking recommendations (Szatkowski, 1979).

12.2.3.2 Mixing

The mixing of these materials should take place at ambient temperatures. The aggregate is used in a damp state and the binder is stored in a tank with sufficient agitation to prevent settling of the emulsion; this is not a serious problem for the stable emulsions normally used in this process. The tanks should have sufficient heating to prevent frost damage in winter. Because the aggregates are added damp to the mixer, there must be good control over the water added to maintain the correct total fluid content needed by the material for proper workability. This may be done by:

- pre-mixing the aggregate and water to a standard moisture content prior to adding the aggregate to the mixer; or
- continuously monitoring the moisture content of the aggregate and adding the necessary additional water through a separate spray-bar in the mixer.

It is also necessary to use the correct combination of grading and fluid content so that there will be no binder drainage during handling, storage and laying. Cement addition has been advocated as an additional ingredient that would increase the modulus of the material, but adding cement to a mixture containing water automatically brings problems of shelf-life. The shelf-life is probably longer than the two hours permitted by the concrete specifications, but it is unlikely to be much longer because the cement will start abstracting water for hydration almost immediately on addition to the mixture. Both this process and the stiffening effect of the setting mortar would considerably reduce the workability of the mixture and, therefore, reduce the density obtained by a given amount of compaction and, hence, the measured engineering parameters at the end of the maintenance period.

Therefore, for any additive used other than normal aggregates and emulsions, the supplier should demonstrate compliance with the requirements both at the shortest storage time recommended and at the longest.

The production plant can be relatively simple, more akin to a concrete plant than an asphalt plant, because there is no requirement for heating and drying the aggregate. Mixing time should be sufficient to properly mix the material, but it is likely that prolonged mixing will start to break the emulsion, certainly in hot weather, and this must be avoided.

12.2.3.3 Storage

As with all asphalt materials, cold-laid materials depend on being adequately workable in order to obtain the good level of compaction necessary to maximize life, load spreading and deformation resistance. As has been shown above, either a flux or water is used to impart this workability and these additions then evaporate off to leave a material that is adequately 'strong' to carry traffic. Therefore, it follows that, in order to maintain workability throughout the storage life of a material, it must be stored in such a way as to prevent evaporation. All materials, the cold mixed ones particularly, need protection from rain which would wash the binder from the aggregate because, of course, an emulsion is fully miscible with water. Protection from frost is also very important as freezing breaks an emulsion thus destroying workability.

It has been found that cold mixed materials after two weeks in storage have much lower densities for the same compaction level compared to new material. Alternatively, higher levels of compaction would be needed to achieve the same density. This is caused by drying out of the material thus reducing the fluid content below the optimum level. Therefore, it is important to take precautions to reduce this loss.

When, as is normally the case, storage is out of doors, the material should be kept in as compact a heap as possible and, ideally, in a three-sided bay. The bay should have a hard base which is resistant to the wear and damage caused by any loading shovel used to handle the material. The back and sides should be high enough for the material not to overtop them when a new delivery is tipped or the loading shovel is working and they should be strong enough to withstand inevitable contact by the delivery vehicle or loading shovel. The bay should be constructed in such a manner that water falling on it drains away. For long term bays it is worthwhile building a concrete slab slightly above ground level and sloping by 2–3% towards the open side.

The exposed surface of the material should be completely covered with a waterproof sheet and, when frost is possible, at least a second sheet and, preferably, a quilt should also be used; in prolonged periods of continuous frost, no reasonable amount of protection will prevent the outer layers freezing which will destroy cold mixed materials. Even if the material is stored under cover, it should still be sheeted to prevent evaporation although naturally frost will be less of problem. Fluxed materials are in general not damaged by cold weather, but low temperatures render them unworkable although the workability is regained as they warm up. Storage bays must always be emptied and cleaned out before tipping a new load.

Care should be taken with material while in transit between storage depot and the work site and prior to placing in the highway. Although insulated vehicles are not needed, the material should be sheeted to prevent evaporation or damage from rain. Where a site is large enough for a full load to be delivered direct from the supplier, as many of the above principles as possible should be adhered to. It is important that the material is tipped onto a clean base in order to avoid contamination and unnecessary wastage. Care must be taken to avoid picking up material from under the heap when moving material to the work site. Of course, the heap should be covered to prevent evaporation and weather damage.

Laying and compaction of cold-laid materials should be carried out by experienced people. Material should not be laid in frosty weather. Cutback materials lose workability by temperature effects and emulsion based materials are damaged by frost. Once any of these materials has been laid and fully compacted, weather is not normally a problem. However, the cold mixed emulsion-based materials will take a long time to gain sufficient strength to carry traffic if it is raining or the humidity is high. For example, for surface dressing emulsions there is an upper limit of humidity of 80% above which work must stop (CSS, 1994). Heavy rain will wash the binder from the aggregate if the emulsion has not completely broken.

Limited trials have been carried out to assess the ability of cold mixed materials to be machine laid; these were successful but extensive work on this aspect is still awaited before it is used routinely. The materials are routinely laid by machine in France and Ireland, although there are not the specified engineering parameters which have to be achieved in those countries.

Good compaction is absolutely essential in order to maximize the life of the material. This is just as true for cold materials as for hot ones. For the cold mixed materials that have to achieve required minimum engineering properties, it is even more important because:

- the tensile modulus is heavily dependent on the density achieved; and
- the highway authority has the right to check the engineering parameters reached at the end of the guarantee period.

It is highly uneconomic if, as a result of slight lack of compaction at the time of laying, the work has to be redone; and, of course, the guarantee period starts again from the time of redoing the work. Poor compaction will also increase rutting due both to inherently less resistance to deformation and to additional compaction under traffic; if a 200 mm thick layer densifies 5% under traffic there will be a 10 mm rut produced which will be exacerbated if the lower layer materials also densify under traffic. In order to maximize the density for a given compaction effort, it is important to maintain the workability of the material and hence the emphasis on good storage conditions set out above.

Very little published information is available on the 'strength' gain of cold mixed macadams but some was published in Highways (Robinson, 1995) which showed that the 6 mm dense bitumen macadam made with emulsion had reached a stiffness of 360 MPa at 2 months and 480 MPa at 3 months against a HAUC requirement for 200 pen equivalence of 600 MPa at 2 years.

12.3 GROUTED MACADAM

The basic principle of these surface courses is very simple. A 40 mm nominal layer of open graded bitumen macadam is laid on a sound substrate and then the voids are filled with a cementitious grout which is usually resin modified. The aim is to combine the flexibility of bitumen macadam with the wear resistance, oil resistance and stiffness of a cement bound material. The materials are proprietary and are not covered by any British Standard except that the bitumen macadam should be made and laid in accordance with BS 4987: Part 2 (BSI, 1993b).

The main uses of the materials are in industrial areas and farms. Currently, there are two trade names for grouted macadam, Hardicrete and Salviacim (Tarmac Quarry Products). However, the company no longer carries out any work using the materials despite both having British Board of Agrément Certificates (BBA, 1988, 1989). On the industrial side, it is probable that block paving has taken a large proportion of the market and stronger, tougher asphalt materials have also become available or more acceptable, such as high stone content asphalts (55/10 or 55/14) and stone mastic asphalt (SMA). On farms, it is probable that the cement-based grout had inadequate chemical resistance to the acids in silage and a number of companies offer a low stone content asphalt as a suitable surfacing material in these circumstances; chippings not being added because a smooth surface is required.

The Agrément certificates for both materials are very similar and much of the information is identical, including even the wording. The underlying structure must be adequate for the imposed loading whether it be rubber-tyred, steel-wheeled or tracked vehicles. The materials are certificated for warehouses, cargo handling areas, bus depots, and on airfields (runways, hard standings and maintenance areas).

The open-textured macadam may be 10 mm, 14 mm or 20 mm nominal size and is made with penetration grade bitumen. The grout is a mixture of Portland cement, polymer latex, silica sand, fine mineral aggregates and water; plasticizers and anti-foaming agents may also be added. The grout is spread on the macadam, which must have cooled below 45 °C, immediately after mixing and spread with squeegees and brooms. It is then vibrated with plates or rollers until the voids will take no more grout. The process in any one area must be completed within 30 minutes of mixing the grout. Excess grout is removed by sweeping and the final finish is achieved using a broom or power float depending on the finish required.

The process must be carried out above 3 °C and the weather should be dry. The new surface must be protected from heavy rain for at least three hours although drizzle is not harmful; in hot or sunny weather, it is helpful to the curing process (and hence durability) to keep the surface damp. Pedestrians should be kept off the new surface for at least 12 hours and cars for 24. Full loading should not take place for 1–4 weeks depending on curing conditions and the nature of loading.

12.4 DENSE TAR SURFACING

Dense Tar Surfacing (DTS) was first laid in 1943 and was subsequently developed by the Road Research Laboratory, as it then was, and the then British Road Tar Association, who published the first specification in 1954 (BRTA, 1954). It has changed little subsequently and BS 5273 (BSI, 1975) was essentially only a metrication exercise. Until North Sea gas was discovered and effectively shut down all gas works (except in Northern Ireland) (section 1.1) which produced the majority of it, tar was the major binder used in the United Kingdom. DTS was developed as an answer to rolled asphalt which was gaining market share for the rival bitumen producers, particularly in the heavy-duty surface course area. Tar has two advantages over bitumen and several disadvantages. Dense Tar Surfacing as used today exploits one of tar's advantages; its superior resistance to softening by petrol, diesel fuel, lubricating oil and other petroleum products.

The following disadvantages of tar should be appreciated before using DTS (section 3.2):

- Tar is a known carcinogen, although the tar industry claims that the currently available tars from steelworks coke ovens and smokeless fuel plants, which are produced using a 'low' temperature process, are much less dangerous in this respect than the old 'high' temperature tars produced in gas works. Nevertheless, tar is banned in a number of member states in the European Union.
- Tar is very much more temperature sensitive than bitumen and, hence, has a much narrower range of working temperature. For example, 54 EVT tar (the usual grade) has a maximum mixing temperature of 105 °C and a temperature for completion of rolling of 60 °C, a range of 45 degrees while the equivalent figures for 100 pen asphalt surfacing are 175 °C and 75 °C, respectively, with a range of 100 degrees.
- Tar is much more reactive than bitumen and hardens much more both during mixing and subsequently on the road. The maximum mixing temperature must not be exceeded under any circumstances because then the hardening that takes place could make the material uncompactable.
- Tar also oxidizes and, thereby, hardens to a greater degree and more rapidly than bitumen in the road. Hence, it is extremely important to compact the material sufficiently to make it impermeable to air and water, otherwise the surfacing will fret, ravel and crack very quickly producing early failure. As an example of hardening, the highest stiffness measured by this author on a Nottingham Asphalt Tester (at standard temperature and load rate) was 22 GPa on a dense tar roadbase 28 years old, a similar level to the underlying lean mixture. In the past when open textured tar surface courses were used, they were surface dressed within 12 months for the same reason.

Because of all the above disadvantages, the usage of DTS is decreasing and now very few suppliers do enough to carry tar routinely. Therefore, if DTS is required, it must be planned well in advance and it is very unlikely that back-up supplies will be available in the event of a break-down. Unless oil contamination is likely to be very heavy, it is probable that a mixture such as 55/10F rolled asphalt surface course will perform at least as well as 50% stone 10 mm DTS when both are well compacted. Local spillages should be cleaned up immediately to prevent softening of either material, tar not being immune from, just more resistant to, softening. The asphalt also would have the advantage that it is very much easier to patch. In areas of very heavy contamination, other, non-asphalt materials surfaces should be considered such as concrete or small element paving, which are outside the scope of this book.

The problems of DTS have been dealt with and the material itself will now be considered. A good summary of the factors affecting DTS is given in Chapter 11 of *Bituminous Materials in Road Construction* (RRL, 1962). Very little basic work has been done since, so the information given over 30 years ago is still current.

BS 5273 (BSI, 1975) describes eight basic variants of DTS – any combination of two stone contents, two aggregate sizes and two sand types: crushed rock or natural sand. Although BS 5273 allows 10 mm material to be laid an average of 30 mm thick, it is recommended that 40 mm is normally used because of the narrow temperature range available the time for compaction is very short and, even for 40 mm layers, it is only just sufficient in good conditions. Unless the trafficking is very severe, when the use of crushed rock fines may be necessary to obtain sufficient resistance to deformation, it may be prudent to use the generally more workable natural sand fines. Because tar based materials stiffen rapidly with age, any deformation that will occur is much more likely to happen in the first summer rather than subsequently.

The main use of DTS is in vehicle parking areas and aircraft hard standings where, therefore, traffic speeds are low; hence, texture depth is not necessary in these situations. For these areas, 50% stone content material can be used which will help in compaction as there is no delay while chippings are spread and workability does not change significantly over a stone content range of 30–55%. Chippings would normally be used on 35% stone DTS to give texture depth. Thirty-five per cent stone material has been used without chippings and proved to be the most durable of the options but it did become very smooth.

The recommendation that bitumen based mixtures should not be laid in inclement weather conditions is regularly flouted – problems only showing up several years later with premature fretting at the joints and other abnormally early signs of failure. Flouting these recommendations with tar based mixtures carries a very high risk of failure in the first winter. It is strongly recommended that DTS is only laid in good weather conditions. As an example, for a 40 mm layer with ambient temperature of 10 °C and a wind speed of 4 m/s at 2 m height (equivalent to 10 knots at 10 m), there is less than 7 minutes for compaction of DTS compared to over 12 minutes for 100 pen asphalt in the same conditions (Nicholls and Daines, 1993). The time, in the same reasonable conditions, for 35 mm thick DTS is 5½ minutes and, at 30 mm thickness, it is only 4 minutes. It has been suggested (Brown, 1980) that at least 10 minutes is required for full compaction of rolled asphalt and, even allowing for the fact that chippings are not used, the time available for compacting even a 40 mm layer is very short.

Because there is very little temperature drop available for working, material should not be left in the hopper to await the arrival of another load of material but it should be placed onto the road and the run finished off and compacted to this point and a joint cut because, otherwise, a badly compacted area will result, leading to rapid failure. Other precautions to reduce temperature loss should be considered, such as:

- insulating the paver hopper;
- running the paver relatively slowly so that the rollers can keep up;
- keeping the hopper full by tipping a new load on top of the previous one as soon as is practicable; and
- leaving the delivery vehicle sheeted during tipping.

12.5 DELUGRIP

Although universally known as 'Delugrip', its full name is 'Delugrip Road Surfacing Material' and is often referred to as 'Delugrip RSM' in the literature. For simplicity, the single word will be used in this section. The material was developed from research carried out at Birmingham University in the late 1960s and early 1970s led by Dr G. Lees (Lees, 1970). Also involved were a number of researchers employed or funded by Dunlop Ltd. The two main aims of the research were to define a rational method of designing the aggregate grading and to develop a knowledge and an understanding of the interaction between tyres and the road surface (Williams et al., 1972). For a good overall summary and early experience on roads, see McKellar (1981).

As a result of these researches, Delugrip was developed to optimize the grip between tyre and road in all weather conditions, as well as to reduce the noise generated by the interaction. A patent was applied for in 1971 and granted in 1975 (Patent Office, 1975) quoting Arthur Roger Williams and Geoffrey Lees as the inventors; a subsequent patent was also granted. Originally the patent was owned by Dunlop but, when the business was taken over by a Japanese tyre company, the patent was acquired by the original inventors who set up a company, Leascon Surfacing Technology Ltd, to license the material. A number of producers in this country and abroad hold licences and it is still regularly used.

The essential factor which separates Delugrip from other asphalt surfacings is the fact that it is a planned mixture of two aggregates with significantly different resistance to abrasion, typically in the ratio of 2 : 1. The grading of the material is designed for the particular aggregates, layer thickness and surface texture required. The fundamental principle is that the grading is not based on a recipe but is based on the packing properties of the selected aggregate components to give a mixture with a controlled void content and surface texture. Once the target grading has been designed and the binder content determined, production tolerances are decided upon but they are often closer than for British Standard materials. A typical example for one particular combination of aggregates and one particular site is given in Table 12.2.

Delugrip is not a particularly easy material to lay. In general terms, it is less workable than rolled asphalt surface course and is not tolerant of hand working. It needs good compaction in order to achieve the

Table 12.2 Typical Delugrip grading

Sieve size	Proportion passing (%) Target	Tolerance
14 mm	100	–
10 mm	97	±3
6.3 mm	81	±7
3.35 mm	54	±5
2.36 mm	48	±10
600 μm	20	±4
212 μm	13	±3
75 μm	9	±2
Binder	6.5	±0.3

Note: This is not a recipe specification and is only correct for the combinations of aggregates used, for the particular thickness laid and for the surface texture required.

intended design; measuring this parameter after completion would be good practice. Because thickness is used as a parameter in the design, the substrate must be at least up to the binder course tolerance requirements in the *Specification for Highway Works* (MCHW 1). If necessary, shaping should be carried out. Delugrip has been laid over a wide range of thicknesses; for example, in the first trial of the material on the A462 at Essington, it was laid 1½ in (38 mm) thick and, on a subsequent trial on the A4 at Hammersmith, it was laid ¾ in (19 mm) thick. The designs for the material would have been different from each other.

Lees *et al.* (1977) claims that, for the same aggregate, the sideway force coefficient at 50 km/h (SFC) of Delugrip is 10 points better than rolled asphalt and chippings. On the comparison sites, the materials were only six months old, which is old enough to expose the aggregate but not old enough to achieve the long-term stable SFC, which normally takes about two years. SFC values from a longer term (six-year) study on the A4 is complicated by the fact that different aggregates were used in the Delugrip and the rolled asphalt and chippings, but the Delugrip started with an SFC of 90 when new and dropped to 65 at six years old and the rolled asphalt dropped from 53 to 45 over the same period. (The SFC measurements were taken with SCRIM but not using the current procedure, which gives lower values than the ones given here even allowing for the factor of 100 used to differentiate between the two styles (approximately multiply by 0.78).) The A4 carried 50 000 vehicles per day over the three lanes with 14% heavy goods vehicles.

Delugrip is not available off the shelf for two reasons:

- the designs are site specific and time must be allowed to carry them out; and
- two different aggregates are used so that the production unit may well need to import at least one of them.

In the early days (1975-80), it was frequently used on highways but the requirement for 1.5 mm on high-speed trunk roads on new material (MCHW 1) has reduced its popularity in this area. Delugrip does not achieve this requirement but a deeper texture does develop with time. This developed texture does not always reach 1.5 mm (sand patch) but it is usually well above the maintenance warning level of 0.5 mm (DMRB 7.3.2) measured with the high speed road monitor (HRM). The texture is less likely to be worn away than with macadams and surface dressing because of differential wear being part of its philosophy. When compared to rolled asphalt and chippings, it is difficult to be definitive as the long-term texture of rolled asphalt depends on:

- the rate of wear of the chippings;
- the rate of weathering and wear of the asphalt; and
- the resistance to penetration of the chippings under trafficking.

The main uses of Delugrip over the more recent past is on:

- airfields, both taxiways and runways;
- motor racing circuits, including Silverstone and Mallory Park;
- vehicle proving grounds, including the Motor Industries Research Association (MIRA) and the Lucas Girling test track; and
- container storage areas.

Racing cars can generate very high lateral forces on corners and very high longitudinal forces on braking, several times the weight of the car. Therefore, it is important for the surface course (not necessarily Delugrip) to be laid several weeks ahead of a major meeting so that stearic hardening can take place in the bitumen in order to maximize the stability of the material and, hence, its resistance to cornering and braking forces.

It has been shown in a number of studies that road surfaces made with smaller aggregates generate less tyre noise than surfaces made with larger aggregate. This is a separate phenomenon from the reduction in noise given by porous asphalt which absorbs noise after its generation (section 7.3.3). Delugrip has been shown to produce 3–4 dBA less noise than rolled asphalt with 20 mm chippings (Walker and Oakes, 1979). In the laboratory, it has been demonstrated that the rolling resistance of a tyre on Delugrip is less than for a tyre on lateral brushed concrete, rolled asphalt and chippings and surface dressing (Leascon, 1995).

12.6 BITUMEN BOUND AGGREGATE

It is current practice for combined surface water drains and ground water drains to be placed adjacent to the edge of the paved shoulder. It is not unusual for vehicles to run off the paved area onto the type B filter media, with which the drains are filled to ground level, dislodging

40 mm aggregate, often in considerable quantities. Originally, these drains were backfilled with 40 mm natural aggregate which, because the particles were smooth and rounded, lacked aggregate inter-lock and vehicles often sank into these drains and the difficult process of extraction scattered considerable quantities of large aggregate onto the carriageway. This problem was reduced when the requirement for the top layer of aggregate to be crushed rock was introduced. However, the problem of loose aggregate remained, particularly where vehicles travel close to the drain, such as in motorway contraflows situations, on all purpose roads with 1 m marginal strips and on the inside of bends on slip roads having a significant degree of curvature.

If a piece of this 40 mm aggregate flies up under the action of traffic, it can inflict substantial damage to a vehicle. It can easily break glass in windscreens or headlights and can put a considerable dent into sheet metal. It is obviously potentially very dangerous and, on busy main roads and motorways, it is difficult, dangerous and/or expensive to clear the hazard away. It would obviously reduce the problem considerably if the top of the drainage material was bound in some way and in such a manner as not to be seriously damaged by occasional overrunning. The material should be sufficiently durable in order that the original problem does not return and sufficiently permeable, say with a permeability coefficient better than 10^{-2} m/sec. Conventional type B filter material has been measured at more than twice this figure.

Various attempts have been made to alleviate the problem, mainly by spraying the top of the aggregate with bitumen emulsion, which is a palliative at best because it is not particularly durable and not resistant to traffic induced forces. Airfields have a similar problem with the added hazard of ingestion into jet engines generating considerable hazard or damage. The Ministry of Defence specification for rolled asphalt and macadam for airfield pavements (MOD, 1995) contains a material specification that would be suitable for highway works.

The material in the MoD specification is called '28 mm pervious macadam' but this is likely to be changed for highway use to 'Bitumen bound aggregate'. The grading is shown in Table 12.3.

No fine aggregate is added and the amount of aggregate passing the 14 mm sieve is to allow for the naturally occurring dust and filler and either hydrated lime or ordinary Portland cement used as an anti-stripping agent. These last may be substituted by stearine amine or other wetting agent provided that binder drainage is not increased.

The binder can be either 200 pen bitumen or modified binder. The binder content is designed using the binder drainage test, as used for porous asphalt with minor modifications. The binder content used is the design binder content from the drainage test with a maximum of 4.5%. The minimum binder content permitted is 3.0% for 200 pen bitumen or 4% for modified binders. There is also a recipe mixture using 3.4% of

Table 12.3 28 mm pervious macadam

BS sieve	Proportion passing (%)
37.5 mm	100
28 mm	80–90
20 mm	10–30
14 mm	4–6

200 pen bitumen for crushed rock and 4.3% for slag. In all cases, the manufacturing tolerance for the binder is ± 0.3%.

Because the materials made with modified binders have a thicker binder film than the 200 pen materials, they will be more durable; in this situation, the nature of the modifier will have little or no effect on performance. It would be sensible for these higher binder content materials to be used in all new work and in maintenance work of significant size. If the experience of 20 mm porous asphalt is relevant, and this is likely, then suitable modifiers to give the required binder retention include natural rubber latex or powder (section 3.5.2) and cellulose fibres (section 3.5.5). Other modifiers, like SBR and SBS (section 3.5.2), are supplied ready mixed to the asphalt manufacturers and would not be viable for less than about 500 tonnes laid in a few days, which is equivalent to about 2 km of drain laid 100 mm thick. The natural rubber and fibres are added at the mixer on a batch by batch basis so is available like any other hot asphalt material.

As with all asphalt mixtures, it is necessary to place the materials while they are still hot and to compact them properly to obtain maximum service life. Because drains are alongside roads, there is an opening for someone to design a small machine with a hopper wide enough to receive hot material from a delivery vehicle and, with some form of lateral conveyor to a simple screed, to place the bitumen-bound aggregate to the correct line and level. Compaction should be carried out while the material is still hot and a number of small rollers are likely to prove suitable. For small areas, a single drum pedestrian vibrating roller is likely to be suitable but, if the laying is mechanized, it will be necessary to use a double drum vibrating roller having the width of the laid material in order to keep up.

An alternative approach (Roadtex, 1994) uses shredded vehicle tyres of a suitable grading and binder content. A draft specification clause has been produced with two alternative gradings, which are given in Table 12.4.

The binder content may seem high but rubber has a relative density of about a third of that of rock aggregate. The binder is polymer modified and fibre reinforced. The material is mixed on site and is placed by lateral chute from the mixer which is lorry mounted. The material is hand raked into place and rolled with a pedestrian roller.

Table 12.4 Asphalt bound chopped/shredded rubber

Sieve size	Proportion passing (%)	
	5/70	5/71
63 mm	100	100
37.5 mm	75–100	85–100
20 mm	0–25	25–50
10 mm	0–5	0–20
Binder	30 ± 5	30 ± 5

The claimed advantages are that the process uses a recycled product that is otherwise difficult to dispose of and, if the material does get broken up and dislodged, the particles are light enough and soft enough not to cause impact damage to vehicles.

A provisional specification for permeability is a minimum of 0.3 sec^{-1} using the same falling head permeameter as is used for porous asphalt. This is equivalent to a permeability coefficient of very approximately 2×10^{-2} m/sec.

12.7 RE-TEXTURING

The processes described under this heading are not asphalt materials but they are surface treatments normally, but not necessarily, carried out on asphalt materials in order to modify the surface characteristics. Some of the processes, such as bush hammering, were developed initially for concrete surfaces. Texturing methods currently in use fall into three main groups:

- mechanical impact;
- cutting or grinding; and
- fluid action.

The methods are not mutually exclusive and some processes combine elements from more than one group.

In general, all the methods are carried out by specialist contractors and the final effect depends greatly on the skill and experience of the contractor, and especially the operator on site. Different surfaces, or even different examples of the same type of surface, react differently to the same treatment. These variations can be due to:

- binder content and hardness;
- aggregate type, hardness and toughness;
- bonding of binder to the aggregate; and
- various mixture characteristics including:
  - air void content,

- aggregate grading, and
- adhesion of binder to the aggregate.

The same piece of equipment has several different variables, such as:

- speed of travel;
- pressure of water;
- size of grit;
- temperature of air; and
- speed of rotation of flails.

The factors given above are not an exhaustive list by any means, but they do indicate why the outcome of surface treatments depends so much on knowledge of the process.

A brief summary of the main types of re-texturing follows based on research (Roe, 1995) which is on-going to improve the understanding of how process and material variables affect the final result.

Bush hammering
Exposes new aggregate surfaces, thus improving micro-texture, but macro-texture is often reduced by the removal of high spots.

Shot blasting
Exposes new aggregate surfaces, thus improving micro-texture, and does not reduce micro-texture. In some circumstances when used more aggressively, it can remove some matrix, thereby increasing macro-texture.

Grooving and grinding
Increase macro-texture but have little effect on micro-texture.

Longitudinal flailing
Exposes new aggregate surfaces but has little effect on macro-texture because the formation of grooves by the flails is balanced by the removal of high spots.

Orthogonal flailing
This is an aggressive treatment forming grooves in two directions, approximately along and across the road. It considerably increases macro-texture and exposes new aggregate surfaces over a large proportion of the treated area. However, it can remove chippings from the surface and crack others in situ which subsequently become lost from the surface.

Carbonizing by hot compressed air
On its own is not normally considered a re-texturing process because it is designed to remove binder by burning it off. It can be used on fatted-up surface dressing to expose aggregate that has been pressed into and

covered by the binder. Care must be taken to avoid setting the surface alight (a possibility if there is a great deal of binder available) because this can damage the surface for a considerable depth and is, of course, potentially dangerous.

Gas burners (also known as infra red heaters)
These are sometimes used for the same purpose as hot compressed air, but the method is losing favour because it is difficult to control and can fairly readily set the road alight. The latter is dangerous to operators and to the general public because of both the fire hazard and the production of toxic fumes from polymer-modified binders, which are becoming very frequently used in surface dressing (section 1.1.1).

High pressure jetting
Using water at varying pressures up to about 1000 bar has gained favour for removal of excess binder from the surface dressings and for the removal of matrix from rolled asphalt and chippings. In both these cases, macro-texture is increased and, in the case of fatted surface dressings, the aggregate is exposed so that the skidding resistance is restored. The method does not affect the surface of the aggregate except to clean it.

It is important to use an appropriate method for the surface course concerned and the outcome required. For example, flailing or grooving is unsuitable for surface dressing and high-pressure jetting has little effect on high stone content surfaces like close-textured bitumen macadam or some of the recently introduced thin surfacings, although it does remove the surface film of binder which may be a means of avoiding early life skidding problems on materials with thick surface binder films. There is little point in using a method that only improves macro-texture on areas of road with lower speed traffic or using a method that does not leave adequate texture, either by increasing it or not removing what is already there, on a higher speed road.

None of the methods can improve the low-speed skidding resistance, which depends on micro-texture, of a road in the long term to a higher level than that which the aggregate in the surface is capable of providing. However, it can improve it in the short term by exposing fresh, angular, aggregate surfaces, although these will be polished by the action of the traffic until it is reduced to the long-term skidding resistance. This usually takes between one and two years.

Depending on the cause of low macro-texture and the speed at which the cause reasserts itself, this can be improved with an appropriate treatment for periods ranging from a few months to the life of the surface. For example, on a first surface dressing over a hard substrate, fatting-up may occur simply because too much binder has been used and a single water jetting treatment to remove the excess binder would probably solve the problem for the life of the dressing. On the other hand, if the dressing was on a 'tar sick' road, that is one with innumerable surface dressings going back into the mists of time, then jetting would remove excess binder temporarily but the problem is likely to recur as soon as there is a spell of hot weather.

The big advantage that all these methods have over other surface treatments, and particularly over veneer treatments, is that they are, to all intents and purposes, weather independent. Fog or heavy snow might stop them but this would be for safety reasons rather than technical ones. This means that they can be implemented immediately a problem occurs and thus restore the road to a safe, or at least a safer, condition until a longer-term treatment can be carried out. When used as an interim measure, it may be feasible to treat only those parts of the area which need it most, such as the wheel-tracks, in order to keep cost down. A long-term solution would be more likely to treat a whole lane width.

It is essential when considering an appropriate treatment to involve the specialist contractor at as early a stage as possible in order to assess the likely outcome of any particular treatment. It may only be possible to give a general idea of the result of treatment and hard and fast specifications may be counter-productive. It would also be useful to check with a contractor what other work he has done on similar sites and whether they were done using the same operator and equipment. As many as possible of the different processes should be considered in order to arrive at the optimum process to deal with the particular problem on the particular site. If a range of sites with different problems are being done at the same time, it may not be possible to use the same process or even the same contractor for all them.

12.8 COLOURED MATERIALS

At first sight, the concept of coloured asphalt materials is slightly bizarre. In an ideal world one would not start with a black material if one were trying to produce another colour but there are ways and means of producing a whole range of coloured surfaces, some being much easier and cheaper to produce than others.

Coloured materials are used for a large range of purposes from high prestige areas, like The Mall in London which traditionally has a red surface, to rural footways surface dressed using white limestone chippings to enhance safety on an unlit road because it is more easily seen in the dark both by pedestrians and motorists. A major use is to separate areas on the ground with separate functions, such as roadways and parking areas in a large car park or bus lanes, from the rest of the road.

By far the most common colour for asphalts and macadam surfaces is a brownish red. This is because the red pigment used, basically an iron oxide, is by far the cheapest pigment available at about £300 per tonne while the next cheapest, green, is about £2200 a tonne and is not such a strong pigment as the red so that more has to added. Between 2 and 5% of the red (by weight of mixture) is used and between 3 and 7% of the green. These amounts vary depending on the amount of bitumen in the mixture, the strength of colour required and how 'black' the bitumen is. There are a few sources of relatively light coloured bitumen available and, if a large area is to be laid, it is probably worthwhile obtaining one of them. Nevertheless, the light bitumen is more expensive than normal bitumen and has to be obtained in full tanker loads, so that its use will be economically doubtful for less than about 300 tonnes of coated material, equating to about 3000 m^2 of surfacing laid 40 mm thick.

For colours other than red and green, it is usual to use a pale-coloured or colourless resin binder, as is used in thermoplastic white and yellow lines. These resins are significantly dearer than bitumen but do permit mixtures using much less pigment and, with red and green, much brighter colours ensue as the black (or dark brown or grey) of the bitumen does not detract from them.

It must be appreciated that, regardless of the binder and pigment used, the ultimate colour of a mixture depends almost entirely on the aggregate used because the surface layer of pigmented binder is worn away. Therefore, the aggregate in any coloured material should be chosen with care. Road Note 25 (RRL, 1960) gave the sources of the different coloured aggregates which were available at the time in Great Britain and most of the sources given are still extant. Sources are available for red, green, pink, buff, white and various shades of grey. The nearest to blue is a grey slag with some inclusions of blue minerals.

The only strongly coloured aggregates with polished stone values (PSV) (section 2.2.4.4) greater than 60 are green. Some pink quartzites are close to 60 PSV but most of the others are between 50 and 58. The best red is in the low 50s and many limestones are lower still. Therefore, if coloured materials are required on heavily trafficked routes, the aggregate should be chosen with care.

As the aggregate is eventually the dominant colour, a good choice of aggregate and a degree of patience would yield a coloured surface for only the additional cost of importing the aggregate. Even the need for patience can be done away with if the surface is high pressure jetted to remove the surface binder film (section 12.7).

When The Mall was resurfaced in the early 1990s, red pigmented asphalt was used and the pre-coated chippings were coated in clear resin rather than bitumen so that a good red colour was visible immediately. The chippings were from Harden Quarry in Northumberland, which has a very strong red colour, because of which they have been used all over the country and have even been exported as far afield as Japan. The disadvantage with Harden is the low PSV, which rules them out of many sites but, because heavy vehicles are banned from The Mall, the source can be used reasonably safely.

By far the cheapest, and often the most effective, way to obtain a coloured surface is to surface dress the area concerned using a conventional binder and chippings of the required colour. Because of the close packing of the chippings, the binder is all but invisible. Mention has already been made of white limestone on footways, which had been in place about five years at the time of writing and was still effective. The only discoloration was due to moss growth on a lightly trafficked footway, which would have affected any other surface in the same way. Another use was in a country park where the underlying soft rock was being eroded by vehicles using the unsurfaced tracks to reach parking areas. The tracks were surface dressed using chippings which were available from an ironstone quarry in the same stratum a few miles away. Although the chippings did not conform to BS 63: Part 2 (BSI, 1987) and were soft and of unknown (but low) PSV, the work was successful in that the erosion of the tracks was considerably reduced without changing the appearance of the park.

Care should be taken in producing and laying coloured materials; it will probably be necessary to put a batch through the mixer which will be scrapped when changing from black to colour in order to clean out any residual black materials. The discarding of a load is particularly necessary if it is only possible to load the delivery vehicle through a storage hopper rather than direct from the mixer. Similarly, the first tonne or so through the paver after laying black material will also often have to be scrapped because, however scrupulous the cleaning, there is always some black left behind. Care must be taken to ensure delivery vehicles are clean and will not stain the coloured materials; similarly for all rollers, shovels, rakes and other tools that will come into contact with the surface. Joints, kerbs and ironwork should be painted with a material of a similar colour to the surface.

Because of the quantity of pigment in the mixture, the overall filler content is quite high, particularly in macadams, and, therefore, the mixtures will be less workable that the equivalent black one unless the binder content is increased. Hence, compaction is even more important (if that is possible). It is also important to ensure that mixture proportions vary as little as possible (particularly the ratio of pigment to bitumen), temperature control must be good and as little hand raking done as possible as any of these affect the finished texture and, therefore, the apparent colour. This is so with black materials but is much more obvious with coloured ones. Also, it will probably be necessary to increase the binder content of the mixture above the British Standard value in order to maintain adequate coating and durability because of the amount and the finely divided nature of the pigment.

When new, coloured materials can look attractive. However, unless care is taken with maintenance, they will quickly lose their appearance.

Compared to their black cousins, coloured surface courses show oil stains much more, so any spillages should be cleared away before they can soak in and stain the material in depth. It may be considered inappropriate to use colours where vehicles routinely park. For example in a car park where it is intended to use a normal and a coloured material to differentiate between roadways and parking bays, it is better to use the colour for roadways and black for the parking areas rather than the reverse.

If a coloured surface needs patching, as it will sooner or later, care should be taken to obtain material as near as possible to the original because, although it will stand out initially, it will fairly rapidly blend in. If an attempt is made to match the current colour, the new material will age much more rapidly than the old and will become more and more distinct.

12.9 MISCELLANEOUS MATERIALS

This section describes a number of specialist materials which, in terms of tonnage laid, form a very small part of the market. They cover a range of uses, most of them off highway, but have been included to demonstrate the range of performance criteria that can be met by asphalt materials, most of which could have application, directly or indirectly, on the highway. As can be seen from the examples below, asphalt materials are 'flexible' enough to be used to solve a wide range of surfacing problems.

12.9.1 Vehicle noise test track material

Vehicle noise test track material is specified almost entirely in terms of the performance of the surface. It is an ISO standard (ISO, 1992) and is part of the international standard for the measurement of vehicle emissions and, in this particular case, the emission of noise. The grading is specified by means of three Fuller Curves specifying the desired target grading and the upper and lower bounds of the target with subsidiary manufacturing tolerance on stone content (greater than 2 mm), sand content (63 μm to 2 mm), filler content (less than 63 μm) and binder content. The binder content is fixed at (5.8 ± 0.5) %. The Fuller Curves are given by Equation 1.1. The maximum aggregate sizes of 6.3 mm, 8 mm, and 10 mm give the upper bound, target and lower bound grading curves, respectively, for the material.

Stone and sand contents have tolerances of ± 5% and filler content a tolerance of ± 2%. The coarse aggregate must be crushed rock with a PSV (section 2.2.4.4) of at least 50 and at least 45% of the sand must be crushed. The binder is unmodified straight run bitumen and the hardest binder from 50, 70 or 100 pen is used consistent with local common practice and obtaining the necessary resistance to deformation. The void content measured from cores shall be less than 8% or, if it fails on that score, the sound absorption coefficient using the impedance tube method shall be less than 0.1 (a<0.1); this latter test may be used in lieu of void content anyway. The patch texture depth shall be at least 0.4 mm at the time of laying, at the start of noise testing and must be checked annually to ensure that it remains above that level. If it does not, at least the tracked area must be resurfaced. The material must be laid at least 30 mm thick. It would appear to be a fairly difficult specification to achieve in all its requirements and more than one attempt has had to be used to successfully complete a complying test track on at least two occasions.

12.9.2 55/6F rolled asphalt surface course

A specific problem on some farm accommodation bridges over the M6 was met by modifying a mixture from BS 594: Part 1 (BSI, 1992). The bridges had to be re-waterproofed and then resurfaced with 45 mm of asphalt material. The thickness was limited by the dead load capacity of the bridge structure. This was too thin to allow for a red sand carpet layer that protects the waterproofing and a surface course that was thick enough and tough enough to resist normal farm traffic and the passage of farm animals (cows, sheep, horses and, in one case, deer). Because of the nature of the traffic, the surface is normally obscured by wet detritus. The bridge engineer would not allow an aggregate size larger than 6 mm to come into contact with the waterproofing so a 55/6F rolled asphalt surface course with 100 pen binder was used laid 45 mm thick in a single layer; the specification being exactly the same as for 55/10F except that the maximum aggregate size was reduced. This was laid successfully and, after about two years, shows every sign of giving a reasonable life.

12.9.3 Sports playing surfaces

A number of suppliers make a 6 mm surface course suitable for tennis courts and similar sports areas. They are similar to 6 mm medium textured macadam in BS 4987: Part 1 (BSI, 1993a) but are made to tighter tolerances with as little oversize as possible. It is important for the playing surface to be consistent over the whole area so that the surface behaves in a predictable manner, for example balls should bounce in the same way anywhere on the surface. These surfaces are usually laid with no falls for drainage so, in order to ensure surface drying takes place as speedily as possible, the surface must be somewhat porous to avoid water being retained on it. A similar material could be used on footways, where level constraints prevent the surfacing being laid to adequate falls for quick drainage.

12.10 REFERENCES

Manual of Contract Documents for Highway Works, Her Majesty's Stationery Office, London.
    Volume 1: Specification for Highway Works (MCHW 1).

References

Design Manual for Roads and Bridges, Her Majesty's Stationery Office, London. HD 29/94 *Structural Assessment Methods* (DMRB 7.3.2).

British Board of Agrément (1988) *Hardicrete Heavy Duty Surfacing. Agrément Certificate No 88/1974*, British Board of Agrément, Watford.

British Board of Agrément (1989) *The Salviacim Process for Industrial Paving. Agrément Certificate No 89/2248*, British Board of Agrément, Watford.

Brown, J.R. (1980) *The Cooling Effects of Temperature and Wind on Rolled Asphalt Surfacings*, Department of Transport TRRL Report Supplementary Report 624, Transport and Road Research Laboratory, Crowthorne.

British Road Tar Association (1954) *Dense Tar Surfacing*, London

British Standards Institution (1975) *Dense Tar Surfacing for Roads and Other Paved Areas. BS 5273: 1975*, British Standards Institution, London.

British Standards Institution (1987) *Road Aggregates: Specification for Single-Sized Aggregate for Surface Dressing. British Standard 63: Part 2: 1987*, British Standards Institution, London.

British Standards Institution (1989) *Bitumens for Building and Civil Engineering; Part 1, Specification for Bitumens for Roads and Other Paved Areas. BS 3690: Part 1: 1989*, British Standards Institution, London.

British Standards Institution (1990) *Soils for Civil Engineering Purposes; Part 4, Compaction-Related Tests. BS 1377: Part 4: 1990*, British Standards Institution, London.

British Standards Institution (1992) *Hot Rolled Asphalts for Roads and Other Paved Areas; Part 1, Specification for Constituent Materials and Asphalt Mixtures. BS 594: Part 1: 1992*, British Standards Institution, London.

British Standards Institution (1993a) *Coated Macadam for Roads and Other Paved Areas; Part 1, Specification for Constituent Materials and for Mixtures. BS 4987: Part 1: 1993*, British Standards Institution, London.

British Standards Institution (1993b) *Coated Macadam for Roads and Other Paved Areas; Part 2, Specification for Transport, Laying and Compaction. BS 4987: Part 2: 1993*, British Standards Institution, London.

Catt, C.A. (1994) *Pavement Construction and Structural Maintenance Policy 1994*, Planning and Transport Department, Warwickshire County Council.

County Surveyors Society (1994) *Highway Authorities Standard Tender Specification for the Tender Year ending 31 March 1996, section 4, Surface Dressing (Works)*, Cheshire County Council, Chester.

House of Commons (1991) *New Roads and Streetworks Act*, Chapter 22, Her Majesty's Stationery Office, London.

Highway Authorities and Utilities Committee (1992) *Specification for the Reinstatement of Openings in Highways, A Code of Practice*, Her Majesty's Stationery Office, London.

International Standards Organisation (1992) *Motor Vehicle Emissions. L371/1-31, Annex VI test track specifications*, International Standards Organisation.

Leascon Surfacing Technology Ltd (1995) *Delugrip Road Surfacing Materials*, Leascon Surfacing Technology Ltd, Stratford on Avon.

Lees, G. (1970) The rational design of aggregate gradings for dense asphaltic compositions, *Proc Assoc of Asphalt Paving Technologists*, Feb., Kansas City.

Lees, G., I.D. Katekdha, R. Bond and A.R. Williams (1977) The design and performance of high friction dense asphalts, *Transportation Research Record 624*, National Research Council, Washington DC.

McKellar, K.S. (1981) An approach to the design of skid resistant road surfaces, *Recent Developments in Highway Surfacing Seminar*, Nov., Institution of Municipal Engineers, London.

Ministry of Defence (1995) *Defence Works Functional Standards: Hot Rolled Asphalt and Coated Macadam for Airfield Pavement Works*, Her Majesty's Stationery Office, London.

Nicholls, J.C. and M.E. Daines (1993) *Acceptable Weather Conditions for Laying Bituminous Materials*, Department of Transport TRL Project Report 13, Transport Research Laboratory, Crowthorne.

Patent Office (1975) *Patent Specification, Road Surfacing Materials. Patent No 1393885*, Her Majesty's Stationery Office, London.

Road Research Laboratory (1960) *Sources of White and Coloured Aggregate in Great Britain (1959)*, Department of Scientific and Industrial Research Road Note 25, Her Majesty's Stationery Office, London.

Road Research Laboratory (1962) *Bituminous Materials in Road Construction*, Her Majesty's Stationery Office, London.

Roadtex (1994) *Safedrain Rubber Topping for Filter Drains*, Uckfield, East Sussex.

Robinson, H. (1995) Cold arm of the law, *Highways*, Jan./Feb.

Roe, P. (1995) Re-texturing, *Surface Treatments Seminar*, Dec.,Transport Research Laboratory, Crowthorne.

Smith, R. (1995) *HAUC and Cold Laid Materials*, paper given to West Midlands Branch of Institute of Asphalt Technology, June 1995.

Szatkowski, W.S. (1979) Rolled asphalt wearing courses with high resistance to skidding, *Rolled Asphalt Road Surfacings*, Institution of Civil Engineers, London.

Walker, J.C. and R.D. Oakes (1979) The reduction of tyre road interaction noise, *Proc Aviation, Surface Transportation and Plant Noise Symposium*, Jan., Dallas.

Whiteoak, D. (1990) *The Shell Bitumen Handbook*, Shell Bitumen, Chertsey, Surrey, pp 45–57.

Williams, A.R., T. Holmes and G. Lees (1972) Toward the unified design of tire and pavement for the reduction of skidding accidents, *Society of Automotive Engineers, Automotive Engineering Congress*, Jan., Detroit.

CHAPTER 13

Recycling materials

A. Stock, Consultant

13.1 CONCEPT AND HISTORY

13.1.1 Introduction

This book has been prepared to provide information on bituminous surfaces. In recent times, the industry has seen some significant changes in that the approach to specifying surfacing materials has changed significantly; innovative materials and treatments, most noticeably thin surfacings, are being accepted more readily than they would have been just a few years ago. Performance based specifications are being introduced, currently for sites where the combination of traffic and environmental conditions make unusual demands on the paving materials. These changes are a response to the need to provide improved levels of service to road user, as well as the important goal of providing value for money.

Recycling asphalt is not directed at providing a further series of specialist mixtures, but at incorporating reclaimed material into asphalt mixtures. This will ensure that cost savings can be realized through utilization of material which would otherwise represent a cost in terms of disposal. Hence, some information on recycling processes and the characteristics of asphalt mixtures produced with reclaimed material is an essential part of this book.

13.1.2 Terminology

Asphalt recycling processes are marketed under a large number of names. This may be somewhat bewildering to professionals who are not familiar with asphalt recycling, but it is a problem which can be overcome with ease. All recycling processes fall into one of four possible categories as follows.

- **Hot recycling** is a process in which reclaimed asphalt is combined with appropriate materials, at normal asphalt mixing temperatures, to produce an asphalt mixture.

- **Cold recycling** is a process in which reclaimed asphalt is combined with appropriate materials, at ambient or near ambient temperatures, to produce an asphalt mixture.
- **In-situ recycling** processes the material in situ, through plant which is in motion. In-situ recycling takes place directly on the road.
- **In-plant recycling** is a process in which the material is processed after removal from the site to a stationary plant which has been set up to produce asphalt.

It follows that both the in-situ and in-plant processes can be either hot or cold, and no matter what name is used to describe any particular process it will be either hot or cold, in situ or in plant.

13.1.3 The development of recycling

13.1.3.1 General

Recycling is probably as old as paving. It is very likely that material which has been removed from a road and would have been dumped has paved areas, such as drives of car parks, for enterprising or lucky individuals as the material became available. This kind of informal recycling is still continuing today, and long may it continue.

It seems likely that asphalt recycling, in a form that would be formally recognized today, started in India and Singapore in the early 1930s. Information on this early work can be found in articles written by N.H. Taylor (1978, 1981). The process used was of the hot in-plant type using standard batch mixing equipment built in the mid 1920s. Taylor reports that the recycled surfaces provided satisfactory performance for 25 years, and were then recycled for a second time. It would appear that, if Mr Taylor attempted to persuade the UK industry that recycling was a viable option in England when he was developing the process in India, he was not successful.

In the late 1930s, the Road Research Laboratory was experimenting with a cold in-situ recycling process, but implementation of this work was interrupted by the Second World War. However, the process was implemented following trials in 1948, and is still in use. This cold in-situ process is known as the 'retread' process, and has always been the preserve of a few specialist contractors. It continues to be used to treat a substantial area of pavement in the United Kingdom each year. Hence, it can be concluded that recycling has been around for many more years than people customarily believe, and that the United Kingdom had an early start in this area of paving technology.

In the 1970s, there was an upsurge of interest in recycling, probably as a result of the high bitumen prices at the time. The contemporary literature indicates that the North American industry was leading development at the time while, in Europe, France appeared to be paying particular attention to developing various processes.

13.1.3.2 Hot in-situ recycling

At the end of the 1970s, hot in-situ recycling was introduced in the United Kingdom with the importation of equipment and technology from the United States of America. The first process simply involved heating, scarifying and re-profiling the surface, generally to a depth less than 30 mm, followed by an overlay of conventional surfacing material. The process was given the name 'repave'. Some counties found that repaving provided a suitable surface treatment and added it to their list of maintenance options and, in 1982, the Department of Transport issued HD/7/82 (Department of Transport, 1982) for the 'In-situ recycling repave process'. This process was developed very quickly to involve the addition of new material to that which had been heated and loosened from the surface, and mixing the two together. This process has become known as 'remix', but has not achieved the popularity of the repave process. Both these processes require specialized and dedicated equipment, and are only available through a limited number of contractors.

13.1.3.3 Hot in-plant recycling

Hot in-plant recycling was also developing, notably in continental Europe and North America, at the same time as the hot in-situ recycling. However, it was not until the end of 1991, or the beginning of 1992, that hot recycling found its way into the specifications for major UK highways, when mixtures containing up to 10% of reclaimed material could be used in layers other than the surfacing without testing. The specification permitted greater percentages to be used provided that the mixes were tested and found to be acceptable. In a revision to the specification in 1994, the maximum quantity of reclaimed material which could be included without testing was increased to 30%.

The coincidence of the interest in recycling in the late 1970s and the step change in plant for producing asphalt, from batch type plant to continuous drum mixing, was unfortunate for two reasons. The first was that retrofitting of the components required for recycling in a drum mixer was very costly. This was a great disincentive to asphalt producers who had switched to this type of plant. The magnitude of the problem of retrofitting recycling attachments is exemplified by the fact that virtually all drum mixers currently manufactured have the key components fitted at the factory, regardless of whether the purchaser intends to produce recycled mixtures. The second reason was that the high profile of the debate about the blue smoke, and the understandable effort put into publicizing the solutions developed for new drum mixers. This focused attention on this type of plant and created the impression that it was necessary to purchase a drum mixer in order to produce recycled asphalt mixtures. This diverted attention from the relative ease of recycling in a batch plant.

13.1.3.4 Cold recycling

As already noted, a cold in-situ recycling process was introduced in the early 1950s into the United Kingdom. There have been many more recent efforts to develop cold recycling for a variety of reasons. In Germany, where disposal of old tar-bound paving materials in landfill sites is very expensive, encapsulation of plainings of tar-bound material in a bitumen emulsion-based mixture which is then used in pavement construction is a preferred option for disposal of this material.

One significant step forward was the development of specialist plant for cold in-situ recycling based on successful soil stabilization equipment. This plant can recycle to a significant depth, depending on the hardness of the material to be recycled, and has been used successfully with appropriate bitumen emulsions, as well as with foamed bitumen. It is worth noting that mixtures containing significant quantities of reclaimed material produced with foamed bitumen gained full approval under the Highway Authorities and Utilities Committee (HAUC) for use as a material for pavement reinstatements. At the time of writing (late 1996), this product is the only one approved for use on both carriageways and footways. As materials for reinstatement have to meet stringent requirements, it is surprising that this obviously successful material has not found wider use in paving operations.

A further development is based upon cold-planing equipment. Using a cold planer as the basic platform, plant which removes the surface and carries it to an internal mixing chamber where it is mixed with an appropriate emulsion, and then placed has been developed. It appears that similar, but apparently independent developments of this type of plant have been taking place in England and Australia. However, while this equipment is being used, it does not appear to have a significant share of the paving market as yet.

13.1.3.5 Cold in-plant recycling

Cold in-plant recycling appears to be virtually non-existent. The reinstatement market, which uses a very large volume of bituminous material, but in relatively small quantities, is an obvious target for cold in-situ recycling. However, the current technology does not appear to be successful with respect to meeting the requirements of this market.

13.1.3.6 Comments on recycling developments

The concept of working in situ is very attractive as there are significant potential efficiencies arising from savings in transport and time. However, there are difficulties associated with the process, be it hot or cold.

Given the historic approach to quality control in the United Kingdom, which is based upon testing samples of material for composition, it does seem surprising that this process was at the forefront of the recycling processes promoted by the paving industry in the United Kingdom. Careful quality control would require sampling the section to be recycled at frequent intervals to establish its composition. In the event of significant variation in the composition of the existing asphalt, which could be caused by the existence of openings or the fact that the section to be recycled included pavement surfaced at different times, then the production of a consistent composition from the combination of reclaimed and new material can be very difficult. It would involve delivery of the appropriate quantity of a mixture of the relevant composition to the machine whist it is in the location requiring that composition. In practice, this is not possible. This problem is, of course, eliminated if the quality of the pavement is judged by criteria which are not based on composition. However, good long-term performance of a surface is more likely to be achieved if the surfacing material has consistent properties and, therefore, composition. Hence, this process will continue to face problems in relation to use on heavily trafficked roads. Also, given the limited depth which can be scarified economically by heater-planers, the treatment of surfaces showing terminal wheel-track rutting by a hot process would be difficult.

13.2 EQUIPMENT FOR RECYCLING

13.2.1 In-plant recycling

The growth of interest in recycling in the 1970s and early 1980s coincided with the development of drum-mix asphalt coating plant. This coincidence was unfortunate with respect to the development of recycling as it introduced a practical problem associated with the mixing process. Drum mixers produce asphalt by a continuous process which combines the aggregate drying with mixing in a specially designed drum. The aggregates are introduced to the drum in the vicinity of a burner, which provides the heat to dry the aggregate and to bring them to mixing temperature. Early experiments with recycling in this type of plant, in which the reclaimed material was added with the new aggregate, indicated that the very high temperatures in the vicinity of the burner, which are of the order of 1500 °C, caused unacceptable emissions from the plant. These emissions, which became known as blue smoke, were the result of the formation of very small particles generated when the bitumen on the reclaimed material either boiled or was burnt due to the very high temperatures in the vicinity of the burner flame. Various systems were investigated by plant manufacturers, the one eventually adopted most widely being to introduce the reclaimed material about halfway down the drum through a ring specially and cleverly designed openings in the drum. In some equipment, a circular heat shield was also fixed inside the drum, across its centre, to ensure that there could be no direct contact with the burner flame.

Use of batch plant for recycling totally eliminates the blue smoke problem because the reclaimed material is added directly to the mixer box and the coincidence of the interest in recycling and the step change in plant for producing asphalt was unfortunate for two reasons. The first was that retrofitting the components required for recycling to new drum mixers was very costly. This was a great disincentive to asphalt producers who had switched to this type of plant. The magnitude of the problem of retrofitting recycling rings is exemplified by the fact that virtually all drum mixers have the key components fitted at the factory, regardless of whether the purchaser intends to produce recycled mixes. The second reason was that the high profile of the debate about the blue smoke, and the understandable effort put into publicizing the solutions developed for new drum mixers focused attention on this type of plant and created the impression that it was necessary to purchase a drum mixer in order to produce recycled mixtures. This diverted attention from the relative ease and low modification cost associated with recycling through a batch plant.

Comparing recycling through drum and batch plant, the drum mixer does have some advantages. In a batch plant, the reclaimed material is raised to mixing temperature by heat transfer from the new aggregate and bitumen added. This imposes a practical limitation of the quantity of reclaimed material which can be incorporated in a mixture without seriously affecting throughput, or requiring that the new aggregate be heated to a temperature which is excessive. A figure of 50% has been suggested as the maximum proportion of reclaimed material which can be incorporated in a mixture in a batch plant. This will clearly vary between installations, but a figure of 30% seems to be a reasonable working limit. In contrast, it is possible to mix 100% reclaimed material in a drum mixer. However, this is not advisable and is very unlikely to be commercially viable because recycling at this level requires significant increases in dwell time in the drum. Again, specific limits for the quantity of reclaimed material to be incorporated into a mixture will vary between installations, but it seems that operating costs increase when more than 70% reclaimed material is included and that 50% is a reasonable operating figure for this type of equipment.

Figures 13.1 and 13.2 show diagrammatically how recycling may be carried out through batch and drum mix plants. Very little if any work is known to have been done to develop plant specifically for cold in-plant recycling.

13.2.2 In-situ recycling

In early 1980s, some German equipment was imported which extended the capability of the heater-planer systems with an option to collect the

Equipment for recycling

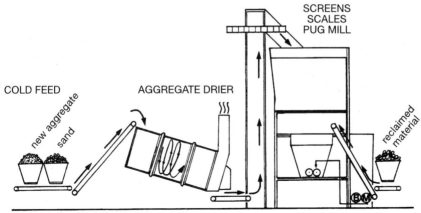

a. Standard batch plant with reclaimed material added to superheated aggregate at the pug mill

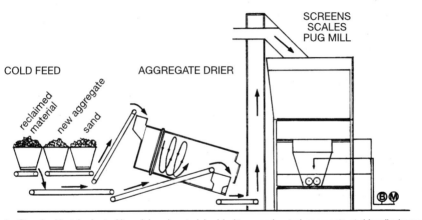

b. Standard batch plant with reclaimed material added to superheated aggregate at drier discharge

Key: B – New bitumen
      M – Modifying/rejuvenating agent

Fig. 13.1 Schematic of two options for recycling in batch plants

heated and scarified material into a windrow, drop a fresh mixture on top of it, mix the two together and then spread and compact it. This process was given the name re-mix, and was trialed on some sections of motorway.

During this period, hot in-situ processes appeared, with two rival items of equipment known as the 'Cutler' and the 'Jim Jackson' machines being imported from the United States of America. The Cutler machine appeared to be the most versatile, and the Wirtgen Company from Germany produced equipment based upon the same principles as the Cutler. The Wirtgen machinery combines hot scarification with placement

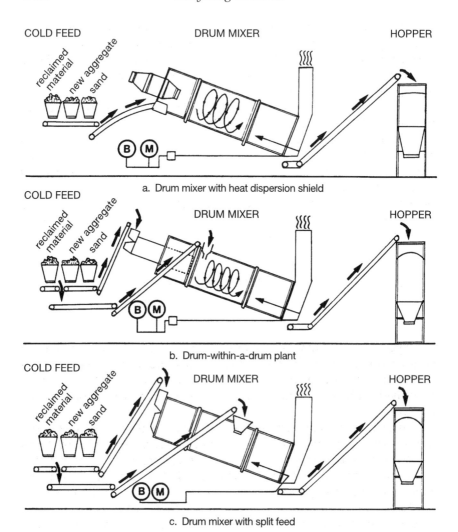

Fig. 13.2 Schematic of three options for recycling in drum mix plants

and, in effect, places a hot scarifier between the receiving hopper and the paving screed of a conventional paver. The machine can place new material as an overlay to hot scarified surface, can combine the hot scarified material with new mixture and then pave with the recycled mixture.

At this time, two companies (one of which had imported a Cutler machine, the other whose equipment appeared to be based on the Jim Jackson machine) were prominent in pioneering these processes in the

United Kingdom. While both companies carried out the same process, their equipment was very different in style. These machines heated and scarified the surface material, reprofiled it as necessary, and either placed an overlay themselves, as with the Cutler, or were followed by a conventional paver placing the overlay.

Figures 13.3 and 13.4 show diagrammatically how the hot processes 'repave' and 'remix' can be performed.

There are some truly mobile hot mix coating plants in existence. These are in effect a drum mixer on wheels or tracks. This is an exciting development which does not appear to have been tried in the United Kingdom. Such plant has been in existence for many years, and some of the recent developments have led to plant which is quite compact and is more likely to be usable in the generally restricted working areas common in the United Kingdom.

13.3 RECYCLED ASPHALT AS A COMPONENT MATERIAL

13.3.1 Recycled asphalt

As recycling has become acceptable in many countries throughout the world, it is not surprising to find that the data published relating to the performance of recycled asphalt mixtures is positive, indicating that they do provide adequate performance in comparison with non-recycled mixtures.

However, there may still be some suspicion that material which has been removed from the surface of a road must be in some way so defective that it cannot possibly be reused and, therefore, must be disposed of rather than recycled. As this view indicates a lack of understanding of the competence of paving materials, and of the way in which pavement structures function, it is appropriate to reproduce data from investigations of recycled asphalt mixtures. This information also underpins mixture design for recycled mixtures, which will be of increasing importance as performance based specifications become more common.

Fig. 13.3 Material flow in a repaving machine

Fig. 13.4 Material flow in a remixing machine

Therefore, this section will review some of the published data to provide the background information which, it must be presumed, convinced specifying authorities that recycled materials are in general, and within prescribed limits, fit for use in pavements. Emphasis will be placed upon a few comprehensive studies and where possible those relevant to the United Kingdom.

13.3.2 Reclaimed binder and rejuvenation

The binder in an asphalt mixture is a very important component because it provides the cohesion in the mixture. It is by far the most costly component, and also one upon which suspicion frequently falls when the performance of mixtures is called into question. Therefore, the properties of reclaimed bitumen are of considerable importance to the performance of recycled mixtures.

Whiteoak (1990) has summarized the changes which take place in bitumen during mixing, placement and service, identifying the four principal mechanisms of hardening, namely:

- oxidation;
- loss of volatiles;
- physical hardening; and
- exudative hardening.

Whiteoak shows that the majority of hardening takes place during mixing and placement, with further hardening taking place during the service. A rule of thumb widely used in mechanistic pavement design (Brown *et al.*, 1980) is that the penetration is reduced to 65% of its original value during mixing and placement. Whiteoak reports data from a study in which the hardening of bitumen in an asphalt mixture was measured after 15 years' service. This study showed that the hardening is dependent on the voids content of the mixture as well as the time since the start of processing. Therefore, the issue which has to be addressed in the design of mixtures for recycling is whether the changes which have taken place during ageing can be reversed effectively, and how best to do this.

Recycled asphalt as a component material

There are a number of papers which address the issue of reversing the effects of ageing in bitumen, and a number of proprietary products are available for rejuvenation. A comprehensive study of rejuvenating agents (Holmgreen et al., 1980) examined most products which were available at that time. The study found that the agents available were capable of restoring aged bitumen to a desired consistency, but that different agents did produce binders with a different temperature susceptibility. It was also found that there could be some problems relating to the compatibility between the aged bitumen and the rejuvenating agent. A comprehensive study of several proprietary recycling agents blended with material reclaimed from five different sites (Kallas, 1984) confirmed these results. Kallas also conducted compositional analysis tests on the blends which confirmed that there were no alterations in composition due to chemical interactions between the aged bitumen and the recycling agents for blends produced with a conventional blending chart.

An investigation of the properties of bitumen reclaimed from UK pavements (Stock, 1985a) also investigated issues related to rejuvenation. An additional facet of this study was to investigate the effects of hot scarification, of the sort carried out during hot in-situ recycling processes. Table 13.1 shows some results obtained from this study.

This data clearly shows that, for the sites studied, the overall effects of heating prior to scarification are less than the effects of processing bitumen through a conventional coating plant. Of greater relevance is that there is a wide range of values for the recovered penetration, indicating that it is not possible to make general assumptions concerning the hardness of the binder in the pavement surface. Therefore, it is necessary to measure the properties of the binder in reclaimed asphalt.

These results are plotted on a graph of penetration against softening point in Figure 13.5. Also plotted on this figure are the specification limits for penetration and softening point. This shows quite clearly that, with

Table 13.1 The properties of recovered binder and effects of scarification

Site number	Effect of scarification			
	Penetration (dmm)		Softening point (°C)	
	Before	After	Before	After
1	50	34	53	59
2	39	27	56.5	60
3	25	25	59.5	60
4	35	25	56	59
5	22	20	60.5	60.5
6	26	19	58.5	61
7	16	17	not measured	not measured
8	36	26	not measured	not measured
9	23	21	59	60
10	66	64	50	49.5

the exception of the hardest binder (with a penetration of 19 dmm after hot planing) for which both penetration and softening point are available, all the samples of bitumen conform to the requirements for one or another paving grade. The fact that all the recovered bitumen samples conform to the requirements of the bitumen specification has very important implications in relation to rejuvenation requirements, namely that they only need to have their grade adjusted to that required for a particular mixture.

Modern bitumen production is often based upon the manufacture of a hard grade and a soft grade of bitumen, which are then blended to produce the grades required for delivery. For the grades used in the United Kingdom, 'rejuvenation' can be successfully achieved by the well-known process of blending different grades of bitumen, the softer grade being added during mixing in the asphalt plant. Flux oils are also perfectly effective, and may be preferred if the mixture design does limit the scope for increasing the binder content. Figure 13.6 shows a blending

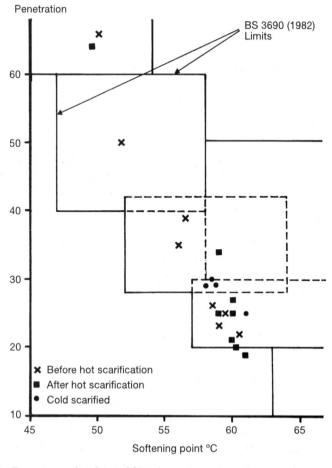

Fig. 13.5 Properties of reclaimed bitumen

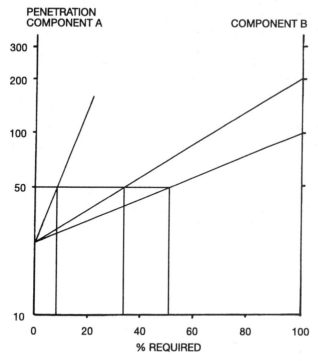

Fig. 13.6 Bitumen blending chart

chart, which is simply constructed by plotting the logarithm of penetration (or viscosity) on the left and right Y-axes, which are joined by a naturally scaled X-axis with values from 0 to 100. The relative proportions of a blend to achieve any penetration can be read off the X-axis.

Not withstanding the simplicity of achieving the grade of bitumen required by utilization of the well established process of blending a soft grade of binder or flux oil with that in the reclaimed material, the softening agent must be one which is compatible with the bitumen. For example, use of reclaimed engine oil will not produce a suitable binder.

13.4 HOT RECYCLED MIXTURES

13.4.1 Performance requirements

The key technical question associated with recycled asphalt mixtures is 'will they perform'. As noted previously, the answer must be yes because the *Specification for Highway Works* (MCHW 1) permits the use, within prescribed limits, of hot recycled mixes, and they are widely accepted as paving materials in other countries throughout the world. However, performance based specifications are being developed within the United

Kingdom, and so it is appropriate to summarize the information available with regard to the performance of recycled asphalt mixtures. While there is a reasonable volume of information in the literature, only that contained in a few comprehensive studies will be reproduced.

The current approach to the performance of asphalt mixtures recognizes that they must have resistance to cracking, deformation and ageing. Dynamic stiffness has also become recognized as an important characteristic of asphalt mixtures because it is an essential parameter for the structural analysis of pavements, which is a requirement in relation to efficient structural design. This section will review the data available against these individual requirements.

The approach which has generally been adopted for these studies is to formulate mixtures containing reclaimed material so that they are identical in terms of aggregate grading, binder content and binder grade for a range of reclaimed material contents. The binder grade is usually obtained by adding a softening agent in an appropriate quantity determined from a simple blending procedure.

13.4.2 Resistance to deformation

One method for the assessment of resistance to permanent deformation is the creep test. This approach has an advantage over the wheel tracking test in that it is capable of generating data which can be used in design procedures for pavements, such as the Shell design procedure (Shell International, 1978).

Table 13.2 shows the creep data from a study in which the reclaimed material came from two different sources, one being hot planed, the other cold planed (Stock, 1985b). In addition, two different softening agents were used, one being a 200 pen bitumen, the other a flux oil. The mixtures were produced to a rolled asphalt recipe, and the comparison was made on the basis of the ratio of the applied stress to the total strain after 60 minutes of loading.

This data is reproduced in graphical form in Figure 13.7 which shows that there is very little variation in the creep stiffness of the mixtures as the quantity of reclaimed material in the mixture is increased. It is also evident that the creep stiffness of the mixtures containing reclaimed material is similar to that of the control mixture. Therefore, this data suggests that, provided care is taken with proportioning the mixture, the creep stiffness is unaffected by the type of planing, be it hot or cold, or the type of softening agent used, be it flux oil or 200 pen bitumen.

13.4.3 Resistance to cracking

Resistance to cracking may be measured by carrying out fatigue tests, or by measuring the propagation of cracks through a sample of asphalt mixture. Recycled mixtures have been examined with both of these techniques.

Table 13.2 Creep stiffness of recycled mixtures

Reclaimed material (%)	Recovery process	Softening agent	Number of samples	Creep stiffness
0 (Control)	None	None	7	55
20	Hot	Flux oil	5	60
40	Hot	Flux oil	3	48
60	Hot	Flux oil	5	53
83	Hot	Flux oil	4	44
20	Hot	200 pen	4	43
40	Hot	200 pen	4	45
60	Hot	200 pen	3	48
83	Hot	200 pen	1	31
20	Cold	Flux oil	2	54
40	Cold	Flux oil	3	45
60	Cold	Flux oil	5	64
76	Cold	Flux oil	3	65
20	Cold	200 pen	4	51
40	Cold	200 pen	4	55
60	Cold	200 pen	5	62
76	Cold	200 pen	5	55

Fatigue tests have been carried out on laboratory prepared beams tested under constant stress in a four point bending system at temperatures of 25 °C and 5 °C (Brown, 1984). The study encompassed reclaimed material from three different sites, two different types of new aggregate (crushed gravel and crushed limestone) and two softening agents (a soft grade of

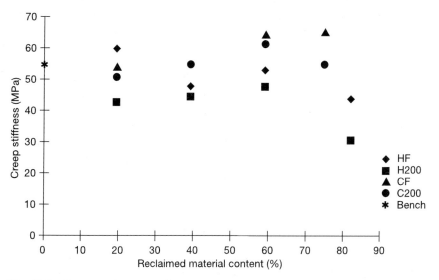

Fig. 13.7 Comparative creep data

bitumen and a proprietary recycling agent), and, therefore, may be regarded as quite comprehensive. The mixture design procedure used in this study ensured that the gradation of the combination of reclaimed and new aggregate had closely similar gradings. However, because the Corps of Engineers Marshall mixture design procedure was used to determine the binder content for the mixtures, the binder contents in the eight mixtures tested varied from 5.4% to 7.1%, which will have an effect on the fatigue life. The results show that, when the specimens were loaded to give an initial strain of 100 microstrain, the life to failure was approximately one million cycles. There are two mixtures from the 16 tested that were judged to deviate significantly from this norm, one at 25 °C and the other at 5 °C. Nevertheless, Brown concluded that overall the recycled mixtures were comparable with the benchmark mixtures having no reclaimed material.

A study of the propagation of cracks through mixtures containing various quantities of reclaimed material has been carried out on a recipe rolled asphalt mixture (Sulaiman and Stock, 1995). The material for test was mixed in a small Bomag recycler, placed in a special large mould and compacted with a pedestrian roller. Test samples were cut from this slab and tested for crack propagation. Tests were carried out at temperatures ranging from –5 °C to 35 °C . This necessitated the use of three different methods of analysing the results because the mixture behaved:

- as an elastic material at low temperatures;
- as an elasto-plastic manner at intermediate temperatures, and
- in a manner requiring creep crack growth analysis at 35 °C.

The results of these tests are presented in terms of a crack growth parameter which is shown in Tables 13.3, 13.4 and 13.5. The results from the limited range of tests that were carried out under creep conditions have been compared on the basis of a regression of the rate of change of the crack characterization parameter, C^*, with respect to time. Table 13.5 shows the coefficients of a regression of this rate of change plotted against C^*.

The data for linear elastic behaviour is shown in Figure 13.8. The larger the absolute value of the Crack Seed Index, the better the resistance to cracking, so crack resistance is greatest at the highest rates of loading and at the lowest temperatures. For the most part, the quantity of reclaimed material in the mixture has very little effect on the crack growth, the effects of temperature and loading rate being much more significant. While there is a perceptible decrease in crack resistance at –5 °C when the reclaimed material content is increased from 50% to 70%, this is of the same order as the effect of reducing the rate of loading from 10 Hz to 1 Hz, and much less than the change which occurs when reducing the loading rate from 1 Hz to 0.1 Hz. Thus, when this type of mixture is used in a road, the operational conditions will have a much greater effect on the performance of a mixture than the addition of extra reclaimed material.

Table 13.3 Crack susceptibility parameters for the linear elastic data

Test temperature (°C)	Rate of loading (Hz)	Quantity of reclaimed material (%)	Number of tests	Crack speed index
–5	10	0	12	–3.643
		30	9	–3.310
		50	9	–3.320
		70	9	–3.080
	1	0	9	–3.467
		30	9	–3.223
		50	9	–3.170
		70	9	–2.957
	0.1	0	6	–1.985
		30	9	–1.915
		50	9	–1.851
		70	6	–1.788
5	10	0	9	–2.500
		30	9	–2.475
		50	9	–2.388
		70	9	–2.417
	1	0	6	–2.000
		30	9	–1.972
		50	9	–1.953
		70	6	–1.930
	0.1	0	6	–0.823
		30	9	–0.921
		50	9	–0.913
		70	6	–0.830
15	10	0	9	–1.761
		30	9	–1.735
		50	9	–1.754
		70	9	–1.714
	1	0	9	–0.965
		30	9	–0.949
		50	9	–0.987
		70	9	–0.978

Figure 13.9 plots similar data from the tests in the elasto-plastic region. While the ordinate has a similar title to that in Figure 13.8, the two crack speed indices are not directly comparable. Once again, the data shows clearly that the proportion of reclaimed material in the mixture has very little effect on the resistance to cracking of the formulation.

Finally, Figure 13.10 plots the data from the creep crack growth tests. This data confirms that crack propagation is not affected by the quantity of reclaimed material in the mixture.

Table 13.4 Crack growth in the elasto-plastic region

Quantity of reclaimed material (%)	Test temperature (°C)	Rate of loading (Hz)	Number of specimens	Crack speed index
0	15	1	3	0.642
		0.1	3	0.377
	25	1	3	0.342
		0.1	3	0.110
30	15	1	3	0.601
		0.1	3	0.398
	25	1	3	0.405
		0.1	3	0.126
50	15	1	3	0.654
		0.1	3	0.351
	25	1	3	0.377
		0.1	3	0.080
70	15	1	3	0.634
		0.1	3	0.433
	25	1	3	0.378
		0.1	3	0.110

To summarize the data on crack propagation, there is some indication from both the fatigue tests and the crack growth tests that, at very high reclaimed material contents and at low temperatures, there is a very small tendency for the mixtures to be more crack-susceptible. However, this effect is much smaller than variations in physical conditions, and so, for practical purposes, is negligible in a properly designed mixture.

13.4.4 Durability

13.4.4.1 Concept

While durability is a concept with which engineers are familiar, it has proven to be very difficult to quantify the durability of paving materials. As

Table 13.5 Crack growth under creep conditions

Quantity of reclaimed material (%)	Regression parameters	
	A	n
0	0.2512	0.734
30	0.2531	0.782
50	0.2503	0.729
70	0.2644	0.764

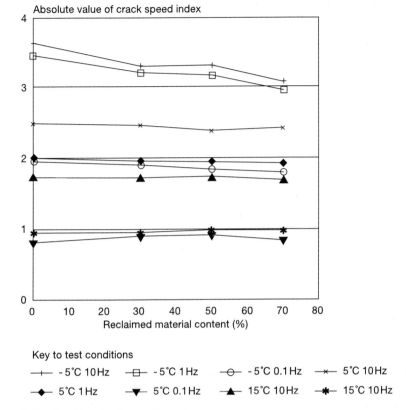

Fig. 13.8 Cracking in the elastic region

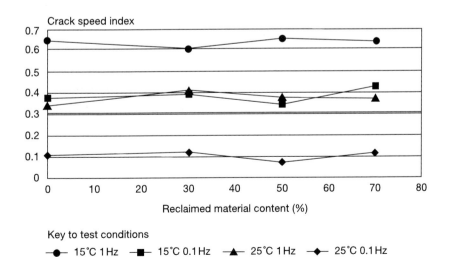

Fig. 13.9 Cracking in the elasto-plastic region

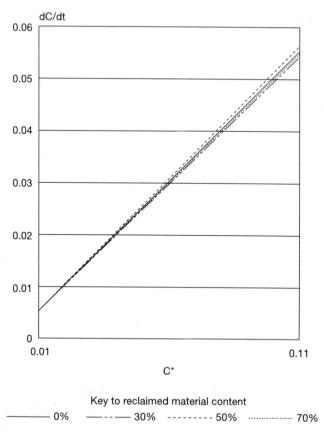

Fig. 13.10 Cracking under creep conditions

a result of this difficulty, a number of tests have been developed, none of which has achieved universal acceptance. Among the tests which have been developed to assess the durability of bitumen, it is probably true to say that the rolling thin film oven test (RTFOT) is probably the most widely accepted, although it only assesses the effect of exposing bitumen to the effects of elevated temperatures, as experienced in an asphalt mixing plant. The other form of testing which is relatively widely used is a water immersion test on a sample of mixture, usually compacted with a Marshall hammer, to check for the effects of water on a mixture. This assessment is based upon a comparison of a mechanical property, such as Marshall stability or dynamic stiffness, before and after a predefined soaking regime.

13.4.4.2 RTFOT study

Two parameters from a programme with the RTFOT on blends of reclaimed bitumen and two softening agents (a soft bitumen and a

Table 13.6 Ageing tests on blends of reclaimed asphalt cement and softening agents (Brown, 1984)

Binder	Retained penetration (%)	Viscosity ratio
New AC – 20*	56.8	3.2
Site 1 plus 3.5% AC - 5*	68.6	7.1
Site 1 plus 3.9% agent	40.3	5.4
Site 2 plus 5.1% AC - 5*	64.0	4.3
Site 2 plus 2.6% agent	44.3	3.2
Site 3 plus 5.5% AC - 5*	58.0	6.2
Site 3 plus 3.8% agent	45.0	3.1

* Some North American bitumens are produced to a viscosity grading specification, these products are designated 'AC grades'.

proprietary rejuvenating agent) (Brown, 1984) are shown in Table 13.6. The retained penetration is the difference between the penetration of the binder before and after RTFOT, expressed as a proportion of the before penetration; the Viscosity ratio is also calculated from before and after tests.

The results in Table 13.6 show that the blend of aged and new AC-5 bitumen has better resistance to ageing than the new AC-20 grade. It is also noticeable that the blends using the recycling agent show significantly less resistance to ageing, suggesting that this agent, which is not identified, is not appropriate for these blends. Therefore, it may be concluded that blends of aged and new bitumen are resistive to ageing.

13.4.4.3 Water immersion tests

A series of water immersion tests were studied (Brown, 1984) using measurements of the Marshall stability before and after soaking to evaluate the resistance to moisture damage. The Corps of Engineers criterion to judge mixture quality requires a retained stability of no less than 70% of the initial stability. The study used two different new aggregates to adjust the gradation of the reclaimed material. The test programme indicated that the type of new aggregate was far more significant in relation to the moisture sensitivity of the mixture than the presence of reclaimed material. Hence, recycled mixtures can provide adequate resistance to moisture, but care is necessary in order to ensure that corrective aggregate with good resistance to moisture is used.

The effect of moisture, combined with freeze-thaw conditions, on recycled mixtures, including with anti-stripping agents, has also been studied (Kallas, 1984). The effect of the anti-stripping agents was discernible, but short-term ageing of the recycled mixtures did not affect the results of the moisture damage tests.

A large study of recycled mixtures, which included examination of material placed on 14 different sites as well as laboratory studies (Dhalaan, 1982), used seven different softening agents. The evaluation was based upon measurements of tensile strength before and after soaking and also indicated that recycled mixtures show similar resistance to moisture damage to mixtures which do not include any reclaimed material.

13.4.4.4 Stiffness

The results of a significant programme of stiffness tests on rolled asphalt mixtures were produced in the same way as the samples used in the cracking tests described in section 13.4.3 (Sulaiman and Stock, 1995). The tests were carried out at temperatures of −5 °C, 5 °C, 15 °C and 25 °C, and the results are reproduced in Table 13.7 and shown graphically in Figure 13.11.

The results show that:

- the stiffness of the mixture is independent of the quantity of reclaimed material in the mixture; and
- the temperature susceptibility of the mixture, as indicated by the rate of change of stiffness with temperature, is also independent of the quantity of reclaimed material.

13.4.4.5 Summary

The weight of evidence is that recycled mixtures have similar performance to mixtures which contain no reclaimed material. However, there

Table 13.7 Results of dynamic stiffness tests

Temperature (°C)	Frequency (Hz)	Stiffness (MPa)			
		0% material content	30% material content	50% material content	70% material content
−5	10	14 294	15 958	15 466	17 863
	5	12 995	14 664	13 083	14 339
	1	9 473	9 938	11 066	11 491
	0.1	6 467	6 785	7 066	7 443
5	10	8 590	9 553	9 498	10 030
	5	7 810	8 450	8 359	9 460
	1	4 131	5 450	4 971	5 334
	0.1	2 165	2 121	1 975	2 947
15	10	6 498	6 901	7 076	7 303
	5	4 798	4 917	5 438	5 844
	1	2 080	2 565	2 505	2 191
	0.1	554	820	605	624
25	10	3 276	3 434	3 407	3 603
	5	2 118	1 818	1 832	2 316
	1	586	611	696	640
	0.1	148	168	167	237

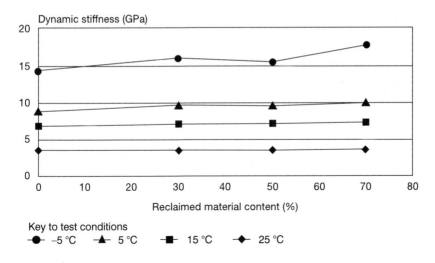

Fig. 13.11 Stiffness of mixtures measured at 10 Hz

is also sufficient evidence to show that this cannot be taken for granted, and that care is necessary to ensure that the relevant softening agent is selected, with the evidence indicating that a soft grade of bitumen is the option which carries least risk. Also, if a new aggregate is selected which would not provide good performance in a mixture without reclaimed material, it will have a detrimental effect on the performance of the recycled mixture.

13.5 COLD RECYCLED MIXTURES

There is very little information on the properties of cold recycled mixtures in the literature. *The Journal of the Association of Asphalt Paving Technologists* is a very good indicator of topics which are considered to be important, and a scan of the papers published between 1986 and 1996 has not unearthed any which report measurements of cold recycled mixture properties. Also, *Bituminous Mixtures in Road Construction* (Hunter, 1994) neglects to mention cold recycled mixtures in the sections discussing paving materials.

However, cold in-situ recycling has been in use in the United Kingdom for many years, through the 'repave' process, and the industry literature contains many articles describing:

- projects which have been carried out successfully;
- developments in materials, such as the use of foamed bitumen; and
- developments in equipment, such as the purpose built machinery which carries out the whole process.

Therefore, it is surprising that there is virtually no information on the properties of materials produced with these processes.

While there were a number of eloquent supporters of cold recycling at the 1990 Seminar held by the British Aggregates & Construction Materials Industries (BACMI), the trade association for the paving industry, who described successful projects, the use of cold recycling remains limited and low-key. Reference can be found in the engineering press to quotations, one example being a statement to the effect that a particular proprietary type cold mixture 'can regularly achieve between 4000 MPa and 6000 MPa'. However, there do not appear to be reports in the public domain which report systematic investigations of the material. It seems likely that the general conservatism in the United Kingdom concerning the use of cold mixtures acts as an additional inhibition to an industry which is generally reluctant to embrace asphalt recycling.

13.6 MIXTURE DESIGN

13.6.1 Hot mixtures

Successful mixture design for hot recycled mixtures is straightforward, and is based upon the overwhelming evidence that mixtures containing reclaimed material exhibit the same properties as mixtures containing no reclaimed material provided that the following conditions are met within the recycled mixture:

- the aggregate gradings are the same;
- the total binder content is the same;
- the corrective binder/softening agent, when combined with the reclaimed bitumen, produces a binder which is the same grade as that which would be used in the mixture with no reclaimed material.

Inevitably, the design of a recycled mixture involves an analysis of the reclaimed material. It is necessary to measure the binder content, and to recover some bitumen in order to determine its penetration; it is also necessary to determine the grading of the aggregate in the reclaimed material. From this information, it is possible to determine the gradation of additional aggregate to ensure that the combination meets the target, and also the quantities of softening agent and new binder to meet grade and content requirements. The process is shown as a flow chart in Figure 13.12.

13.6.2 Cold mixtures

There is no reason to believe that the process described above would not work on cold mixtures, although it has not been verified. The major

Mixture design

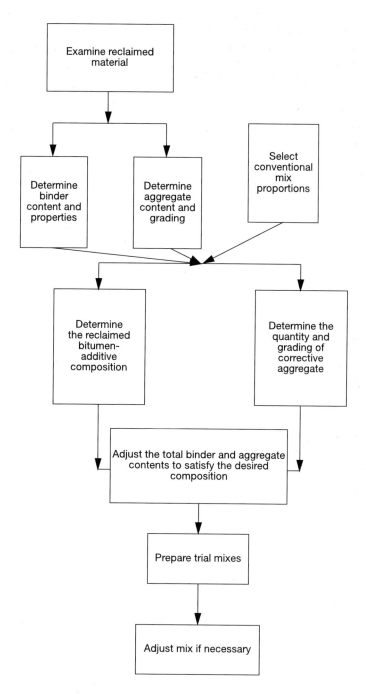

Fig. 13.12 Flow chart for the mixture design procedure

difficulty facing this approach is the lack of use of cold mixes in the United Kingdom, which means there is no 'conventional' design to copy.

While the process of matching the composition of existing mixtures is an effective approach to the design of mixtures containing reclaimed material, it is not necessarily the approach which will make the most cost-effective use of the reclaimed material that is available to a coating plant. The progressive introduction of performance orientated specifications will, hopefully, offer competent operators with the opportunity to optimize the use of reclaimed material to produce mixtures which provide the required levels of performance.

13.7 IMPLEMENTATION OF RECYCLING

13.7.1 Introduction

Notwithstanding the ready availability of proven equipment for the production of recycled mixtures, the considerable volume of positive high-quality data from the evaluation of material in the literature, and that many other countries have embraced the process, recycling has not been introduced in the United Kingdom with any enthusiasm.

The US Army Corps of Engineers, the largest engineering organization in the world, include asphalt recycling processes in their procedures, and have used these mixtures successfully on airfields, when high-quality materials are essential, and the risks of damage to aircraft dictate that a cautious and conservative approach is taken to the selection of paving materials.

Within the United States of America, recycling is commonplace and rarely warrants notice any more. It is so well enshrined in the industry practice that there is, and has been for over a decade, a specific trade association for contractors engaged in recycling. The Asphalt Institute, which is funded by bitumen producers, has issued publications in its manual series covering both hot (Asphalt Institute, 1986) and cold (Asphalt Institute, 1983) recycling. Also, recycling has achieved a level of acceptance within many European countries which makes it no longer a subject for comment or debate.

Given this background, this section will discuss the opportunities for recycling in the United Kingdom and consider the issues surrounding its implementation.

13.7.2 Opportunities for recycling

There are approximately 50 000 km of motorway, trunk and principal roads in the United Kingdom. Assuming that these roads, which were designed for a 20-year life, are reconstructed on a 40-year schedule, then somewhere between 1 and 2 million tonnes of asphalt are reclaimable each year. This will include some 60 to 120 thousand tonnes of bitumen. At a reclaimed material content of 70%, studies have shown that savings in the region of £10/tonne of mixture are reasonable in the United Kingdom. Hence, utilizing two million tonnes of reclaimed material in this manner represents a saving of £20 million.

If the remaining 300 000 km of the road network are included in the calculation, the opportunity is enhanced considerably. For example, there is a huge length of road used principally by local traffic and tourists. The ride and level of service provided by these roads is frequently very poor, and the usual form of maintenance is surface dressing. Many of these roads would benefit greatly from surface recycling, and a cold process would probably be the most effective. While this treatment would cost more than a surface dressing, it would restore the riding quality of the pavement as well as seal the surface, which will in turn reduce the longer-term maintenance costs. Benefits to the local economy would accrue through an improved level of service, and also from increased tourism resulting from the better service.

It may be argued that recycling will reduce the demand for stone and bitumen and, therefore, is not in the interests of the paving material industry. However, the saving mentioned above would purchase about one million extra tonnes of coated material, which would represent an increase of about 5% in the output to the industry, which would presumably be welcome.

Recycling also offers the opportunity to reduce costs during reconstruction work. Material removed from pavements during reconstruction is almost invariably transported off site, and if dumped attracts a significant charge. As hot in-plant recycling is currently the most controllable process, a well-organized job can improve utilization of transport by loading lorries which have delivered new asphalt with the reclaimed asphalt for the return trip. This reclaimed material can be stored at the coating plant and become available for incorporation in some recycled mixture. This improves vehicle utilization and also avoids charges for the disposal of the removed material, and can represent a significant overall improvement in efficiency.

All coating plants create waste. There are inevitable reject loads, not necessarily due to poor control at the mixing plant, but due to uncontrollable factors such as weather conditions and plant breakdowns. Again, recycling at the coating plant represents an opportunity to utilize these materials avoiding transport costs and fees for dumping in appropriate sites. There is at least one supplier of coated material in the United Kingdom who has found this approach sufficiently attractive to collect material which is considered to be 'waste' from neighbouring coating plants and recycle it. If this was not a cost-effective procedure, it would not be performed. The pressures created by the need to complete major road maintenance and reconstruction projects with the minimum disruption to the road user have generated an interest in recycling because on large projects, facing great cost and time pressure, it can make the difference between success and failure.

13.7.3 Economics and energy

In the United Kingdom, the Department of Energy have monitored energy consumption during paving operations utilizing both hot and cold recycling processes. In both cases, they have concluded that there are significant energy savings as a result of utilizing recycling processes.

A separate study (Servas *et al.*, 1987) measured the savings resulting from recycling, and compared them with estimates. The results are reproduced in Table 13.8, and confirm that, to a very large degree, estimated cost savings can be achieved. This is particularly reassuring because this study was undertaken when recycling was being introduced, that is when the people involved were at the start of their learning curve, and were quite properly being cautious with their pricing.

13.7.4 Implementation of recycling

It is appropriate to ask why there is not more widespread utilization of recycling given that:

- recycled asphalt mixtures are of acceptable quality;
- the highly respected Department of Energy audit team has measured significant savings in energy consumption; and
- studies of savings have shown that estimated savings are to a significant degree realized in practice.

It is clear that, within the industry, there is not a receptive atmosphere towards recycling. For example, the author received a letter from a well-known contractor who explained that, on finding out that the Highways Agency specification of the early 1990s was permitting the addition of 10% reclaimed material, he had sold his recycling plant.

Table 13.8 Cost savings resulting from recycling

Proportion of reclaimed material (%)	Estimated saving	Actual saving (from completed project)
	Proportional cost of new asphalt (%)	
20	11	10
50	24	24
75	36	29

Some of the very confused thinking relating to recycling asphalt, but underlining the generally negative attitude towards it was summarized neatly by Mathew Pettipher (1994) of the *New Civil Engineer*. Mr Barry Neville, campaigns manager for the BRF, was quoted as saying:

- If you want good roads you need the best materials and the only way to be sure of getting them is to use primary aggregates.
- With the arrival of whole life costing and 44 t lorries it would be false economy to build roads not of the highest standard because we would just have to repair them in five years.

These statements from the influential pressure group are not substantiated by any hard facts, because the asphalt produced with reclaimed material is indistinguishable from that produced with all new material. British Aggregates & Construction Materials Industries' (BACMI) economist Jerry McLaughlin is quoted as saying that:

> Cost is a factor but not as much as the client's perception about the quality of recycled materials. Aggregates are not a large part of a project cost but they are very important in terms of quality,

confirming the prejudicial view that secondhand is substandard. In truth, the requirements for road building aggregates are so stringent that stone in the paving mixtures does not degrade. It is only at the tyre/surface interface that stone polishes and wears, so the reclaimed aggregates found in asphalt are generally as good as new. Maurice Webb, speaking on behalf of the Federation of Civil Engineering Contractors (FCEC) has said:

- A lot of Contractors suggest to local authorities that road planings should be re-used. But they are classified as unfit because of the bitumen content.
- The DoE claims to have no problem with bitumen but the 6 Mt to 7 Mt of planings lifted by Contractors every year continues to go to waste.

It seems that the FCEC believe that planings should be recycled into some use other than asphalt, because there is no other reason to reject stone coated with bitumen. Unfortunately, this is a misguided and even irrelevant proposal, and certainly ranks as a major missed opportunity as it disregards the most costly component, the bitumen.

Having spoken at many meetings of learned societies and participated in many discussions on implementation of recycling, the author is convinced that the material supply industry does not believe that recycling is commercially viable. However, it is essential that there is no confusion between a process which yields economic benefits or benefits to the community, because this is undoubtedly true where recycling is concerned, and one which yields commercial benefits, i.e. improved return on investment, which is what industry requires.

It is possible that industry believes that the spending authorities would simply retain all the savings which may accrue from recycling, and return them to some central fund from which they may be re-assigned to non-highway cost centres, or even returned to taxpayers. This is of no benefit to the material supply industry because they will be contributing to a reduction in their own business. Recycling is not like the introduction of new products, such as the range of thin surfacings currently in vogue, because it does not necessarily generate a new market, and a spending authority which simply pockets all the saving instead of re-investing a significant proportion of it will inevitably stifle innovation. According to the *BACMI Statistical Year Book* for 1995, there were 350 coating plants producing 37.7 million tonnes of asphalt, which gives an average production of about 108 000 tonnes of asphalt/plant. According to Table 13.8, recycling at 10% reclaimed material content yields a saving of about 10%. Assuming that a contractor can keep all that saving, which could be about £2.5/tonne, he or she will have to produce 40 000 tonnes of mixture containing this quantity of reclaimed material in order to generate £100 000, which is probably representative of the lowest start-up cost for recycling. This is, of course, a very crude sum and does not include finance and other costs associated with this sort of development. Given the overall luke warm reception of recycling, and the fact that the one or two million tonnes of planings which may become available each year only represent 2.5% of the total production of asphalt, the average plant operator can only expect to produce 2000–3000 tonnes of mixture containing reclaimed material each year. It will take a long time to reach 40 000 tonnes at that rate, and so the decision to sell the recycling plant is perhaps not surprising. It is certain that competent investors are looking for better and quicker returns.

This is, of course, a worst-case scenario, because reclaimed material is available in commercially significant quantities in some areas, and the quote above from the FCEC does suggest that significantly more reclaimed material is available, but it does indicate the gap between the strong economic case for recycling and the weak commercial one.

If the Industry is truly committed to providing value for money for the road user, a way must be found to overcome this problem. In the United States of America, the Federal Highways Agency set up 'Demonstration Project 39'. This project ran for a limited period and provided funds for some of the one-off costs faced by contractors who were setting up to produce recycled mixtures for the first time. Because the United States of America is one of the strongest advocates of a free market economy, this recognition of the need to underwrite some start-up costs, in order that a significant economic gain can be made, is an example which could be of value to others.

13.8 REFERENCES

Manual of Contract Documents for Highway Works, Her Majesty's Stationery Office, London.
Volume 1: Specification for Highway Works (MCHW 1).
Asphalt Institute (1983) *Asphalt Cold-Mix Recycling*, The Asphalt Institute Manual Series No. 21 (MS-21), The Asphalt Institute, Maryland, USA.
Asphalt Institute (1986) *Asphalt Hot-Mix Recycling*, The Asphalt Institute Manual Series No. 20 (MS-20), The Asphalt Institute, Maryland, USA.
Brown, E.R. (1984) *Evaluation of Properties of Recycled Asphalt Concrete Hot Mix*, Technical report GL-84-2, Final Report, US Army Engineer Waterways Experiment Station, Vicksburg, Mississippi.
Brown, S.F., A.F. Stock and P.S. Pell (1980) The structural design of asphalt pavements by computer, *The Journal of the Institution of Highway Engineers*, March, Institution of Highway Engineers, London.
Department of Transport Roads and Local Transport Directorate (1982) *Departmental Standard HD/7/82 Specification for Road and Bridge Works. In-situ Recycling: The Repave Process*.
Dhalaan, M.H. (1982) *Characterization and Design of Recycled Asphalt Concrete Mixtures Using Indirect Tensile Test Methods*, PhD Thesis, The University of Texas at Austin, USA.
Holmgreen, R.J., J.A. Epps, D.N. Little and J.W. Button (1980) *Recycling Agents for Bituminous Binders*, Report No. FHWA-RD-80, Federal Highways Administration, National Technical Information Service, Springfield Virginia, USA.
Hunter, R.N. (ed.) (1994) *Bituminous Mixtures in Road Construction*, Thomas Telford Services Ltd, London.
Kallas, B.F. (1984) *Flexible Pavement Mixture Design Using Reclaimed Asphalt Concrete*, Research Report No. 84-2, Oct., The Asphalt Institute, Maryland, USA.
Pettipher, M. (1994) Quality block to recycling campaign, *New Civil Engineer*, 21 Apr., Thomas Telford Services Ltd, London, p. 6.
Road Research Laboratory (1962) *Bituminous Materials in Road Construction*, Her Majesty's Stationery Office, London.
Servas, V.P., M.A. Ferreira and P.C. Curtayne (1987) Fundamental properties of recycled asphalt mixes, *Proceedings of the 6th International Conference on the Structural Design of Asphalt Pavements*, July, Ann Arbor. Michigan, USA.
Shell International Petroleum Company Ltd (1978) *Shell Pavement Design Manual*.
Stock, A.F. (1985a) Bitumen recovered from highways and its use in recycled mixes, *Highways and Transportation*, May, The Institution of Highways and Transportation, London.
Stock, A.F. (1985b) Structural properties of recycled mixes, *The Journal of the Institution of Highway Engineers*, March, Institution of Highway Engineers, London.
Sulaiman, S.J. and A.F. Stock (1995) The use of fracture mechanics for the evaluation of asphalt mixes, *Journal of the Association of Asphalt Paving Technologists*, 64, March, 500–33.
Taylor, N.H. (1978) Life expectancy of recycled asphalt paving. In L.E. Wood (ed.), *Recycling of Bituminous Pavements, ASTM STP 662*, The American Society of Testing Materials, 3–15.
Taylor, N.H. (1981) Recycling of asphalt paving in Singapore and India, *Construction Industry International Conference on Road Surfacing Re-Cycling Technology '81*, Imperial College, London.
Whiteoak, D. (1990) *The Shell Bitumen Handbook*, Shell Bitumen, Chertsey.

CHAPTER 14

Summary

J.C. Nicholls, Transport Research Laboratory

14.1 SELECTION OF TYPE OF SURFACING MATERIAL OR TREATMENT

The properties of the various surfacing materials and surfacing techniques are described in the relevant chapter with a crude summary given in Table 14.1 (Nunn *et al.*, 1997). The choice of which material(s) or treatment(s) is (or are) the most appropriate for a particular site requires a rational policy. There is rarely a unique solution, and selection will involve consideration of:

- the properties required of the road surface after work is completed;
- the local availability of constituent materials;
- the local experience of the relevant options;
- the suitability of laying the options (particularly surfacing treatments) at the time of day and in the expected weather conditions when the work is to take place;
- any physical constraints imposed on the new surfacings or its construction; and
- the relative cost of the various options.

The main decisions in deciding on the application of a surfacing material or treatment, and those responsible for making them, will generally be led by the following questions:

- Why?   The client (or engineer on his or her behalf) should have a policy for when construction or maintenance is needed, and required levels of properties such as skid resistance, texture depth, ride quality, waterproofing.
- Where?   The client (or engineer on his or her behalf, particularly for maintenance) will assess where construction or treatment is needed according to the adopted criteria.
- Who?   The engineer will select potential contractors dependent on their expertise with the material or treatment required (the selection may be problematic when the option(s) include alternative proprietary processes; the possible solutions include performance contracts or parallel tenders).
- What?   The engineer has to decide what to use but, if more than one option is permitted, the contractor will make the final decision.

Table 14.1 Performance of surfacing materials

Material		Suitability for re-profiling	Deformation resistance	Resistance to cracking	Spray reducing	Noise reducing	Skid resistance	Texture depth	Initial cost	Durability	Speed of construction	Quality of ride
Rolled asphalt		✓✓✓✓	✓✓✓	✓✓✓✓	✓✓✓	✓✓✓	✓✓✓✓	✓✓✓✓	✓✓✓✓	✓✓✓✓✓	✓✓	✓✓✓
Thick wearing course	Porous asphalt	✓✓✓	(✓✓✓✓✓*)	✓✓✓✓✓	✓✓✓✓✓	✓✓✓✓✓	✓✓✓✓	✓✓✓✓✓	✓✓✓	✓✓✓	✓✓✓	✓✓✓✓✓
	Asphalt concrete / dense bitumen macadam	✓✓✓	✓✓✓✓	✓✓✓	✓✓✓	✓✓✓	✓✓✓	✓✓✓	✓✓✓✓	✓✓✓	✓✓✓	✓✓✓✓✓
	Mastic asphalt / gussasphalt	✓✓✓	✓✓	✓✓✓✓	✓✓	✓✓✓	✓✓	✓✓	✓✓	✓✓✓✓	✓✓	✓✓✓
	Stone mastic asphalt	✓✓✓✓✓	✓✓✓✓	✓✓✓✓	✓✓✓	✓✓✓	✓✓✓	✓✓✓	✓✓✓	✓✓✓✓	✓✓✓	✓✓✓
Thin wearing course	26–39 mm thick	✓✓✓	✓✓✓✓	✓✓✓	✓✓✓	✓✓✓	✓✓✓	✓✓✓	✓✓✓✓	✓✓✓	✓✓✓	✓✓✓✓
	18–25 mm thick	✓✓✓	✓✓✓	✓✓✓	✓✓✓	✓✓✓	✓✓✓✓	✓✓✓	✓✓✓✓	✓✓✓	✓✓✓✓	✓✓✓✓
	< 18 mm thick	✓✓	✓✓✓	✓✓✓	✓✓✓	✓✓✓	✓✓✓✓	✓✓✓	✓✓✓✓	✓✓	✓✓✓✓	✓✓✓✓
Veneer treatment	Surface dressing	n/a	n/a	✓✓✓	✓✓✓	✓	✓✓✓✓	✓✓✓✓✓	✓✓✓✓✓	✓✓	✓✓✓✓✓	†
	High-friction systems	n/a	n/a	✓✓✓	✓✓✓	✓✓	✓✓✓✓✓	✓✓✓✓	✓✓	✓✓✓	✓	n/a
	Slurry surfacing	✓✓‡	n/a	✓✓	✓✓	✓✓	✓✓✓	✓✓✓	✓✓✓✓✓	✓✓	✓✓✓	n/a

✓ = least advantageous to ✓✓✓✓✓ = most advantageous

Notes

* The deformation of rolled asphalt can be enhanced by selection, as for clause 943 of the *Specification for Highway Works*.
† The quality of ride for surface dressing will depend on the design of surface dressing, the aggregate size(s) employed and the evenness of the substrate
‡ Slurry surfacing can give a useful improvement to the profile of the type of surface to which it is applied, for which this rating is appropriate – for other types of surfacing, it may not be appropriate.

suitability for re-profiling	The suitability of the material to be used for regulating or re-profiling an existing surfacing.
deformation resistance	The ability of the material to resist the effects of heavy traffic to create ruts in the wheel-paths during hot weather.
resistance to cracking	The ability of the material not to crack or craze with age, particularly in cold weather and in areas of high stress.
spray reducing	The ability of the material to form a surfacing which minimizes the amount of water thrown up by the wheels of passing traffic into a driver's line of sight in wet conditions.
noise reducing	The ability of the material to form a surfacing which reduces the noise generation at the tyre/surfacing interfaces and/or increase the noise absorbed.
skid resistance	The ability of the material to form a surfacing which can achieve a high mid-summer SCRIM coefficient.
texture depth	The ability of the material to form a surfacing which can achieve a high texture depth, with particular reference to the requirement for high-speed trunk roads of 1.5 mm using the sand-patch method.
initial cost	The initial cost to supply, lay and compact an area with the material.
durability	The ability of the material to remain in place and retain its other properties with time while exposed to traffic and the weather.
speed of construction	The time required between closure and re-opening of the road when surfacing it with the material.
quality of ride	The ability of the material to form a surfacing which gives a driver a comfortable ride.

Source: Nunn *et al*., 1997.

- **When?** The engineer will specify the period during which the work is to be carried out and may impose constraints on the time of day or the day of the week; the contractor may need to ensure that the work will be carried out in the appropriate season for the material or treatment.
- **How?** The contractor needs to be competent in the option, but the engineer also needs to appreciate the time that the road will be occupied in order to allow for traffic diversions, road closures, etc.

Generally, the use of surface treatments will be preferred to overlaying or replacing the surface course because of the cost, although these are generally more dependent on the weather conditions.

14.2 SURFACE TREATMENTS

A simple flow chart for selecting an appropriate surface treatment based on the need for certain important properties is shown in Figure 14.1 (Catt, 1995). However, there are often other matters that need to be considered (such as the time of year if the work has to be done quickly), so that there may be a more appropriate alternative. Generally, one of the branches has to be followed while the other can be followed irrespective of the answer (such as if texture depth is not required, treatments either providing or not providing texture can be used but, if texture is required, the treatment must provide texture).

It is important to note that all the treatments are weather sensitive to a greater or lesser extent, in particular with regard to minimum temperature requirements and wet roads. A crude guide to the season in the United Kingdom is given in Table 14.2.

14.3 SURFACING MATERIALS

When considering the selection of a 'thick' surfacing material (i.e. excluding veneer and the thinner thin surfacings), the choice will depend on the properties required, which are summarized in Table 14.1 and reviewed below.

Table 14.2 Season for surface treatments

Treatment		Preferred season	Extended season
Conventional surface course		March to mid-November	All year
Thin surface course	Ultra thin layers	Mid-April to September	Mid-March to October
	Thin layers	March to October	All year
Surface dressing	14 mm	Mid-May to mid-July	May to July
	10 mm	May to July	Mid-April to mid-August
	6 mm	May to mid-August	Mid-April to mid-September
Slurry surfacing		April to mid-September	March to October
High-friction surfacings	Chemical set	April to September	March to October
	Thermoplastic	All year	All year
Re-texturing		All year	All year

Surfacing materials

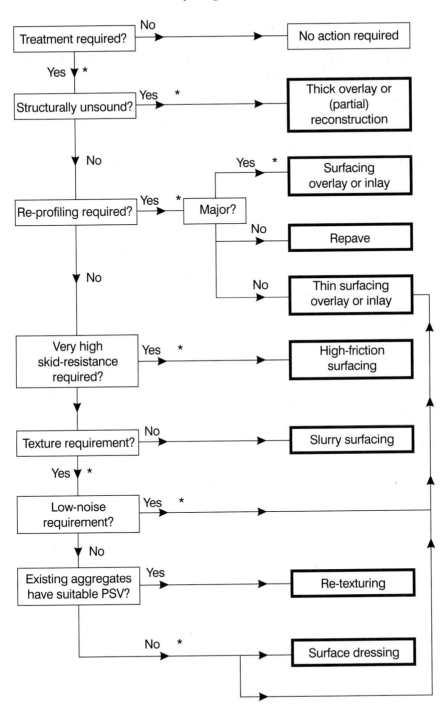

* Criteria other than those given here can override the decision to send the choice down this branch

Fig. 14.1 Flow chart for choice of treatment

14.3.1 Regulating ability

All the materials, other than the veneer coats, can provide regulating ability, the extent being basically dependent on the depth of the layer. The uniformity of the final profile may also be inversely affected by the maximum nominal size of aggregate in high stone-content mixtures, the maximum nominal size usually being related to the depth of the layer. Therefore, the need for regulation and profile will determine the thickness of layer rather the type of mixture.

14.3.2 Skid-resistance

The skid-resistance, as measured by mean summer SCRIM coefficient, is dependent on the polished stone value of the exposed aggregate. Therefore, all mixtures with exposed aggregate can be designed to give a skid-resistant surfacing; the obvious exception is gussasphalt and, more particularly, mastic asphalt. However, some mixtures with thick binder films, such as porous asphalt, stone mastic asphalt and several of the proprietary thin surfacings, may require to be trafficked before the aggregate becomes sufficiently exposed to provide the skid-resistance.

14.3.3 Texture depth

Materials in which the texture is below the screeded surface, colloquially known as 'negative textured' (such as porous asphalt, stone mastic asphalt and thin surfacings), provide, or can be designed to provide, a good texture depth. Porous asphalt has particularly good texture depth. Rolled asphalt with pre-coated chippings will also give good texture by the application of the chippings. However, high stone content rolled asphalt will generally develop texture with trafficking removing excess mortar on the surface. Macadam and asphalt concrete provide limited texture depth, while gussasphalt and mastic are relatively smooth unless chippings are applied or a crimping roller is used.

14.3.4 Permeability

The approximate permeabilities of the surfacing materials are given in Table 14.3 (Daines, 1995). These show that the binder-rich materials are, as would be expected, the most impermeable and therefore suitable when water penetration is deprecated.

14.3.5 Noise

The noise generated on surfaces in which the texture depth is above the screeded surface, colloquially known as 'positive textured' (such as rolled asphalt with pre-coated chippings and surface dressing), increases with

Table 14.3 Relative permeability of surfacing materials

Material	Typical voids content (%)	Approximate water permeability (m/sec)
Mastic asphalt	<1	<10^{-11}
Rolled asphalt (30% stone)	2–8	10^{-11}–10^{-10}
Rolled asphalt (55% stone)	2–6	10^{-11}–10^{-10}
Asphalt concrete	3–5	10^{-10}–10^{-8}
Close-graded bitumen macadam	4–7	10^{-8}–10^{-5}
Open-graded bitumen macadam	12–20	10^{-8}–10^{-3}
Porous asphalt	15–25	10^{-4}–10^{-2}

the texture depth, and therefore the safety. This is not the case with 'negative textured' materials, which are also generally quieter than 'positive textured' materials. Of all the materials, porous asphalt is the quietest with the pores absorbing noise.

14.3.6 Spray and glare

The extent of spray and glare generated on the surfaces is not generally considered in deciding which of the materials are to be used, but can tilt the decision if all other things are equal. Porous asphalt, designed to remove the water that causes spray and glare, is the best material, while thin surface course materials and stone mastic asphalt have shown some advantage. However, of the impermeable materials, greater texture depth will generally improve the spray- and glare-reducing properties.

14.3.7 Physical constraints

The various physical constraints that need to be considered include:

- adequate access on the site for the equipment, such as loading the chipper for rolled asphalt or to allow laying in echelon with porous asphalt;
- the load, and hence maximum layer thickness possible, on under-bridges; and
- the maximum thickness when maintaining existing kerbs or headroom to over-bridges.

All these constraints are very much site-specific.

14.4 COMBINATION TREATMENTS

An alternative is to provide a new surface course layer in a material which does not have all the required properties and then overlay it with a suitable surface treatment to provide the missing properties. This has to be done in some circumstances, such as the need for high skid-resistance which can only be provided by a high-friction system.

14.5 RELATIVE ECONOMIC BENEFITS

The relative cost of the various materials and treatments will be dependent to a certain extent on the availability of component materials. However, the ranking is essentially in order of binder content, with gussasphalt and mastic asphalt the most expensive and macadams generally the cheapest. The exception is the high-friction surfacings, which are more costly with their specialist binders. The other consideration in ranking materials is whether the comparison is by area or by volume. Thin surface course materials are generally more expensive than surfacing materials laid at full depth when compared by volume or weight, but tend to be less expensive in terms of area covered because they are laid thinner. The economic benefits need to be compared on a site-by-site basis.

14.6 REFERENCES

Catt, C.A. (1995) Choices for the maintenance engineer, *Surface Treatments Seminar*, Dec., Transport Research Laboratory, Crowthorne.

Daines, M.E. (1995) *Tests for Voids and Compaction in Rolled Asphalt Surfacings*, Department of Transport TRL Project Report 78, Transport Research Laboratory, Crowthorne.

Nunn, M.E., A. Brown, D.J. Weston and J.C. Nicholls (1997) *Design of Long-Life Flexible Pavements Modulus*, TRL Report 250, Transport Research Laboratory, Crowthorne.

APPENDIX A

Draft UK Specification for Stone Mastic Asphalt Surface Course

GENERAL

1. Stone mastic asphalt shall comply with the general requirements of BS 4987 for coated macadam, the Specification for Highways works and the specific requirements of the following sub-clauses.

MATERIALS

AGGREGATES AND FILLER

2. Coarse aggregate shall be crushed rock or crushed slag complying with Clause 901.

3. When tested in accordance with the procedures of BS 812, the coarse aggregate shall additionally have the following properties:
    Polished Stone Value – not less than 45, or as specified in Appendix 7/1.

    Ten Per Cent Fines Value – not less than 180 kN when tested in a dry condition, or as specified in Appendix 7/1.

    Maximum Aggregate Abrasion Value – not more than 12, or as specified in Appendix 7/1.

    Maximum Flakiness Index – for the coarse aggregate only, 30 per cent, or as specified in Appendix 7/1.

4. Fine aggregate shall comply with Clause 901 and shall comprise crushed rock, crushed slag or crushed gravel fines, which may be blended with not more than 50 per cent natural sand.

5. Added filler shall be hydrated lime, crushed limestone or Portland Cement, in accordance with the requirements of BS 594: Part 1 and shall be at least 2 per cent by mass of total aggregate.

BINDER

6. Unless specified otherwise in Appendix 7/1, either a modified binder or, alternatively, bitumen with a stabilizing additive, shall be used, at the choice of the Contractor. Modifiers include any material added to or blended with the base bitumen.

7. Bitumen shall comply with BS 3690: Part 1 and shall have a nominal penetration of 50 or 100, unless specified otherwise in Appendix 7/1.

8. If a modified binder is used, the base bitumen, before modification, shall comply with BS 3690: Part 1, and shall have a nominal penetration of 50 or 100 or 200, unless specified otherwise in Appendix 7/1.

9. The choice of bitumen grade or type of modified binder shall be notified to the Engineer before the commencement of work.

STABILIZING ADDITIVE OR MODIFIED BINDERS

10. When bitumen complying with BS 3690: Part 1 is used as the binder, at least 0.3 per cent by mass of total mixture of stabilizing additive shall be used to ensure binder drainage does not occur during transport and handling. Stabilizing additives shall be cellulose or mineral or other suitable fibre.

11. Proposals to use a bitumen and stabilizing additive, or a modified binder, shall be submitted to the Engineer, complete with all details, including binder drainage test results, manufacturer's recommendations for addition or means of incorporating any stabilizing additives or modifiers, homogeneously, without segregation, into the mix.

12. Before agreeing the use of an additive or modified bitumen, the Engineer shall be satisfied it has proved satisfactory in use under circumstances, similar to the Contract, elsewhere or that it has undergone appropriate performance trials. For the purpose of this sub-clause, documented evidence of use and trials of the additive or modifier, in any member state of the European Economic Area, will be acceptable.

13. Where information on use or trials is inadequate or lacking, in the opinion of the Engineer, trials may be required to be undertaken before agreeing the use of the additive or modifier.

Draft UK Specification for Stone Mastic Asphalt Surface Course

Table A1.1 Aggregate grading

BS sieve size	Per cent by mass of total aggregate passing	
	Nominal size	
	14 mm	10 mm
20 mm	100	
14 mm	90–100	100
10 mm	35–60	90–100
6.3 mm	23–35	30–50
2.36 mm	18–30	22–32
75 μm	8–13	8–13
Binder (% by mass)	6.5–7.5	6.5–7.0

MIXTURE

14. The target aggregate grading and target binder content proposed by the Contractor shall fall within the envelope formed by the limits given in Table A1.1, unless agreed otherwise by the Engineer before the commencement of work.

15. Adjustments may be required to the above binder content ranges to account for the varying density of slag aggregates should these be used.

16. The Contractor shall demonstrate the properties of the proposed mixture, at the target composition, by preparing loose mixture and compacted specimens in accordance with the general requirements of BS 598: Part 107. The loose mixture and compacted specimens shall comply with the requirements of sub-Clause 17. and 19. below.

17. When tested at the target composition, the loose mixture shall demonstrate not more than 0.3 per cent binder drainage, by total mass of mixture, at a temperature of 175 °C. The test shall be carried out using the apparatus and general principles stated in Clause 939. The drainage shall be calculated as:

$$Binder\ drainage = \left[\frac{(W_2 - W_1)}{(1100 + B)}\right] \times 100\%$$

Where B is the initial mass of binder in the mixture, W_1 and W_2 are the mass of tray and foil before testing and tray and foil and drained binder after testing and the mass of combined aggregate before addition of binder was 1100 g, all as stated in Clause 939.

18. Three compacted specimens shall be manufactured at the target composition and the air void contents of these shall be measured by the procedure described in ASTM D 3203 (or DD 228 – Methods for determination of maximum density of bituminous mixtures), using:

    (a) the maximum density of the mixture, obtained using the theoretical maximum specific gravity of the loose mixture, determined in accordance with ASTM D 2041 and converted to relative density using the appropriate correction factor.

    (b) the bulk density of the specimen, determined in accordance with BS 598: Part 104: Clause 4, as the bulk density required by ASTM D 3203, except the specimens shall not be coated in wax.

19. At the target composition, the air void content of the mixture shall be within the range 2–4 per cent.

MIXING

20. Stone mastic asphalt shall be mixed in accordance with the requirements of BS 4987: Part 1, such that an homogeneous mixture of aggregate, filler, bitumen and, when used, additive, is produced at a temperature of 150–190 °C. At the time of mixing, the coarse aggregate shall be in a surface dry condition.

TRANSPORTATION

21. Stone mastic asphalt shall be transported to site in double-sheeted or tented and sealed ridge sheeted insulated vehicles.

22. To facilitate discharge of stone mastic asphalt, the floor of the vehicle may be coated in accordance with the requirements of BS 594: Part 2, Clause 4.3. When a coating is used then, prior to loading, the body shall be tipped to its fullest extent, with the tailboard open, to ensure drainage of any excess. The floor of the vehicle shall be free from adherent bituminous materials or other contaminants.

SURFACE PREPARATION

23. Existing surfaces shall be prepared in accordance with the requirements of BS 4987 and the Series 700. Tack coat shall be K1-40 cationic bitumen emulsion complying with BS 434: Part 1. It shall be spray-applied at a rate of 0.3–0.5 L/m^2 to completely cover the surface

and shall be allowed to completely break before the stone mastic asphalt is laid.

24. Where necessary, or when required by the Engineer, existing surfaces shall be regulated in accordance with the requirements of Clause 907.

25. Unless raised prior to surfacing, iron-work and reflecting road studs shall be located for lifting and relaying after completion of surfacing works. Gullies shall be covered prior to surfacing.

LAYING

26. Unless required otherwise, stone mastic asphalt shall be laid and compacted in accordance with the requirements of Clause 901, to the thickness stated in Appendix 7/1

COMPACTION

27. Stone mastic asphalt shall be compacted immediately, to practical refusal, using at least two steel-wheeled rollers, with a minimum mass of 6 tonne, per paver. One roller shall be a tandem drum roller.

28. The tandem drum roller shall operate directly behind the paver, while the other roller shall be used for completion of rolling and the removal of all roller marks.

SURFACE TEXTURE

29. When stated in Appendix 7/1, the texture depth of the surfacing shall be in accordance with the requirements of Clause 921 after compaction.

COMPLIANCE OF MIXTURE

30. The agreed mixture shall be that obtained following completion of mixture design and the agreement of a target binder content and target aggregate grading for the mixture.

NOTE FOR GUIDANCE: The agreed mixture is that obtained after the Contractor demonstrates a mixture which complies with the above requirements, and then proposes that mixture to the Engineer for agreement.

Table A1.2 Tolerances for aggregate grading 14 mm and 10 mm size

BS test sieve	Tolerances for aggregate grading in per cent by mass of aggregate passing BS test sieve
14 mm	± 5
10 mm	± 10
6.3 mm	± 8
2.36 mm	± 7
75 μm	± 2

32. When tested in accordance with the methods of BS 598, the sampling and testing tolerance for binder content shall be ± 0.6 per cent.

DETAILS TO BE SUPPLIED

33. The Contractor shall supply all the details required in this Clause to the Engineer before commencement of work under this Clause and when requested during the work.

34. The Contractor shall supply the Engineer with test certificates stating the properties of the materials used. Samples of emulsion tack coat, modified or unmodified bitumen, additive or mixed bituminous materials from the pavement surface or other suitable sampling point shall also be supplied to the Engineer by the Contractor when so instructed by the Engineer.

APPENDIX B

Draft UK Specification for Thin Surface Course Systems

1. Thin wearing course systems shall comply with sub-Clauses 2 to 26 of this Clause and the requirements of Appendix 7/1.

2. Thin wearing course systems shall have a British Board of Agrément Roads and Bridges Certificate. In the event that no such Certificates have been issued, thin wearing course systems shall have Departmental type approval.

AGGREGATES AND FILLER

3. Coarse aggregate shall be crushed rock complying with Clause 901 and BS 63: Part 2: Table 2, unless agreed otherwise by the Engineer before commencement of work.

4. When tested in accordance with the procedure of BS 812, the coarse aggregate shall additionally have the following properties:

    Polished Stone Value (PSV) – as specified in Appendix 7/1.

    Ten Per Cent Fines Value (TPV) – not less than 180 kN, or as specified in Appendix 7/1.

    Maximum Aggregate Abrasion Value (AAV) – not more than 12, or as specified in Appendix 7/1.

    Flakiness Index (I_F) – not more than 25 per cent.

5. Fine aggregate shall comply with Clause 901 and shall be either crushed rock fines or natural sand or a blend of both. Fine aggregate shall be added as required to suit the particular system.

6. Filler shall be crushed limestone complying with the requirements of BS 594: Part 1. Filler shall be added as required to suit the particular system.

Table B2.1 Aggregate grading

BS sieve size	Per cent by mass of total aggregate passing	
	Nominal size	
	14 mm	10 mm
20 mm	100	
14 mm	80–100	100
10 mm	50–80	55–100
6.3 mm	15–50	20–55
5.0 mm	10–45	15–55
2.36 mm	8–45	15–45
1.18 mm	7–30	10–35
600 μm	5–25	7–30
300 μm	5–20	5–25
75 μm	0–15	0–15

7. When sampled and tested in accordance with the procedures of BS 598: Parts 100, 101 and 102, the aggregate grading shall fall within the envelope formed by the limits given in Table B2.1, unless agreed otherwise by the Engineer before the commencement of work.

8. The design and selection of aggregates, filler and bitumen proportions shall be the responsibility of the Contractor, who shall supply the necessary details to the Engineer for information only.

BINDER

9. The binder shall be petroleum bitumen complying with BS 3690: Part 1. The penetration of the bitumen shall be grade 70, 100 or 200 penetration, as selected by the Contractor, unless stated otherwise in Appendix 7/1. A polymer may be added, as selected by the Contractor.

10. The choice of bitumen grade and the penetration and softening point of the modified or unmodified binder shall be notified to the engineer before the commencement of work.

11. When sampled and tested in accordance with the procedures of BS 598: Parts 100, 101 and 102, the binder content of the surfacing material shall be in the range 3.5 to 7.5 per cent, by mass of total mixture.

12. Where appropriate to the system, the target binder content shall be determined by the binder drainage test in Clause 939, except that the range to be tested shall be amended to suit the grading of the

aggregates proposed for use. The target binder content determined in the laboratory may be adjusted to suit the mixing plant and the aggregate type which is used, subject to plant trial and delivery distance. The adjusted binder content shall be notified to the Engineer prior to delivery and shall not be lower than that specified above. The tolerance on sampling and testing for binder content shall be ± 0.3 per cent.

TACK COAT

13. Tack coat shall be a hot-applied cationic bitumen emulsion complying with BS 434: Part 1, with a minimum bitumen content of 38 per cent. To suit the particular system, it may be modified with a polymer. The choice of tack coat shall be notified to the Engineer before commencement of work.

SURFACE PREPARATION

14. Existing surfaces shall be cleaned using steel brooms and suction sweeping or other appropriate means. The surface may be moist but not wet; standing water shall not be present. All mud, dust, dirt and other debris and organic material shall be removed.

15. Where necessary or required by the Engineer, existing surfaces shall be regulated in accordance with the requirements of Clause 907, in advance of laying surfacing material to this clause.

16. Unless raised prior to surfacing, iron-work and reflecting road studs shall be located for lifting and relaying after completion of surfacing works. Gullies shall be covered prior to surfacing.

17. Where possible, existing road markings shall be removed.

MIXING

18. The material shall be mixed in accordance with the requirements of BS 4987: Part 1, such that an homogeneous mixture of aggregate, filler and bitumen is produced at a temperature of 150–180 °C.

TRANSPORTATION

19. Mixed materials shall be protected from contamination and undue heat loss by being transported to site in sheeted lorries. To facilitate

discharge of the materials, the floor of the lorry may be coated with the minimum of light vegetable oil or liquid soap or other non-solvent solution. When such coating is used, the lorry body shall be tipped to its fullest extent with the tailboard open to ensure drainage of any excess, prior to loading. The floor and sides of the lorry shall be free from adherent bituminous materials or other contaminants before loading the surfacing material.

LAYING

20. Tack coat shall be spray-applied, in accordance with the requirements of the SHW Series 900, at a rate selected by the Contractor and notified to the Engineer before the commencement of work, to completely cover the surface where the material is to be placed. The particular spray rate shall be dependent on the proprietary system and the porosity of the surface being covered.

21. Bituminous materials shall be applied at a suitable temperature and compacted by at least two passes of a tandem roller, capable of vibration, and with the minimum deadweight of 6 tonnes, before the material cools below 80 °C, measured at mid-layer depth.

SURFACE TEXTURE

22. Where stated in Appendix 7/1, the texture depth of the surfacing shall be in accordance with the requirements of Clause 921 after compaction.

DETAILS TO BE SUPPLIED

23. The Contractor shall supply all the details required in this Clause to the Engineer before commencement of work under this Clause and when requested during the work.

24. Checks shall be made at the end of each working day and records kept, to determine the quantities used of both tack coat and bituminous material.

25. The Contractor shall supply the Engineer with test certificates stating the properties of the materials used. Samples of emulsion tack coat, modified or unmodified bitumen or mixed bituminous materials from either the spray bar or storage tank or the pavement surface or

other suitable sampling point shall also be supplied to the Engineer by the Contractor when so instructed by the Engineer.

GUARANTEE

26. The Contractor shall guarantee the surfacing materials and workmanship for a period of two years from the date of opening the surfacing to traffic. This guarantee shall exclude defects arising from damage caused by settlement, subsidence or failure of the carriageway on which the material has been laid, but shall include for fretting, stripping, loss of chippings and loss of texture to below 1 mm measured by the sand patch method described in BS 598: Part 105.

APPENDIX C

Draft Irish Specification for Hot-Laid Thin Bituminous Surfacings

SCOPE

Hot-laid thin surfacing is here defined as an application of a tack coat of polymer-modified bitumen emulsion followed by a hot bituminous mixture or bitumen coated aggregate. These materials may be proprietary in which the tack coat is modified by a polymer or other material and the bituminous mixture may also be modified by a polymer-modified binder and/or added fillers or fibres.

The tack coat is applied by a paving machine fitted with a spray bar and the bituminous mixture is laid in the conventional manner, all in a single pass.

EQUIPMENT AND MATERIALS

At least two weeks before laying is to commence, the Contractor shall submit, to the Engineer, full details of the materials and mixtures he intends to use.

These details must include the following.

A. Equipment

(i) An integrated paving machine fitted with an appropriate spray bar so that tack coat and bituminous mixture are laid in a single pass.

(ii) Steel wheeled or pneumatic rollers as required.

Draft Irish Specification for Hot-Laid Thin Bituminous Surfacings

B. Materials

(i) Tack coat

(a) Type
The tack coat shall be a cationic bitumen emulsion modified with natural/synthetic polymer/rubber latices with a nominal hydrocarbon content of not less than 65 per cent. The content of bitumen, polymer (also polymer type) and any flux oils shall be stated.

(b) Rate of application
Shall be not less than 0.6 L/m² for a 65 per cent polymer-modified bitumen. If the polymer-modified bitumen emulsion has a hydrocarbon content greater than 65 per cent, the rate of application shall be adjusted to give an equivalent application of polymer bitumen.

(ii) Regulating material

The type of tack coat to be used with the regulating material and type of regulating material must be stated.

(iii) Composition of mixtures

(a) Aggregate grading and type
The maximum aggregate sizes shall be either 14, 10 or 6 mm. The target proportions passing the various test sieves, expressed as per cent by mass of total aggregate, shall be provided by the Contractor. The coarse aggregates used in the mixture shall be crushed rock or gravel complying with the quality requirements of BS 4987. In addition, the resistance to polishing, crushing, abrasion and maximum flakiness index shall be as described in Table C3.1, when tested in accordance with BS 812.

The fine aggregates may be crushed rock or sand or a mixture thereof but shall be subject to the approval of the Engineer. The quality shall comply with the requirements of BS 4987 and with Table C3.1 when tested in accordance with BS 812.

(b) Binder type and content for the mixture
The binder shall be a nominal 100 or 200 penetration grade bitumen. It shall be mainly petroleum bitumen complying with the specification issued by the Department of the Environment/National Roads Authority. The binder may be modified by the addition of not less than 5 per cent of thermoplastic elastomers, polymers or rubbers. The type and the

Table C3.1 Values when tested in accordance with BS 812

Category of site		Aggregate abrasion values	Polished stone value	Flakiness index	Aggregate crushing value
A.					
• All National Primary Roads and other roads carrying high-speed traffic with Annual Average Daily Traffic (AADT) greater than 2500, including urban areas.	Coarse aggregate – retained on 5 mm BS Sieve	10 max.	60 min.	25 max.	20 max.
• Bends of less than 150 m radius on roads not subject to a speed limit of less than 40 mph	Fine aggregate – passing 5 mm BS Sieve	12 max.	45 min.	–	–
• Gradients of 1 in 20 or steeper.					
• Approaches to traffic signal lights					
B. Other roads	Coarse aggregate	12 max.	57 min.	25 max.	20 max.
	Fine aggregate	12 max.	45 min.	–	–

concentration used shall be storage stable when tested in accordance with Appendix G of HA 50/93, 'Porous Asphalt Surface Course' (issued by the UK Department of Transport), except that the duration of the test shall be 120 ± 0.25 hours and the test temperature shall be the maximum mixing temperature. Storage stable shall mean that, under these conditions, the maximum allowable difference in softening point (BS 2000: Part 58) between the contents of the upper and lower third of the container shall be 2 °C.

The binder content, as per cent by mass of total mixture, and aggregate gradings, as per cent by mass of total aggregate, shall be stated. Binder contents and aggregate gradings shall be determined by the Contractor on mixture samples taken during the work. The sampling and testing rate shall be not less than one per 200 tonnes of mixture and a minimum of one sample must be taken and tested each day during work. The test procedures shall be subject to the approval of the Engineer and comply with the sampling and testing procedures specified in BS 598.

(c) Binder–aggregate affinity

Prior to commencing the work, the Contractor shall carry out a standard test to verify that the binder–aggregate affinity is adequate to prevent stripping in the presence of water. The modified version of SHRP method M–001 is recommended in which case the initial absorption value shall be 1 mg of bitumen per gramme of aggregate or greater and, after water addition, the net absorption shall be not less than 70 per cent of the initial absorption. (Details of the modified SHRP method are given in National Roads Authority Report RC372.) Other acceptable methods are ASTM D1075, D1664, and indirect tensile strength tests on mixture specimens as described in NCHRP reports 246 and 274. The modified M–001 method shall take precedence.

(d) Filler type and amount

The filler shall be crushed rock or limestone or hydrated lime or Portland cement or other material approved by the Engineer, and not less than 75 per cent shall pass the 0.075 mm sieve.

The filler content (as per cent by mass of total aggregate) shall be stated.

(e) Other additive type and content

The type of any other additive, e.g. mineral fibre, or glass fibre, or cellulose fibre, etc., and additive content (as per cent by mass of total aggregate) shall be stated.

C. Mixing

(i) At least two weeks before surfacing is to commence, the Contractor shall submit full details of mixing and laying proposals. This shall include all data as described under 'Materials' above, as well as storage of raw materials, and mix temperatures at delivery on site and for rolling. These shall be subject to the approval of the Engineer who may require the Contractor to lay a trial length to demonstrate that the surfacing can be laid to the requirements of the specification.

(ii) The materials, including any added filler, shall be accurately weighed into a mechanical mixer of approved type and thoroughly mixed in such a manner that all particles are completely and uniformly coated. All weighing mechanisms shall be properly maintained within the accuracies recommended by the manufacturer. The Contractor shall, at the request of the Engineer, supply a service certificate showing that the weighing devices have been calibrated against standard loads, according to BS 5750: Part 1, within the six months prior to the start of the contract.

(iii) The aggregate, before coating, shall be surface dry.

(iv) Temperatures of the material delivered on site and at the time of rolling shall be checked and shall be within the ranges agreed with the Engineer.

D. Laying

(i) Surface regularity and crossfall

The finished road surface shall be laid to an agreed line and level and the maximum longitudinal and transverse deviation of the finished road surface shall be ± 10 mm from the agreed level. The number of surface irregularities shall not exceed the maximum permitted for Class B roads, for flexible basecourse surfaces, as specified in the current specification for road works. When tested with a 3 metre straight edge at right angles to the centre line of the road, the maximum allowable depression of the road surface below the straight edge shall be 10 mm.

The minimum crossfall of the finished surface shall be 2.5 per cent.

(ii) Preparation of the site

(a) Prior to laying operations, the Contractor shall carry out any necessary patching and/or other repairs. The extent and type of these repairs shall be agreed with the Engineer. The thin overlay mixture shall not be used for any major pothole or rutting repairs.

(b) Where a bituminous bound material is used for regulation, a tack coat shall be employed.

(c) The Contractor shall remove any loose material, dust and vegetation by suction sweeping and air jets if necessary. Any standing water shall also be removed.

(d) The treatment of existing furniture by the Contractor shall be agreed with the Engineer. The Contractor shall raise any manholes and gullies to a level agreed with the Engineer and all catseyes shall be removed by the Contractor prior to the surfacing.

(iii) Paving

(a) The surfacing shall be laid only between April and November.

(b) The material shall be spread evenly by the paving machinery and the following thickness tolerances shall apply:

Mix maximum stone size	6 mm	10 mm	14 mm
Thickness (mm)	10–20	15–30	20–35

(c) One hundred per cent of the carriageway shall be covered with a uniform layer of surfacing.

(d) During the application, the Contractor will carry out routine sampling and testing to ensure the materials comply with the requirements of the specification. These shall be verified during construction using the procedures specified in BS 598 or according to additional procedures agreed prior to the work.

In particular, the tolerances, for the target mixture, for the proportions passing the various sieve sizes shall be as follows:

sizes above 2.36 mm:	± 4 per cent;
sizes passing 2.36 mm and retained on 0.075 mm:	± 3 per cent; and
sizes less than 0.075 mm:	± 1.5 per cent.

(a) Binder type
Unmodified bitumens shall comply with the requirements of the Department of the Environment/National Roads Authority specifications. Where modified binders are to be used, the Engineer may request additional tests to be carried out.

(b) Binder contents

Shall be determined by the Contractor on mix samples taken during the work. The procedures shall be subject to the approval of the Engineer and generally shall be according to test procedures given in BS 598. The tolerance shall be within ± 0.3 per cent of the target binder content given by the Contractor and must be in the range 4.5 to 6.5 per cent.

E. Rolling

Rollers used for initial compaction shall not cause crushing of the aggregate. Final rolling shall be carried out by a smooth steel wheeled roller of 8 tonne dead weight, which may be a vibrating roller operated without vibration.

PERFORMANCE REQUIREMENTS

The Contractor shall guarantee in writing that the surface will satisfy the following performance specification for skidding resistance, surface texture depth and resistance to disintegration after the completion of the works for the period specified. In the event of failure of the surface to satisfy the performance specification for the said period, the Contractor shall replace the material in accordance with this specification as directed by the Engineer or otherwise rectify the surface to the satisfaction of the Engineer, all at the Contractor's cost.

If rectification other than by 'replacement of the material' is required, the Engineer shall inform the Contractor in writing at the time of tender of the type of rectification that will be required.

(i) Skidding resistance

The skidding resistance shall be measured using either the SCRIM machine or the Portable Skid Resistance Pendulum, as directed by the Engineer and shall conform to the following specification. Skid resistance measurements made by SCRIM machine shall take precedence over measurements made with the pendulum.

(a) SCRIM

Where possible tests shall be carried out at 80 km/h. The mean value of the measurements of Sideway Force Coefficient shall be greater than 0.50 and not more than 15 per cent of the values shall be less than 0.50. No individual value shall be less than 0.45. Where measurements at 80 km/h are not possible, for safety reasons, tests shall be carried out at 50 km/h. Measurements at 50 km/h shall have a mean value of the Sideway Force

Coefficient of not less than 0.55, and not more than 15 per cent of the values shall be less than 0.55. No individual value shall be less than 0.50.

(b) Portable skid resistance pendulum
Measurements shall be made in accordance with Road Note 27. The locations at which the measurements are taken will be chosen by the Engineer as follows:

Transverse lines will be selected at suitable intervals. Five measurements will be made at approximately equal spacings across each line. Ninety-five per cent of the measurements thus made shall be greater than 60.

(ii) Texture depth

Surface texture depth measurements shall be made by the sand patch method and shall comply with the requirements set out below. Locations at which the measurements are made will be chosen by the Engineer and according to BS 598 as follows:

(a) Initial measurement
Measurements shall be made, by the Contractor, directly after completion of the work and before the road has been opened to traffic. They shall be made over 50 m lengths, regularly spaced along the section, and covering not less than one third of the area of the surfacing. On each 50 m lane length, 10 individual measurements shall be made at 5 m intervals along a diagonal line across the carriageway lane width, but not within 300 m of the edge of the carriageway.

(b) Measurements made after periods in service
Measurements shall be made in the nearside wheel lane (i.e. nearest to the edge of the pavement) at 5 m intervals. These shall be carried out on both sides of the carriageway, i.e. in both traffic directions.

Where maximum speed limits are in excess of 50 km/h Where the maximum stone size is 10 or 14 mm, ninety-five per cent of texture depth measurements, after laying, shall be equal to or greater than 1.5 mm. After three years, ninety-five per cent of measurements shall be equal to or greater than 1.3 mm.

Where the maximum aggregate is 6 mm, ninety-five per cent of the measurements of texture depth shall be equal to or greater than 1.2 mm after laying and after three years in service.

In any of the above cases, no individual texture depth measurements shall be less than 1 mm.

Where maximum speed limits are 50 km/h or less For stone sizes of 6, 10, and 14 mm, ninety-five per cent of texture depth measurements after laying shall be equal to or greater than 1.2 mm and no individual measurement shall be less than 1 mm. After three years, ninety-five per cent of texture depth measurements shall be equal to or greater than 1 mm.

(iii) Visual assessments:

After five years in service, visual assessment shall show no loss of the surfacing material due to attrition or debonding or ravelling or surface disintegration or cracking, save for the reflection cracking of any defects from the underlying original foundation.

Index

Abrasion, Cantabro abrasion test, binder modifiers 73–4
Abrasion value (AAV) 29–30
ACMA publications, thin surface course materials 260
Acrylic resin, high-friction surfacings 305–6
Adhesion agents 76–7
Adsorption test (NET) USA 31–2
Aggregate crushing value (ACV) 26–7
Aggregates and fillers 18–46
   abrasion value (AAV) test 29–30, 36
   adhesion of bitumen 31–2
   CEN TC-154 10–11
   classification, rock types, BS 812 18–20
   coarse aggregates, asphalt concrete 153–8
   durability
      in asphalt concrete 155–6
      classification, criteria 155
   fine aggregates, asphalt concrete 157
   frictional wear (MDE France) 29
   performance and specification 32–9
      cleanliness 32–5
      durability 37–9
      hardness 35–7
   porous asphalt 204–5
   properties and methods of test 21–32
      bitumen adhesion 31–2
      dimensions 21–25
      durability 27–31
      dust content 22–3
      polished stone value (PSV) 30
      porosity 25–6
      relative density 25–6
      resistance to crushing 26–7
      resistance to impact 26
      resistance to polishing 30–1
      resistance to soaking 27
      resistance to wear 29–30
      shape 23–5
      size
         sedimentation tests 22
         sieve tests 21–2
      skid resistance (SCRIM) 85–6
      soundness 27–9
      strength 26–31
      water absorption 25–6
   rock types, BS 812 18–20
   rolled asphalt surface courses
      coarse 114–16
      fine 116–17
   shape classification 24
   sieve sizes 10–11
   test methods and procedures 11
Airfields
   asphalt concrete 148–51, 163–8, 175
   bitumen bound aggregate 322–5
   and highways, criteria compared 177
   porous asphalt 203
   recycled materials, US Army Corps of Engineers 360
   slurry surfacing, veneer coats 299
   standards, UK 141–2
      BAA/1993 9
      DWS/1995 9
Anti-strip agents 162
Appendix A, draft UK specification, stone mastic asphalt 373–8
Appendix B, draft UK specification, thin surface course materials 379–83
Appendix C, Irish specification, thin hot laid bituminous surfaces 384–92
Aquaplaning resistance 187–91
   see also Skid resistance

Index

Artificial rock types 18
Asphalt 49–50
  component materials 7–8
  dynamic shear rheometer 153
  dynamic stiffness tests 356
  natural 49–50
  porous 185–218
  recycling 335–65
  terminology 4–5
  types and purposes 112
  see also Asphalt concrete; Rolled asphalt surface courses
Asphalt concrete 139–184
  6mm close grade macadam grading curves 146
  6mm dense surface course 144, 146
  10mm close grade surface course 144–5
  14mm close grade surface course 143–4
  comparison of production limits 167
  limestone aggregate 148
  40mm single course 200 pen granite coarse and fined 145, 147
  40mm single course macadam grading curves 147
  airfield surfacing 148–51
    design requirements 148, 153–8, 175
  applications 141–52
  component materials 152–68
    anti-strip agents 162
    binder modifiers 161
    binders
      estimation of content 171
      properties 160–1, 169
    cleanliness 156
    coarse aggregates 153–7
    durability 155–6
    fillers 158
    fine aggregates 157
    grading 139–40, 175–6
    hydrated lime 158–60
    Macadam asphalt concrete 161
    pigments 162
    shape 156
    skid resistance 155
    strength 155
  concept 139–41
  defined 139
  design 168–76
    binder content 169, 172
    Hveem design 170–2
    Marshall design 170, 176–7
    mixture approval and verification 172–4
  history 139–41
  international gradings and properties 151–2, 154
  production and paving 176–81
    criteria for highways and airfields 177
  properties 141–52
  rollers
    combination 181
    pneumatic type (PTR) 181
    tandem 181
    three point static 181
  standards
    BS 4987 140–1
    particle size distribution (PSD), Fuller curve 6, 139–40
    SHRP binder specification 164
    UK highways and byways 142–8
  voids 173–4
    Asphalt Institute Manual (2) method 174
    BAA/DWS method 174
Asphalt mixtures, relative properties 111
Australia, bitumen, specifications 63
Austria
  asphalt concretes 152
  porous asphalt 203, 204
Avis Technique, France 13

BAA/1993, standards, UK airfields 9
BAA/DWS, volumes calculation 174
Basalt 18
Bauxite, high-friction surfacings 304, 307
Belgium, porous asphalt 204
Binder modifiers 68–78, 120–4
  adhesion agents 76–7
  alternatives, fibre additives 207–8
  asphalt concrete 161
  binder and mixture data for approval of 128
  Cantabro abrasion test 73–4
  characteristics modified 120–2
  chemical 74–6

Index

fibre additives 77–8
generic types 68
mastic asphalt 225–6
polymers 68–74
temperature effects 71–3
types 68–9
types and examples, list 69
Binders 47–79
  anti-strip agents 162
  asphalt concrete 159, 160–1
  bitumen 50–68
  bitumen bound aggregate 323–4
  classification 48
  estimation, centrifuge kerosene equivalent (CKE) 171
  key functional requirements 129
  mastic asphalt 224–5
  natural asphalts 49–50
  porous asphalt 205–8, 205–8
  properties 161
  reclamation, rejuvenation 344–7
  rejuvenation, recycled materials 344–7
  rubber 324–5
  softening point, and road damage 92
  specifications SHRP 64, 164
  stabilometer test 171
  stone mastic asphalt 250
  swell test 171
  tar 47–9
  see also Polymer-modified binders (PMBs)
Bitumen 50–68
  adhesion 31–5
  ageing, tests 63
  binders, mastic asphalt 224–5
  BS 3690 118
  CEN TC 19/SC-1 11–12
  classification
    aromatics 50, 54
    asphaltenes 54
    resins 54
    saturates 54
  cutback 67–8
  deformation response tests 61
  droplet stabilization 65
  emulsion 63–7
    manufacture 65
  history, replacement of tar 3
  penetration grade 11–12, 50–63
  petroleum bitumen 50–3
  polycyclic aromatic compounds (PCAs) 50
  polymer modifiers 68–74, 93
    and temperature susceptibility 72
  production 50–3
    comparison of qualities 50–1
    crude oil composition 52
    schema 51
  properties
    chemistry 53–6
    colloidal nature 55
    content and emulsion viscosity 66
    physical properties 56–62
    temperature effects 60
    visco-elastic creep 61
  reclaimed
    blending chart 347
    properties 346
  SHRP binder specifications 64
  specifications 62–3
  terminology 4–5
  tests
    ageing 63
    Fraass breaking point test 56, 58
    Instron adhesion pull-off test (INAPOT) 31, 34–5
    needle penetration test 11–12, 56–7
    net adsorption test (NET) USA 31
    rolling thin film oven test (RTFOT) 63, 164
    softening point (ring and ball) test 56–7
    Vialit plate test, (LCPC1963) France 31, 33–5
  viscosity 58–62
    of emulsions 66–7
Bitumen macadams, fluxed 309–11
Bitumen-bound aggregate 322–5
  binder 323–4
Bitumen-extended epoxy resin, high-friction surfacings 305–6, 304–6
Bridge decks
  gussasphalt 236–7
  mastic asphalt
    concrete 227–8
    steel 228–30

Index

British Standards Specification
  BS 594
    aggregates, grading 114–16
    binders 118–19
    traffic, and surface stability 113
  BS 598, binders 119
  BS 812, rock types 18–20
  Bs 3690, bitumen 118–19
  BS 4987, asphalt concrete 140–1, 142
  BS 5273, DTS 319
  BS and CEN durability tests 37
Buildability factor 94
Bush hammering, re-texturing 326

Calcined bauxite, high-friction
  surfacings 304, 307
Cantabro abrasion test, binder
  modifiers 73–4
Capillary viscometer, bitumen 60
Carbonizing by hot compressed air,
  re-texturing 326–7
Carcinogenic effects of tars 3, 49
Cellulose 78
CEN durability tests 37–9
CEN TC 19/SC 1, bitumen 11–12
CEN TC 154, aggregates 10–11
CEN TC 227, mixed materials 12–13
CEN TC 317, tar 12
Centrifuge kerosene equivalent
  (CKE), estimation of binders 171
Chemcrete 75
Chipping spreaders, veneer coats
  293–4
Choice of materials (*table*) 367
Choice of treatment (*flow chart*) 369
Chrysolite 78
City of London 97
Civil Engineering Sec.14: Bituminous
  Materials – Aircraft Pavements
  (BAA/1993) 9
Classifications
  bitumen 54–5
  current materials 5–7
Cold-laid materials 309–16
  cold mixed materials 312–16
    mixing 313–14
    storage 314–16
  deferred set macadams 311–2
  fluxed bitumen macadams 309–11
Coloured materials 328–31
Coloured surfaces 114

Comité Européen de Normalisation
  10–13
  see also CEN
COMPARE computer model, whole-
  life costing 99–101
Copper sulphate, in bitumen 75
Crack Seed Index 350
Crude oil compositions 52
Crystalline polymers 70
Cutback bitumen 67–8
Cutler machine, recycling 341–3
Czech republic, porous asphalt 204

DBM *see* Asphalt concrete
De-icing salts 215
Decorative finishes, pigments 162–3
Deferred set macadams 311–12
Deformation 166
Delugrip 320–2
Denmark, porous asphalt 204
Dense tar surfacing 317–20
Development
  Comité Européen de
    Normalisation (CEN) 10–13
  pavement material 8–16
  performance specifications 8–10
  national standards 8–10
  Strategic Highway Research
    Program (SHRP) 14–16, 62–4,
    152, 164
  technical approval schemes 13–14
  see also CEN TC
Durability
  classification of aggregates, France
    38
  coarse aggregates, asphalt concrete
    155–6
  comparative tests 37
  gussaphalt 221–2
  hot recycled mixtures 352–7, 352–7
  recycled materials 352–7
  rolled asphalt surface courses 92–4
  selection criteria, asphalt surfaces
    92–4
  stone mastic asphalt 244
  tests
    British Standards Specification
      37
    CEN tests 37–9
Dust content, methods of test,
  aggregates and fillers 22–3

Dust nuisance, historic 47
Dust quality, assessment, methylene blue test 33–4
Dynamic shear rheometer 153
Dynamic stiffness tests 356

Economics
  and energy, recycled materials 43, 362–4
  high-friction surfacings 307
  porous asphalt 203, 216
  rolled asphalt surface courses 99–101, 134
    COMPARE computer model 99–101
    whole-life costing 99–101
  slurry surfacing 302
  summary 372
  thin surface course materials 274–5
  veneer coats 295–7
Emulsion, bitumen 63–7
Epoxy resin, high-friction surfacings 304–6
Ethylene vinyl acetate (EVA) 63, 163, 164, 206
Europe
  aggregates, sources 39
  asphalt concretes 152
  Comité Européen de Normalisation (CEN) 10–13
  porous asphalt 203–4

Farm accommodation bridges, 55/6F
  rolled asphalt surface course 332
Fatigue cracking 166
Fibre additives 77–8
  alternatives to binder modifiers, porous asphalt 207–8
  characteristics, asphalt 78
  MCHW-1 206–7
  stone mastic asphalt 251
Filler 117–18
  asphalt concrete 158, 160
  defined 40
  porous asphalt 205
  pulverised fuel ash 118
  stone mastic asphalt 250
  see also Aggregates and fillers
Flailing, longitudinal flailing re-texturing 326

Flint 18–19
Flow chart, choice of surface treatments 369
Fluxed bitumen macadams 309–11
Footways 113–14
Fraass breaking point test, bitumen 56, 58
France
  asphalt concretes 152
  Avis Technique 13
  bitumen, specifications 63
  durability classification of aggregates 38
  frictional wear (MDE) 29
  micro-Deval co-efficient (MDE) test 29–30, 36
  porous asphalt 204
  thin surface course materials 261
  Vialit plate test 31, 33–5
Fuller curve, ideal PSD 6, 139–40

Gabbro 19
Gas burners, re-texturing 327
Germany
  asphalt concretes 152
  bitumen, specifications 63
  freeze/thaw (F/T) test 28
  impact test, aggregates 26
  porous asphalt 204
  stone mastic asphalt 261
    specification 252
Gilsonite, binder modifier, mastic asphalt 225
Glare and illumination
  porous asphalt 196–9
  rolled asphalt 197
Glass wool 78
Grading
  aggregates, BS 594 114–16
  asphalt concrete 139–40, 175–6
    international gradings and properties 151–2, 154
  stone mastic asphalt 375
Granite 19
Gritstone 19
Grooving and grinding, re-texturing 326
Ground radar plot, delamination, coring 178
Grouted macadam 316–17
Gussasphalt 231–7

Gussasphalt (continued)
  aggregates, specification and origin 224, 233
  application 232–7
  binders 224–5
  manufacture 222, 231–2

HAPAS (Highway Authorities Product Approval Scheme) 14, 267
Hardness, assessment 35–7
Heavy goods vehicles (HGVs)
  parks, static load and surface material 94–5, 113–14
  road damage 92
    surface irregularities 82
  wheel tracking test 93
High pressure jetting, re-texturing 327
High stone content rolled asphalt 132
High-friction surfacings
  acrylic resin 306
  bitumen-extended epoxy resin 305–6
  calcined bauxite 307
  concept and history 302–4
  economics 307
  epoxy resin 306
  history 302–4
  maintenance 307
  polyurethane resin 306
  properties and applications 304–5
  surface dressing binders 306–7
  thermoplastics 306
  veneer coats 302–7
High-Speed Road Monitor (HRM) 83
Highway Authorities Product Approval Scheme (HAPAS) 14, 267
Highway Authorities Standard Tender Specifications (CSS, 1996) 9
Highways
  and airfields, criteria compared 177
  asphalt concrete gradings and properties, USA 154
  and byways, asphalt concrete 142–8
  standards 141–2
  UK, porous asphalt 200–3
  see also UK; USA
Highways Agency, Procedure for Evaluating New Materials, UK 13–14, 267, 269

Historical developments 1–4
Hot Rolled Asphalt and Coated Macadam for Airfield Pavement Works (DWS 1995) 9
Housing estates 260
Hungary, porous asphalt 204
Hveem design, asphalt concrete 170–2
Hydrovac process 212

Iceland, freeze/thaw (F/T) test 28
Impact, aggregate impact value (AIV) 26
Infra red heaters, re-texturing 327
Instron adhesion pull-off test (INAPOT) 31, 34–5
International gradings and properties, asphalt concrete 151–2, 154
Ireland, specifications
  thin hot-laid bituminous surfaces 384–392
  thin surface course materials 270–1
Iron oxide 162, 328–9
Italy
  asphalt concretes 152
  porous asphalt, two-layer systems 212

Lime
  asphalt concrete 158–60
  binder modifiers 77

Macadam, John Loudon 2
Macadam asphalt concrete *see* Asphalt concrete
Macadams and asphalts
  28mm pervious 324
  grouted macadam 316–17
  terminology 4–5
Macro-texture (profile depth) 83, 85
Magnesium sulphate soundness value (MSSV) 27–9
Marshall asphalt
  14mm surface course 179–80
  airfield pavement works 9, 149–51
  asphalt concrete 170
Marshall specimen testing
  design asphalt 124–6
  stone mastic asphalt 253
Mastic asphalt 219–37

aggregates
  formulations 233
  specification and origin 224
applications 227–31
  other paving 230
  roofing 230–1
  steel bridge decks 228–30
  tunnels and concrete bridge decks 227–8
  waterproofing 227
binder modifiers 225–6
  gilsonite 225
  polymers 225–6
binders 224–5
  bitumen 224–5
  Trinidad lake asphalt 225
bitumen 224–5
concept 219–21
cost 222
durability 221–2
equipment for application 222–4
history 219–21
manufacture 226
waterproofing characteristics 221
Materials and classifications, current 5–7
MCHW-1 *see* Specification for Highway Works
Methylene blue absorption (MBV) test 23, 33
Micro-asphalts 298–9
Micro-Deval (MDE) coefficient test 29–30, 36
Motor racing circuits 322

National Specifications for Road Bitumens 63
Needle penetration test, bitumen 11–12, 56–7
Net adsorption test (NET) USA 31–2
Netherlands
  asphalt concretes 152
  porous asphalt 204
    two-layer systems 212
Noise levels *see* Tyres
Nottingham asphalt tester (NAT) 84, 127

Particle shape
  contact sensor, computer ratio calculations 24–5

Zingg graph 24–5
Particle size distribution (PSD)
  Fuller curve 6, 139–40
  ideal, equation 6
  sieve and sedimentation tests 21–2
  skid resistance 22
Pavers 181
PCSM *see* Surfacing materials
Performance specifications, national standards 8–10
Permanent International Association on Roads Congress (PIARC) 89
Permeability, relative, surface materials 371
Pervious macadam *see* Porous asphalt
Petroleum bitumen 50–68
  see also Bitumen
Pigments 162–3, 328–31
Plastomers *see* EVA
Polished stone value (PSV) test 30–1
  porous asphalt 200, 202
  skidding resistance 30, 85–6
Polycyclic aromatic compounds (PCAs) 50
Polymer-modified binders (PMBs) 68–74, 93, 163
  elastomers/plastomers 70
  mastic asphalt 225–6
Polymer-modified bitumen
  asphalt improvements 72
  improved rut resistance 71
Polyurethane resin, high-friction surfacings 305–6
Porous asphalt 185–218
  advantages in use 199
  aggregates and filler 204–5
  applications
    airfields 203
    hydraulic applications 199, 202
    overseas use 203–4
  binders 205–8
    fibre additives 207–8
    modified 206
    unmodified 205–6
  concept and history 185–6
  differences in thermal characteristics and conventional surfacings 214–16
  drawbacks 92
  economics 203, 216
  history 185–6

Porous asphalt (continued)
  maintenance and repairs 213–16
  manufacture and laying 208
  performance in service 212–13
  properties 186–99
    aquaplaning and skidding resistance 187–91
    driver comfort and fatigue 194
    glare and illumination 196–9
    macro-texture 96
    spray 194–6
    tyre noise 191–4
  quality control 208
  recycling 213
  site suitability 199–200
  sizes, 20, 10, 14, and 16mm 209–11
  tack coats 208–9
  terminology 185
  two-layer systems 212
  UK highways 200–3
    aggregate quality and cost 200
    aggregate size 201
    compliance with specification 201
    laying and compaction 201
    mixing plant operations 201
    reducing costs 202–3
  very high void mixtures 211
  winter maintenance 214–16
    freezing conditions 214
    receptivity to salt 214
Porphyry 20
Portland cement, adhesion agents 77
Power spectral density 90
Profile depth, macro-texture 83
PSV *see* Polished stone value (PSV) test
Pulverised fuel ash, as filler 118

Quality control, rolled asphalt surface courses 102–6
Quarries and quarrying
  European sources 39
  tonneage, UK sources 39
Quartzite 20

Recycled materials 40–4, 335–65
  asphalt 343–4
  binder rejuvenation 344–7
  bitumen blended chart 347
  cold recycled mixtures 336, 338, 357–60

  design aspects 358
  concept and history 335–9
  confusion in perception 362–4
  design of mixture 358–60
  development 336–9
  economics and energy 43, 362–4
  engineering 42–3
  environment 42
  equipment
    Cutler machine 341–3
    in-plant recycling 339–40
    in-situ recycling 340–3
    Jim Jackson machine 341–3
    repaving/remixing 343–4
  expected development 43
  hot recycled mixtures 337, 347–58
    crack growth 352
    crack propagation studies 350–2
    crack susceptibility parameters 351
    cracking resistance 348–52
    creep condition cracking 354
    creep stiffness 349
    deformation resistance 348
    design 358
    durability 352–7
    elastic region cracked data 353
    in-situ recycling 337
    performance requirements 347–8
    rolled thin film oven tests (RTFOT) 354–5
    stiffness tests 356–7
    water immersion tests 355–6
  implementation 360–4
  in-plant recycling 336
  in-situ recycling 336
  material types 40–2
  MCHW-1 requirements 98
  opportunities, UK 360–2
  properties of reclaimed bitumen 346
  recovered binder and scarification effects 345
  scarification effects 345
  terminology 335–6
  UK 98
Red iron oxide 162, 328–9
Relative density, determination 25–6
Re-texturing, specialized materials 325–8
Road texture, and tyre noise 90, 246
Roads *see* Highways

Index

Rock types
  artificial 18
  basalt 18
  BS 812 18–20
  classification 18–20
  flint 18–19
  gabbro 19
  granite 19
  gritstone 19
  porphyry 20
  quartzite 20
  schist 20
Rock wool 78
Rolled asphalt surface courses 80–138
  0/3mm and 15/10 rolled asphalt 132
  55/6F 332
  applications 80–1, 111–14
  characteristics required 80–107
    durability 92–4
    quality assurance 103–6
    quality control 102–3
    selection criteria 83–95
      site restrictions 94–5
      structural design 83–4
      surface properties 84–92
  component materials 114–29
    binder modifiers 120–24
    binders 92–4, 118–20
    coarse aggregates 114–16
    filler 117–18
    fine aggregates 116–17
    mixed material 124–29
    test requirements 129
  concept 108
  design criteria 126
  durability 92–4
    failure signs 133–4
  economics 99–101, 134
    COMPARE computer model 99–101
    whole-life costing 99–101
  high stone content rolled asphalt 132
  history 108–9
  macro-texture (profile depth) 83
  maintenance 97–8, 133–4
  monitoring existing network 81–3
    surface irregularities, MPN 82
  with pre-coated chippings 129–32
  properties 109–11
    skid resistance 130–1
    surface properties 84–92, 111
    relative properties, various mixtures 111
  sports playing surfaces 332
  tack coats 132–3
  test requirements, summary 130
  uses 95–7
    new works 95–7
    replacement 98–9
  varieties 129–32
  vehicle noise test track material 331
Rolled thin film oven tests (RTFOT) 63, 164, 354–5
Rollers 181, 181
  veneer coats 294
Roman roads 1–2
Roofing applications, mastic asphalt 230–1
Rosin ester, high-friction surfacings 305–6
Rubber, as binder 324–5
Rut resistance
  polymer-modified bitumen 71
  thin surface course materials 265

Schist 20
Schlagversuch impact test 26, 249
SCRIM (sideway force coefficient routine investigation machine) 85–9
  coefficients 245
  Irish specification 390–1
  PSV 85–6
  see also Skid resistance
Season of year, and surface treatments 368
Secondary see Recycled materials
Sedimentation tests, size properties of aggregates and fillers 22
Shear, dynamic shear rheometer 153
Shot blasting, re-texturing 326
Sideway force coefficient routine investigation machine see SCRIM
Sideway force coefficient (SFC)
  influence of traffic 89
  seasonal variation 85–7
  skid resistance 85–9
  skidding accidents, wet conditions 85–7
Sieve sizes 10–11
Sieve tests 21–2
Skid resistance 85–9
  coarse aggregates 155
  high-friction surfacings 302–4

Skid resistance (continued)
  investigatory, different site categories 88
  macrotexture 187–91
  particle size 22
  polished stone value (PSV) test 30–1
  rolled asphalt surface courses 130–1
  RTAs and texture 189
  sideway force coefficient (SFC) 85–6
  stone mastic asphalt 245–6
  thin surface course materials 273–4
  summary 370
  see also SCRIM (sideway force coefficient routine investigation machine)
Slurry sealing 67
Slurry surfacing, veneer coats 297–302
Softening point (ring and ball) test, bitumen 56–7
Soundness, magnesium sulphate soundness value (MSSV) 27–9
Spain, porous asphalt 204
Specialized materials 309–334
  55/6F rolled asphalt surface course 332
  bitumen bound aggregate 322–5
  coloured materials 328–31
  Delugrip 320–2
  dense tar surfacing 317–20
  grouted macadam 316–17
  re-texturing 325–8
    bush hammering 326
    carbonizing by hot compressed air 326–7
    gas burners 327
    grooving and grinding 326
    high pressure jetting 327
    infra red heaters 327
    longitudinal flailing 326
    orthogonal flailing 326
    shot blasting 326
  sports playing surfaces 332
  vehicle noise test track material 331
Specification for Highway Works (MCHW 1) 9, 32–3, 82–83, 202, 268
Specification *see* Performance specifications
Sports playing surfaces 332
Spray tankers 293

Stabilometer test, binders 171
Static loads 94–5
Steel bridge decks, mastic asphalt 228–30
Stiffness, dynamic stiffness tests 356
Stone mastic asphalt 238–58
  composition
    aggregate 249–50
    binders 250
    by mass and volume 242–1
    coarse aggregate 249–50
    fibres, cellulose/mineral 250–1, 254
    filler 250
    polymers 251
  concept 240–2
  cost and maintenance 257
  draft UK specification 373–83
    aggregates and filler 373
    binder 374
    compaction 377
    laying 377
    mixing 376
    mixture compliance 377–8
    modified binders 374
    stabilizing additive 374
    supply of clause details 378
    surface preparation 376–7
    surface texture 377
    target aggregate grading 375
    target binder content 375
    transportation 376
  history and development 238–40
  laying 255–7
  Marshall specimen testing 253
  mixing 251–5
  off-highway applications 249
  properties 247–9
    deformation resistance 242–3
    durability 244
    dynamic stiffness 243
    flexibility and fatigue life 243
    noise generation 246
    skid resistance 245–6
    toughness 244–5
  specification 252
  trial mixture testing 252–3
Strategic Highway Research Program (SHRP) USA 14–15, 62–4
  asphalt program 14–15, 62–4, 152
  binders, specifications 64, 164
  SHRP product list 15

Styrene-butadiene-styrene/rubber (SBS, SBR) 163, 164
  see also Polymer modified binders
Sulphur, binder modifiers 74–5
Superpave system, USA, UK 10, 15, 163
Surface dressing (chip seal)
  binders, high-friction surfacings 306–7
  properties 278
  terminology 5
  see also Veneer coats; Thin surface course materials
Surface treatments
  choice *(flow chart)* 369
  choice of season 368
Surface water drainage, and type B filter media dislodgement 322–3
Surfacing materials 366–71
  choice *(table)* 367
  combination treatments 371–2
  noise 370–1
  performance 367
  permeability 370–1
  physical constraints 371
  regulating ability 370
  relative economic benefits 372
  relative permeability 371
  selection 366–8
  skid resistance 370
  spray and glare 371
  texture depth 370
  treatment 366–8
    flow chart 369
    season 368
  see also Veneer coats
Sweden, porous asphalt 204
Switzerland
  porous asphalt 204
  stone mastic asphalt 261

Tack coats 132–3, 208–9
Tarmacadam, origins 3
  see also Macadam
Tars 47–9
  carcinogenic effects 3, 49
  CEN TC-317 12
  dense tar surfacing 317–20
  disadvantages 318
  polycyclic aromatic compounds (PCAs) 50

production 47
viscosity, equiviscous temperature (EVT) 48
Telford, Thomas 2
Ten percent fines value (TFV) 27
Terminology 4–5
Texture depth, summary 370
Thermoplastics, high-friction surfacings 306
Thin film oven test (RTFOT) 63, 164, 354–5
Thin hot laid bituminous surfaces
  Irish specification 384–392
    aggregate grading and type 385
    binder type and content 385–7, 389–90
    binder-aggregate affinity 387
    equipment 384
    filler type and amount 387
    laying 388–90
    materials 384–8
      test values 386
    mixing 388
    other additive type and content 387
    rolling 390
    SCRIM skid test machine 390–1
    skid resistance pendulum test 390–1
    skidding resistance 390–1
    tack coat 385
    texture depth measurements 391–2
    visual assessments 392
Thin surface course materials 97, 259–76
  categories 261–3
    trade names 262
  components
    aggregates and filler 264
    binder modifiers 265
    binders 264
  concept and history 259–61
  definition 259
  draft UK specification 267, 379–83, 379–83
    aggregates and filler 379–80
    binder 380–1
    guarantee 383
    laying 382
    mixing 381

Thin surface course materials (continued)
  stone mastic asphalt 269
  supply of details 382
  surface preparation 381
  surface texture 382
  tack coat 380
  transportation 381–2
 economics 274–5
 Eire provisional specification 270–1
 Highway Agency specifications 265
 maintenance 274
 properties and applications 263–4
  skid resistance 273–4
  temperature stress 127
 specification guidelines
  production tolerance 271–2
  surface finish and texture 272–3
  texture depth and layer thickness 271
 tack coats 274
 varieties
  paver-laid surface dressing 266
  thin polymer-modified asphalt concrete 266–7
  thin stone mastic asphalt 269
  ultra thin hot mixture layers 266
  very thin surfacing layer 266–7
  UK specification 267
 wheel track rutting 127, 265
 see also Veneer coats
Trinidad lake asphalt 49–50
 binders, mastic asphalt 225
Tunnels, mastic asphalt 227–8
Tyres
 noise levels 191–4
  porous asphalt 371
  and road texture 90, 246
  vibration terms and relationships 91
 noise test track material 322, 331
 studding 240

UK
 Department of Transport see Highways Agency
 opportunities for recycled materials 360–2
 porous asphalt 204
 quarries and quarrying 39
 standards for pavement design 112

USA
 volumes, density and effective binder, compacted asphalt 173–4
 see also Marshall; Strategic Highway Research Program (SHRP)

Vehicle noise test track material 322, 331
Vehicle parks, static load and surface material 94–5, 113–14
Vehicles, tyres see Tyres
Veneer coats 277–308
 advantages 278
 aftercare 294–5
 choice of materials 367
 component materials 287–93
  binders 288–9
  bitumen emulsions 291–2
  chippings 287–8
  cutback bitumen (unmodified) 289–90
  modified bitumen emulsions 292–3
  modified cutback bitumen 290–1
 concept and history 277–83
 economics 295–7
 equipment
  chipping spreaders 293–4
  rollers 294
  spray tankers 293
 high-friction surfacings 302–7
 history 277–83
 maintenance 295
 principles
  forces causing/resisting embedment 279
  hardness categories, and temperature 280–1
  local factors 281
 properties and applications 283–7
  double surface dressing 285
  inverted double dressing 285
  multiple-dressings 285–7
  preparation 283
  racked-in surface dressing 285
  sandwich surface dressing 285
  single surface dressing 284
 slurry surfacing 297–302
  airfield surface treatments 299
  component materials 299–301

concept and history 297
contract types 301
economics 302
footway slurry 298
maintenance 301–2
micro-asphalts 298–9
thick slurries 298
thin carriageway slurries 298
specifications 282
surface dressing
   operations (Road Note 39, TRL) 286
   types 284–7
traffic control and aftercare 294–5
see also Surfacing materials
Vialit plate test, (LCPC1963) France 31, 33–5, 72
Visco-elastic creep
   bitumen 61, 122
      moving wheel-load 123–4
      static loading 122
Viscometry 59
   capillary 60
   deformation 60–2
   elastic and viscous moduli 62
   Standard Tar Viscometer 68
Viscosity, bitumen 58–62

Viscosity measurement, tar (STV) 48–9
Voids, compacted asphalt 173–4
Volumes
   BAA/DWS 174
   density and effective binder, compacted asphalt 173–4
   MS2 174

Waste materials *see* Recycled materials
Water absorption
   equation 25–6
   properties of aggregates and fillers 25–6
Water immersion tests 355–6
   recycled materials 355–6
Waterproofing protection, mastic asphalt 227
Wear, aggregate abrasion value (AAV) 29–30
Wheel tracking test 70–1
   heavy goods vehicles 93
   rates, various modifiers 122
   rut resistance, thin surface course materials 127, 265

Zingg graph, particle shape 24–5